CELL AND MOLECULAR BIOLOGY

CELL AND MOLECULAR BIOLOGY

Prakash S. Lohar

Head, Department of Biotechnology
Reader, Department of Zoology
M.G.S.M's Arts, Science and Commerce College
Chopda, Jalgaon Dt.
Maharashtra

MJP PUBLISHERS
Chennai 600 005

MJP PUBLISHERS

© Publishers, 2024   47, Nallathambi Street
All rights reserved   Triplicane
Printed and bound in India   Chennai 600 005

To

My Parents
and
Dadasaheb Dr. Suresh G. Patil

FOREWORD

It gives me great pleasure to write a few words on the effort of Dr. P. S. Lohar to write this book on *Cell and Molecular Biology*. It is a fast growing field of study and research. Cell Biology has been studied extensively in terms of structure and function of all cell molecules, and the pathways these molecules follow within the cell, thus including many aspects of molecular biology.

This book tries to give the basics of cell and molecular biology and contains twelve chapters. The author tries to present important and latest developments in different areas of cell and molecular biology in a lucid and simple way.

Such a book is timely and I would like to congratulate the author and the publishers for this good effort.

God bless.

Rev. Fr. Dr. S. Ignacimuthu, s.j., Ph.D., D.Sc., F.R.E.S.

Director, Entomology Research Institute
Loyola College, Chennai

PREFACE

During my postgraduation in physiology, I had developed a great interest in cell and molecular biology. Even though many years have passed, I still find this subject the most fascinating to explore and I still love spending time in reading about the latest findings in the field. Thus, for me, writing a book on cell and molecular biology provides an opportunity to keep abreast of what is going on in this field.

My primary goal in writing this text is to help generate an appreciation in graduate and postgraduate students of Biotechnology, Zoology, Botany, Microbiology, Biochemistry, Pharmacy, Health and Medical Sciences for the activities in which the biomolecules and microscopic structures present in the cellular world of life are engaged. Cell and molecular biology is an ever-dynamic subject of life sciences forming the core subject in the curricula of most of the Indian Universities. In this comprehensive book, attempts have been made for a concise coverage of every topic from its fundamental aspect to the latest developments in an easily accessible style that will definitely help the students preparing to appear for various competitive examinations. All cell organelles, their structure and functions have been discussed with incredible illustrations, and the molecular aspects of genes are integrated in this book and would also prove to be very useful to researchers and academicians.

I have designed this text to be student-friendly, to be understood and enjoyed by the students, who have only a minimal background in biology and chemistry. The method I have adopted permits the presentation of complex material in a simple and lively manner without causing boredom and monotony. There are twelve chapters in this book, of which first two chapters deal with the cell world. The diversified nature of cells, the location, biological structure and functions of cellular components including cell wall, plasma membrane, nucleus, ribosomes, mitochondria, plastids, endoplasmic reticulum, Golgi complex, lysosome, microbodies, and cytoskeleton are well explained. Since the study of cell functions generally requires the use of considerable instrumentation, the third chapter emphasizes the importance of microscopy and micrometry. Chapter 4 covers the structural and genomic organization of different viral classes with their involvement in AIDS and cancer. Chapter 5 describes bacterial genetics and the role of bacteria, plasmids and cosmids in genetic engineering.

Cellular cycle, cell divisions, and cell death are discussed in chapter 6, followed by a description of eukaryotic chromosomes with the consequences of structural and numerical variations in chapter 7. The chemical nature, structure, and replication of DNA in prokaryotic and eukaryotic cells are covered in chapter 8. After considering the molecular structure and duplication mechanism of DNA, its mutability and repair processes are emphasized in chapter 9. The next two chapters (10 and 11) provide information on transcription, RNA synthesis, RNA processing, genetic code and the mechanism of translation. Finally, chapter 12 focuses on our present understanding of various mechanisms by which gene expression in prokaryotes is regulated. Every chapter is followed by a summary to reflect the depth of the coverage at a glance, and review questions to test the understanding of the topic. In addition to the breadth of coverage, other features that increase the usefulness of the book include a glossary for defining the key terms and an up-to-date list of Nobel laureates in the field of physiology/medicine and chemistry.

I have attempted to provide coordinated and authentic information. However, there may be several possible interpretations of the data included in the book. As a sort of evolution, new insights are made in relation to concepts and techniques in biological science. I welcome suggestions for improvement of this book, which could be incorporated in the next edition of this book. Hope this book will cater to the immediate need of students for effective learning as well as for the basic understanding of the subject for researchers and teachers.

Prakash S. Lohar

ACKNOWLEDGEMENTS

This book is the outcome of shared insights from many people. I would first like to thank Rev. Fr. Dr. S. Ignacimuthu, s.j., Director, Entomology Research Institute, Loyola College, Chennai, for writing the foreword for this book.

I am especially grateful to Hon'ble Dadasaheb Dr. S.G. Patil, Founder President, Mr. Sandeep S. Patil, President, Mr. V.D. Joshi, Vice-President, Dr. Ashok Kadam, Secretary, and other respected members of Management Council and Governing Council of Mahatma Gandhi Shikshan Mandal, Chopda, who have inspired me a lot from the very beginning of my services and have continuously encouraged me to write books, from the two books *Biotechnology* and *Endocrinology: Hormones and Human Health* to the completion of this third book on *Cell and Molecular Biology*.

To prepare the manuscript for this title, I have extensively reviewed hundreds of papers from the original literature of many journals in basic and applied research. I am sincerely indebted to those authors who supplied some of the illustrations in this book.

I wish to express my deep sense of gratitude to Mr. Kailas Patil (MLA), Dr. Mahendra Patil, Dr. Premchand Mahajan, Dr. Mahendra Jaiswal, Mr. Rajesh and Dr. Swati Chavan, Prof. S.B. Chincholkar, Prof. V.L. Maheshwari, Prof. A.Y. Mahajan, Dr. B.B. Waykar, Dr. Sandhya Sonawane, Prof. Shashikant Borse and Prof. S.B. Wasavada for their moral support throughout the synthesis of this book. I owe reverence to Dr. Vikas Gulve (Ayurvedacharya) for his deep understanding of the cell and molecular biology of the plant world and its interactions with the human body that boosted me in an extraordinary way to complete this book.

My thanks are also due to Dr. D.D. Patil, Prof. D.B. Deshmukh, Dr. S.S. Alizad, Prof. S.G. Gujarathi, Prof. L.M. Patil, Prof. P.M. Patil, Shri. V.R. Dewang, Shri. Shirish S. Patil, Shri. V.J. Tillu, Shri. D.M. Patil, Shri. Mangal Thakre, other teaching and non-teaching staff of MGSM's A.S.C. College, Chopda, for providing their active support.

I would like to remain indebted to my family members and well-wishers for their unstinting support for the completion of this endeavour.

On the publishing side, I extend my heartiest thanks to Mr. C. Sajeesh Kumar, Managing Editor, MJP Publishers for his dedicated support to develop this book. I would also like to acknowledge Ms. P. Parvath Radha, Project Editor, for her careful editing and constructive suggestions in the manuscript. Finally, I express my sincere thanks to Mr. J.C. Pillai, Director, MJP Publishers, who gave me the initial push to write the book and also for his consistent enthusiasm in keeping pace with the new developments in the publishing world.

Prakash S. Lohar

CONTENTS

8. DNA—Chemical Nature, Structure and Replication — 243

9. DNA Mutability and its Repair Mechanism — 273

THE CELL—PROKARYOTIC AND EUKARYOTIC

Discovery of cell

Organization and properties of cell

Different classes of cells

DISCOVERY OF CELL

One of the greatest inventions of the 17th century is the light microscope that had been used to unravel a world that would never have been revealed to the naked eye. Robert Hooke (1665) observed the slices of cork tissue under the compound microscope and saw a honeycomb-like arrangement in the texture of cork's thin section, which is composed of small compartments surrounded by the cell wall. He coined the term **cell** for the individual units.

In 1831, Robert Brown observed a spherical body in each cell and called it as a **nucleus**. The cell as a unit of structure and function of all living organisms was not recognized until about 1839 when M.J. Schleiden, a German lawyer-turned-botanist, and Theodore Schwann, a German zoologist, almost simultaneously concluded that all living tissues either of a plant or animal are composed of individual cellular units and they proposed the **cell theory**.

The Schleiden and Schwann's cell theory stated that

❖   all organisms are composed of one or more cell and

❖   the cell is the structural and functional unit of the life.

By 1855, Rudolf Virchow, a German pathologist, added a third tenet to the cell theory that a new cell can arise only by the division of a pre-existing living cell. In modern terms the cell may be defined as a unit of biological activity delimited by a selectively permeable membrane and capable of self-reproduction.

SHAPE AND SIZE OF CELL

The shape of cells is variable as it mainly depends on the functional requirement of the organism and environmental conditions. It may be irregular as in the case of amoeba or fixed. When the shape is fixed or permanent, all types of shapes are found such as spherical, cuboidal, spindle, rectangular, flattened, oval, columnar, flagellated, disc-shaped, etc. Bacteria may have rod, spiral or even comma shape. The single cell body of *Acetabularia* is differentiated into a base, stalk and an umbrella-like cap. The multicellular animals and plants show some regularity in the shape of their cells (Table 1.1).

The size of the cells also varies considerably. The factors affecting the size of the cells are

1.   the ratio between the volume of nucleus to the cytoplasm of the cell,

2.   ratio of cell surface to volume of the cell, and

3.   the rate of metabolism.

Table 1.1 Types of cells, their shape and examples

Name of the cell	Typical shape	Examples
IN PLANTS		
Epidermal	Cuboidal or brick-shaped	Epidermis
Vascular	Elongated	Phloem
Supporting	Elongated with heavy walls	Bast cells
Parenchyma	Spheroidal	Mesophyll, interstitial cells in stem, root, etc.
Algae (Unicellular)	Diverse	*Chlorella*
Unicellular fungi	Diverse	Yeast
Filamentous fungi	Elongated cells	*Neurospora*
Bacteria	Diverse	*E. coli, Vibrio cholerae,*etc.
IN ANIMALS		
Protozoa	Asymmetrical, slipper-shaped, oval, etc.	Amoeba, Paramecium, *Plasmodium*
Spermatozoa	Flagellated cells	Sea urchin sperm
Eggs	Usually spherical	Sea urchin egg, frog egg
Blood cells	Disc-shaped, amoeboid	Erythrocytes, leucocytes
Epithelial	Cuboidal or brick-shaped	Epidermis, glandular lining
Nervous	Cell body with long axon	Sensory and motor neuron
Muscular	Spindle, cylindrical	Smooth, striated muscle
Connective	Spheroidal	Cartilage and bone cells

Some of the organisms are observed under the compound microscope only since they are not visible to naked eyes, some are barely visible, while some are macroscopic. The largest cell is an ostrich egg (diameter, about 20 cm), while the smallest cell size can be found in mycoplasma (0.1 μ diameter). Table 1.2 summarizes the variation in the size of the cells.

Table 1.2   Average sizes of some living cells

Name of the cell	Average size
Giraffe nerve cell	2 m
Ostrich egg	190 × 140 mm
Chick egg	60 × 40 mm
Human ovum	0.1 mm
Amoeba	100 μ (micron)
Sea urchin egg	70 μ
Hepatocyte (Liver cell)	20 μ
Erythrocyte (R.B.C.)	7 μ
E.coli	1.5 × 0.7 μ
Diplococcus pneumoniae	200 × 100 nm

ORGANIZATION OF CELLS

Generally a cell has three distinct regions namely the cell wall, protoplasm and vacuole of which the cell wall and vacuoles are non-living components. The protoplasm is differentiated into nucleus and cytoplasm, each of which is surrounded by a membrane, the nucleus by the nuclear membrane and the cytoplasm by cell or plasma membrane. The cytoplasm can be differentiated into outer fine granular ectoplasm and inner dense endoplasm that encloses various cell organelles. The cell wall is the outermost envelop of a plant cell (Figure 1.1), which is absent in animal cell (Figure 1.2). Internal to cell wall there is a plasma membrane. The chief cytoplasmic cell organelles are mitochondria, plastids, Golgi complex, endoplasmic reticulum, ribosomes, lysosomes, flagella, centrosomes, etc.

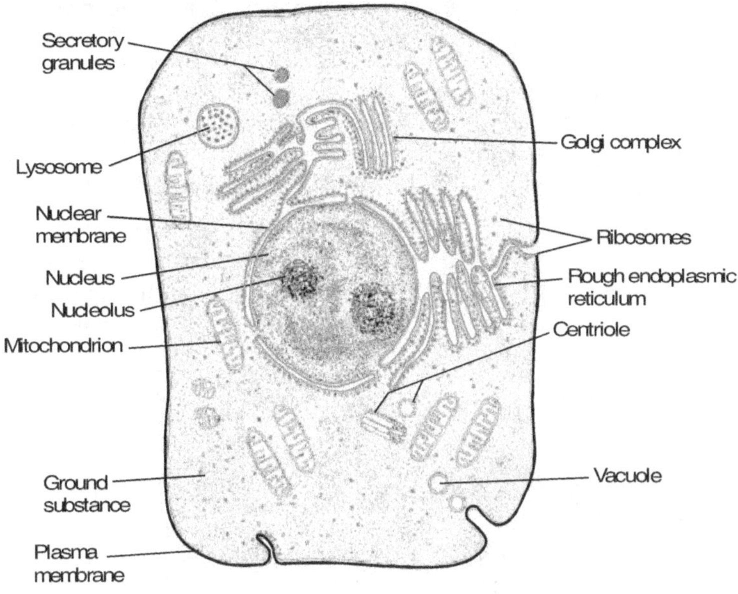

Figure 1.1   Schematic diagram of the animal cell

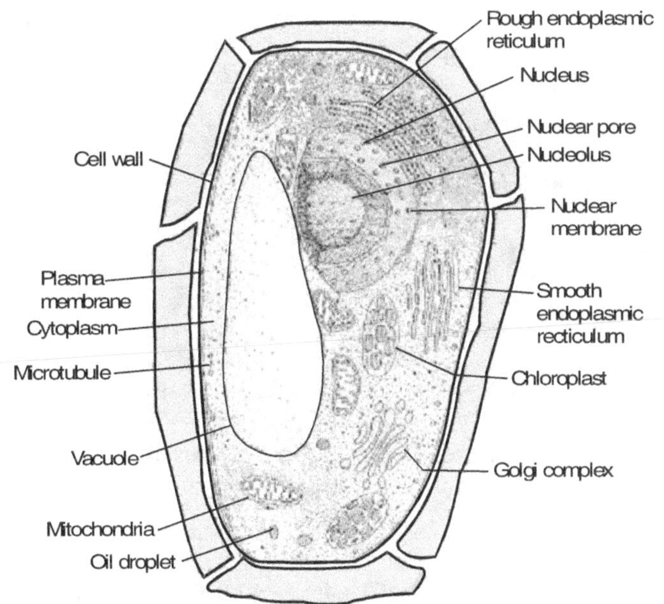

Figure 1.2   Schematic diagram of the plant cell

Some of the fundamental properties of the cells are as follows:

1. There are several types of cells in a multicellular organism. Each type of cell has consistent appearance, so far as its structure and location are concerned within the individuals of a species. Similarly, each type of cell organelle has a consistent composition of macromolecules (Atoms-molecules-polymers-complexes-subcellular organelles-cell). Thus cells are highly complex and organized.

2. Every cell has genetic information that is encoded in its genes (hereditary units). Genes have a) ability to construct cellular structures, b) directions for performing cellular functions and c) program for making the copies of cells by the division of mother cell into daughter cells.

3. Cells require constant input of energy for development and maintenance of complexity.

4. Every cell has the ability to perform chemical transformations in the presence of enzymes.

5. Cells not only perform numerous activities like respiration, movement, digestion, assimilation, transport, etc. but also evoke specific responses to hormones, growth factors, extracellular materials, as well as to substances present on the surfaces of other cells.

6. Every cell has the capability of self-regulation. For example, if the cell is unable to repair its damaged or mutated DNA, such a cell can transform into cancer cell that can destroy the entire organism.

7. Cells have finite lifetime. Once a higher organism reaches its mature size, it stays at that size by carefully regulating the balance between cell division and cell death.

DIFFERENT CLASSES OF CELLS

The electron microscope helped biologists to observe the internal structure of a wide variety of cells. It is apparent from these studies that there are two fundamental classes of cells, **prokaryotic** and **eukaryotic**, which are distinguished by their size, the type of internal structure, or cell organelles present within them. The existence of two distinct varieties of cells, without any known intermediates, represents one of the most fundamental evolutionary discontinuities in the biological world.

It was discovered by 1940 that in some organisms, e.g. typical bacteria, which were classified under the kingdom Monera, the genetic material is not enclosed by nuclear membrane (Figure 1.3). All bacteria consist of **prokaryotic** cells that are structurally very simple cells. While the cells of algae, fungi,protozoa, animals and plants are structurally more complex **eukaryotic** cells since they have membrane-bound nucleus. Some of the fundamental differences between prokaryotic and eukaryotic cells are given in Table 1.3.

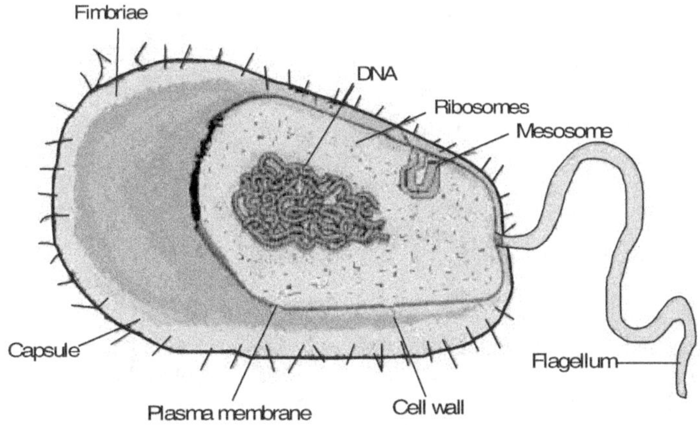

Figure 1.3 A schematic view of a generalized bacterium

Table 1.3 Distinguishing characters between prokaryotic cells and eukaryotic cells

Character	Prokaryotic cell	Eukaryotic cell
Size	Generally small (1–10 μm)	Generally large (5–100 μm)
Group in which they found	Bacteria	Algae, fungi, protozoa, plants and animals
Genetic material	Nucleoid, DNA is circular and naked (without histones)	Nucleus, DNA is combined with proteins (histones)
Chromosome	Single	Multiple
Nucleolus	Absent	Present
Membrane-bound organelles like mitochondria, chloroplast, Golgi complex, endoplasmic reticulum, etc.	Absent	Present
Mesosomes	Present	Absent

(Contd.)

Table 1.3   (Continued)

Character	Prokaryotic cell	Eukaryotic cell
Ribosomes	70S (50S + 30S)	80S (60S + 40S)
Cytoplasmic streaming (cyclosis)	Absent	Present
Exocytosis and Endocytosis	Absent	Present
Cell wall	Non-cellulosic, Presence of peptidoglycan	Cellulosic, Peptidoglycan is absent
Cytoskeleton	None	Complex, with microtubules
Locomotary organelles	Simple fibril	Multifibrilled with "9 + 2" microtubules
Pseudopodia	Absent	Present in some
Membrane bound vacuoles	Absent	Present
Cell division	Amitosis (fission or budding)	Mitosis and Meiosis
Replication of DNA	Single point	Multi point
RNA polymerase in transcription	Single RNA polymerase	Three types of RNA polymerases
Energy metabolism	No mitochondria; Oxidative enzymes bound to plasma membrane	Oxidative enzymes packaged to mitochondria

Prokaryotic Cells

These are primitive cells having nucleus without nuclear membrane that are divided into two major subkingdoms—Archaeobacteria and Eubacteria. The archaeobacteria comprise three groups of ancient bacteria, which are represented by the **methanogens** (bacteria involved in production of methane gas from CO_2 and H_2 gases), the **halophiles** (bacteria living in extreme saline water), and the **thermoacidophiles** (bacteria living in highly acidic hot springs).

The subkingdom Eubacteria consists of all other types of bacteria including the smallest living cells, **mycoplasma** ($0.1\,\mu$ diameter), which are the only prokaryotes without cell wall. These are tiny, irregularly shaped, flexible bacteria

(Figure 1.4) living in close contact with cells of higher eukaryotic organisms and are so small that they can pass through a filter paper. They were at once regarded as disease-causing agents responsible for pneumonia-like respiratory illness (Pleuropneumonia-like organism—**PPLO**) affecting higher vertebrates including humans.

Cyanobacteria (Figure 1.5) are the most complex prokaryotic cells. Formerly they were known as blue-green algae because of their ability to form a blue-green scum on the surface of fresh-water bodies. Many of them are capable not only of photosynthesis, but also of nitrogen fixation.

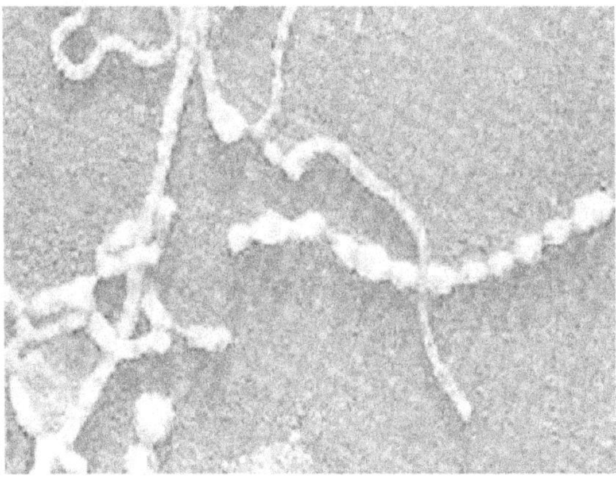

Figure 1.4   Electron micrograph of mycoplasma, the smallest living organism

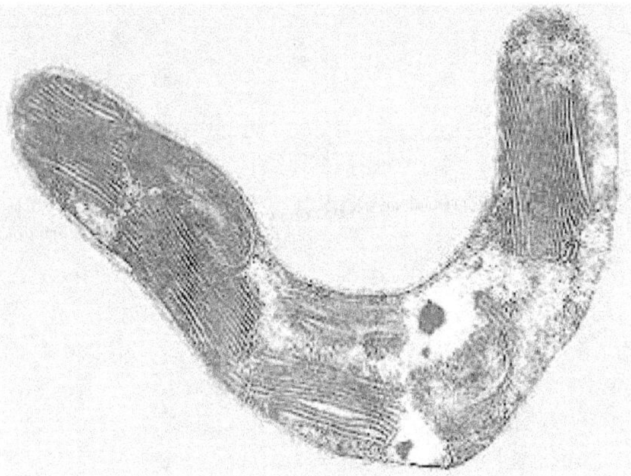

Figure 1.5   Electron micrograph of cyanobacterium (the stacks of cytoplasmic membranes similar to chloroplast of a plant cell)

Eukaryotic Cells

These cells have prominent nucleus and a number of cell organelles, making them complex because an individual cell constitutes a complete, single-celled (unicellular) organism. The formation of highly complex unicellular organisms (e.g. *Amoeba, Paramecium, Vorticella*, etc.) represents one evolutionary pathway, where a single eukaryotic cell performs all the activities including response to the environment, food capturing, ejection of excess fluid, evading predators, etc.

An alternate pathway has been the evolution of multicellular organisms in which different types of specialized cells carry out different activities. Thus there is the division of labour among the cells of a multicellular organism. The process by which a relatively unspecialized cell is modified into a highly specialized cell is called differentiation. As a result of differentiation, different types of cells acquire distinctive appearance and unique materials. For example, disc-shaped red blood cells are filled with a single protein, haemoglobin that helps in transport of oxygen; skeletal muscle cells contain precisely arranged myofilaments composed of contractile proteins, myosin and actin; chondrocytes present in cartilage contain a characteristic matrix made up of polysaccharides and collagen, which together provide mechanical support; the cells present in the digestive tract have the ability to secrete digestive enzyme; the nerve cells release neurotransmitters that help in conduction of impulses; and so forth. Despite the differences in the individual cells with respect to size, shape, number, contents, and functions, the various cells of multicellular plants and animals are made up of similar organelles like mitochondria, endoplasmic reticulum, Golgi complex, lysosomes, plastids, etc. In most of the cells, the number, appearance and location of the organelle is correlated with function of the particular cell.

Viruses

These are disease-causing pathogens, presumably simpler and smaller than bacteria. A virus can be a prokaryotic cell but it does not have any membrane system. Viruses also lack complex enzyme system that is essential for respiration, photosynthesis, etc. They are responsible for dozens of diseases in plants, animals and humans, including tobacco mosaic disease, measles, polio, influenza, AIDS and a few types of cancer (see details in chapter 4). Viruses have different shapes, sizes and constructions (Figure 1.6), but they share some common properties including the following. All of them are obligatory intracellular parasites. Outside the living cell, the virus exists as a particle or **virion** that contains a genetic material made up of either DNA or RNA; the genetic material of the virion is surrounded by a protein capsule, or **capsid** that is made up of specific number of subunits. They are non-cellular pathogens which can only reproduce when present within the living cell.

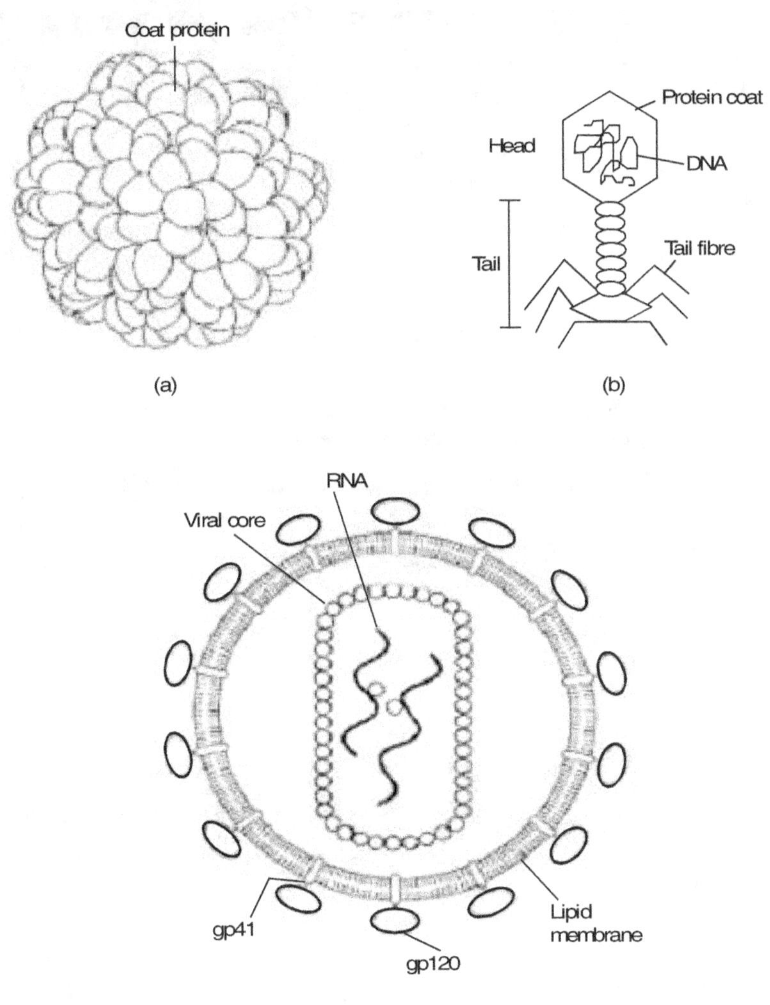

Figure 1.6 (a) Small icosahedral virus (e.g. R17, f2, MS2), (b) Bacteriophage, T4 and (c) Human immunodeficiency virus

SUMMARY

- Robert Hooke (1665) coined the term **cell** for the individual units present in the slices of cork tissue observed under the microscope.

- In 1831, Robert Brown observed a spherical body in each cell and called it as a **nucleus.**

- Schleiden and Schwann(1839)proposed the **cell theory,** which states that all living tissues either of a plant or animal are composed of individual cellular units and the cell is the structural and functional unit of the life.

- Rudolf Virchow (1855) added the third tenet to the cell theory that a new cell can arise only by the division of a pre-existing living cell.

- The shapes and sizes of cells are variable as both parameters depend on the functional, requirement of the organism and environmental conditions.

- The largest cell is an **ostrich egg** (about 20 cm, diameter), while the smallest cell size can be found in **mycoplasma** (0.1 μ diameter).

- Each type of cell has a consistent appearance as far as its structure and location are concerned within the individual of a species. Every cell has genetic information that is encoded in its **genes**.

- Cells require constant input of energy for development and maintenance of complexity. They have the ability to perform chemical transformations in the presence of enzymes.

- Every cell has the capability of **self-regulation**. And the cell has **finite lifetime**.

- Electron microscopic study reveals and there are two fundamental classes of cells, **prokaryotic** and **eukaryotic**, which are distinguished by their size, the type of internal structure, and cell organelles present within them.

- All bacteria are **prokaryotic** cells and are structurally very simple cells. While the cells of algae, fungi,protozoa, animals and plants are structurally more complex **eukaryotic** cells since they have membrane-bound nucleus.

- Prokaryotic cells are primitive cells having nucleus without nuclear membrane and are divided into two major subkingdoms—Archaeobacteria and Eubacteria.

- The archaeobacteria consists of three groups of ancient bacteria—**methanogens, halophiles** and **thermoacidophiles.**

- Eubacteria consists of all other types of bacteria including **mycoplasma** and **cyanobacteria**.

SUMMARY

- Eukaryotic cells have prominent nucleus and a number of cell organelles. In the complex **unicellular** organisms, a single eukaryotic cell performs all the activities, whereas there is the division of labour among the cells of a **multicellular** organism.

- Viruses can be prokaryotic cells but they do not have any membrane system. They are non-cellular pathogens, which can only reproduce when present within the living cell.

REVIEW QUESTIONS

1. Comment on the concept of cell theory.

2. What are the fundamental properties of the cell?

3. How can you distinguish between a prokaryotic and eukaryotic cell?

4. Comment on the shapes and sizes of various cells.

5. What is prokaryotic cell? Add a note on its classification.

6. What are the points of differences between the eukaryotic cell of a unicellular and multicellular organism?

7. Write short notes on

   i. Nucleoid

   ii. Mycoplasma

   iii. Contribution of Rudolf Virchow

   iv. Archaeobacteria

   v. Cyanobacteria

   vi. Amoeba as a unicellular animal

   vii. Virus

CELL ORGANELLES: STRUCTURE AND FUNCTION

Cell wall

Plasma membrane

Cytoplasm

Nucleus and Nuclear membrane

Chromatin structure

Nucleolus

Ribosomes

Mitochondria

Plastid and Chloroplast

Endoplasmic reticulum

Golgi complex

Lysosomes

Cytoskeleton

CELL WALL

It is the outermost boundary in majority of prokaryotic cells and plant cells (eukaryotes). Animal cells do not have a cell wall. Since the chemical composition of prokaryotic and eukaryotic cell wall is different, both types are dealt with separately.

PROKARYOTIC CELL WALL

It is the semi-rigid, complex and non-living component of the cell that surrounds the plasma membrane within the capsule. It is secreted and maintained by the living portion of the cell, called protoplast. Its prime function is to provide shape to the bacteria and protect the protoplasmic inclusions. The thickness of the cell wall rather than its chemical composition is the point of difference between the archaeobacteria and eubacteria. Thickness of the cell wall determines the gram reaction. Gram-negative bacteria (e.g. *E. coli*) have a cell wall thinner than gram-positive bacteria (belonging to the *Bacillus* group).

Chemically, the cell wall in prokaryotes is made up of fibres of heteropolymers called mucopeptides or peptidoglycans (murein)—a complex of disaccharide and polypeptide that forms a lattice to surround and protect the cell. Disaccharide consists of two derivatives of monosaccharide called *N* acetyl glucosamine (NAG) and *N* acetyl muramic acid (NAM). Both NAM and NAG link alternately to form a carbohydrate backbone. Attached to NAM is a tetrapeptide chain having alanine, glutamic acid and either lysine or diamino pimelic acid (DAP). Adjacent tetrapeptides may be directly bound to each other or linked by an oligopeptide bridge to form a cross link between two chains. These cross links make the peptidoglycan layer a giant molecule through which bacteria obtain their structural integrity (Figure 2.1). *E.coli* has a periplasmic space between the outer and inner cytoplasmic membranes, where peptidoglycan forms a periplasmic gel.

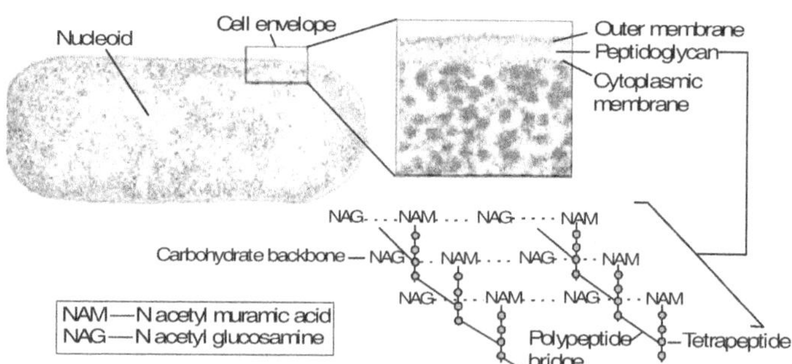

Figure 2.1   Electron micrograph of *E. coli* with its magnified cell envelope showing chemical structure of a peptidoglycan

EUKARYOTIC CELL WALL

The presence of a rigid cell wall is the characteristic feature of a plant cell. Cell wall in plants is made up of non-living matter, which is secreted by the plasma membrane. Plant cell wall performs both morphogenetic and physiological functions, since it forms a rigid skeleton that provides the shape to the cell and also acts as a restraint to cell expansion. A typical cell wall has three different regions—primary wall (thin and elastic), secondary wall (thick and rigid) and middle lamella. The primary wall is generally present in growing cells since it is capable of expansion as the cell grows. Once the cell ceases its growth, the secondary wall is usually formed, as it is thick and inextensible. Usually, cell walls of adjacent cells are separated by a middle lamella (Figure 2.2).

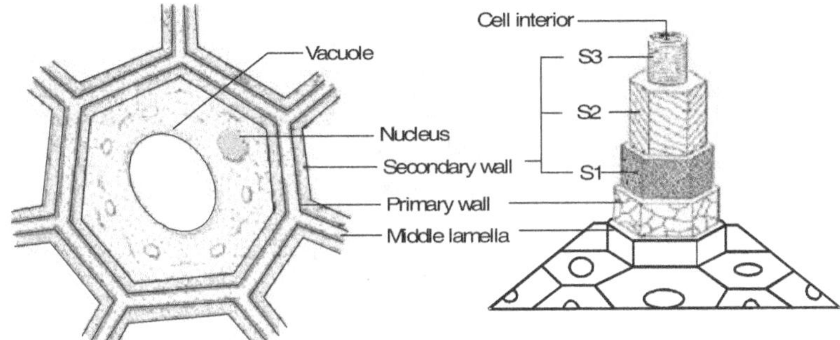

Figure 2.2　Diagrammatic representation of a plant cell showing position of cell wall, various components and orientation of cellulose microfibrils in different layers

Primary wall　It is secreted by the plasma membrane. It is formed in young emerging cells and remains thin and elastic but as the cell approaches maturity, it becomes thick and rigid. Chemically, the primary wall is composed by dry weight of 30–45% cellulose, over 50% other polysaccharides including hemicellulose and pectin, and about 5% glycoproteins rich in hydroxyproline. It is the principal layer of meristematic cells, chlorophyllous cells and true parenchyma.

Secondary wall　It is deposited on the primary wall towards the cytoplasm, once the cell completes its maturation. It is very rigid and does not alter its shape. Chemically, the secondary wall is composed of cellulose, non-cellulosic material and hemicellulose. A typical secondary wall has three distinct layers that are recognized by the different arrangement of microfibrils; these are outer layer (S1), middle layer (S2), and inner layer (S3) as shown in Figure 2.2. In the secondary wall, the microfibrils are more orderly and compact than in the primary wall. The space between the fibrils is usually filled with matrix that contains non-cellulosic materials such as hemicellulose, pectin, xylans, etc. In cellulose, over 10,000 residues of glucose are joined by β1-4 linkage to form a chain. About 100 of such cellulosic chains bundle together to form micelles or elementary fibrils of about 100 Å in diameter (Figure 2.3). Many hundreds of such micelles stack together to form microfibrils having a diameter of 200–300 Å. About 2000 of these microfibrils join to form a macrofibril, which is visible under the light microscope.

The cell wall undergoes chemical changes as it matures, one such change is lignification in which there is deposition of lignin (a phenol-containing polymer and a major component of wood) on the cell wall. As a result, the cell becomes hard, thick and dead. Thus lignification brings about structural and functional changes in the cell.

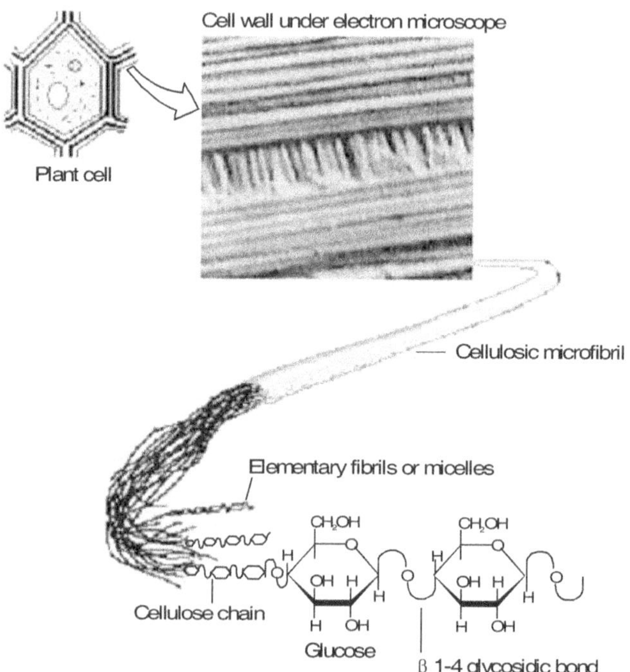

Figure 2.3　Structural elements of cellulosic microfibrils in plant cell wall

In all multicellular plants, there are numerous cylindrical cytoplasmic channels connecting adjacent cells, which are known as plasmodesmata. They are lined by plasma membrane and serve as sites for intercellular communication.

Middle lamella　It is an amorphous and colloidal intracellular substance made up of pectin, cellulose, calcium, magnesium pectate, etc. Middle lamella is present in between adjacent plant cell walls and is formed during cell division.

FUNCTIONS OF CELL WALL

1. The cell wall acts as a physical barrier protecting plant cells from invading pathogens like viruses, fungal spores, or bacteria. Thus they serve as a part of the defence mechanism.

2. It provides a definite shape to the plant cell.

3. Cell wall acts as a structural skeleton for the plant. In growing tissues, both cell wall and turgor pressure provide support to the organs.

4. It presents a barrier to cell expansion during cell division.

5. Some of the polysaccharides present in the cell wall can act as reserve carbohydrates for the plant.

- *Cell wall is the outermost boundary in majority of prokaryotic cells and plant cells whereas it is absent in animal cells.*

- *Bacterial cell wall is the semi-rigid, complex and non-living component of the cell that surrounds the plasma membrane within the capsule.*

- *Chemically, bacterial cell wall is made up of mucopeptides or peptidoglycans (murein)— a complex of disaccharide and polypeptide that includes, N acetyl glucosamine (NAG) and N acetyl muramic acid (NAM) as derivatives of monosaccharide attached to a tetrapeptide chain.*

- *A typical plant cell wall consists of three different regions—primary wall (thin and elastic), secondary wall (thick and rigid), and middle lamella.*

- *The primary wall is secreted by plasma membrane and chemically it is composed of 30–45% cellulose, 50% other polysaccharides including hemicellulose and pectin, and about 5% glycoproteins.*

- *The secondary wall is very rigid and does not alter its shape. It consists of cellulosic chains that bundle together to form micelles or elementary fibrils.*

- *The middle lamella is an amorphous and colloidal intracellular substance that is formed during cell division.*

- *Besides providing a definite shape to the plant cell, cell wall acts as a physical barrier protecting the cells from invading pathogens.*

- *In growing tissues, both cell wall and turgor pressure provide support to the organs and the cell wall presents a barrier to cell expansion during the cell division.*

- *Plant cell wall may provide reserve carbohydrates.*

REVIEW QUESTIONS

1. What is cell wall? Give details of the prokaryotic cell wall.

2. Explain the role of NAM and NAG in the formation of prokaryotic cell wall.

3. Describe the ultrastructure of plant cell wall with its functions.

4. Differentiate between bacterial cell wall and plant cell wall.

5. Comment on the role of cellulose in formation of plant cell wall.

6. Write short notes on:

    i. Mucopeptides

    ii. Elementary fibrils

    iii. Plasmodesmata

    iv. Lignification

    v. Protoplast

PLASMA MEMBRANE

The cytoplasm in both plant and animal cells is surrounded by the plasmalemma or plasma membrane, while the cell membrane is the limiting membrane of the cell organelles like mitochondria, Golgi complex, endoplasmic reticulum, lysosomes, nuclear membrane, etc. The cell membrane is the plasma membrane deposited with a layer of cell cement or the matrix that fills up the gap between the adjacent cells. The plasma membrane and cell membrane are collectively known as biological membranes.

CHEMICAL COMPOSITION OF MEMBRANE

Membranes are mainly composed of lipoproteins—a complex of lipids with proteins. The ratio of protein to lipid varies from 1 : 0.8 to 1 : 4 and it depends on the type of membrane, type of organism, and the type of the cell. In addition to lipoproteins, polysaccharides, DNA and RNA have also been found in certain cases. All these chemical components are held together in a thin film-like pliable sheet of about 75 Å thickness, in three concentric layers. The middle layer is about 35 Å thick and outer layers on either side are 20 Å thick.

Lipids

The most common fats found in the membrane are phospholipids, glycolipids and sterols. All of them are **amphipathic**, i.e., they have both hydrophilic and hydrophobic portions within a molecule.

The three important **phospholipids** are lecithin, choline and cephalin that are made up of glycerol, fatty acids, phosphoric acid, choline and a complex of sphingosine.

Glycolipids contain a sugar (glucose or galactose) in addition to fatty acids and sphingosine. They are either cerebrosides or gangliosides.

Sterols are steroid alcohols having common characteristic cyclopentanoperhydrophenantherene ring in their structure. The common examples of sterols are cholesterols present in animal tissues, phytosterols present in plants and ergosterol present in ergot and yeast.

Proteins

There are three different types of proteins present in the membrane—structural proteins that form the backbone of the membrane and are strongly lipophilic in nature, functional proteins like enzymes that catalyse several physiological reactions, for example, the enzyme glucose 6-phosphatase found in membranes of endoplasmic reticulum and the cytochrome oxidase present in mitochondria, and carrier proteins that bring about the diffusion of substances across the membrane for example the sulphate carrier molecule in the plasma membrane of *Salmonella typhimurium* and glucose-binding factor present in *E. coli*.

The proteins present in the plasma membrane are further differentiated into integral proteins and peripheral proteins. The integral proteins have hydrophobic interaction while the peripheral proteins have hydrophilic interaction with the lipid bilayer.

Other Constituents

Small amounts of polysaccharides, sialic acid, DNA, RNA, and traces of co-enzymes, porphyrins, metal ions, etc. are distributed in certain biological membranes. The presence of DNA has been confirmed in the membranes of mitochondria, chloroplast and plasma membrane while RNA has been found in the membranes of nucleus, endoplasmic reticulum and plasma membrane of certain plant cells.

STRUCTURE OF THE PLASMA MEMBRANE

Several cell biologists attempted to understand the molecular structure of the plasma membrane in various ways and proposed different models for the same. These models are classified into two basic categories: bilayer models and micellar or subunit models. In the bilayer model, the proteins and lipids are arranged in the form of layers, whereas in micellar model, there is the presence of a number of small and similar independent subunits in the plasma membrane. Some of the bilayer models are unit membrane model, Danielli–Davson model, Kavanau's model, etc., while the fluid mosaic model, Benson's model, mosaic-membrane model, Lenard and Singer's model are few examples of micellar models. Some of the models are explained here.

Lipid Bilayer Model

In 1925, E. Gorter and F. Grendal made the first proposal that the plasma membrane might contain a lipid bilayer. They extracted the lipid from human erythrocytes and measured the amount of surface area the lipid would cover when spread over the surface of water. They concluded that the plasma membrane contained a bimolecular layer of lipids, or simply a lipid bilayer. They also suggested that the polar groups of each molecular layer were directed towards the outside of the bilayer (Figure 2.4).

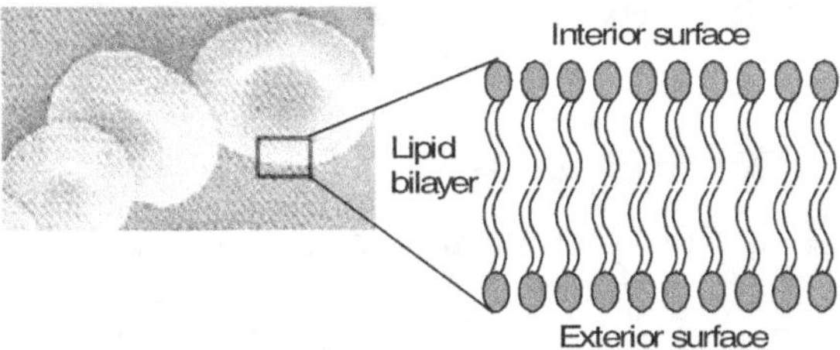

Figure 2.4 Plasma membrane of human erythrocytes showing lipid bilayer model

Unit Membrane Model

It was proposed by Robertson (1957) after studying the membrane structure of Golgi complex, endoplasmic reticulum and nuclear envelope with the help of electron microscope. He found that all biological membranes appear to be made up of lipid layer having the thickness of 35 Å and a dense band of proteins of 20 Å thickness present on either side. Two outer electron dense layers of proteins separated by the lighter middle layer of phospholipids together form a unit membrane (Figure 2.5).

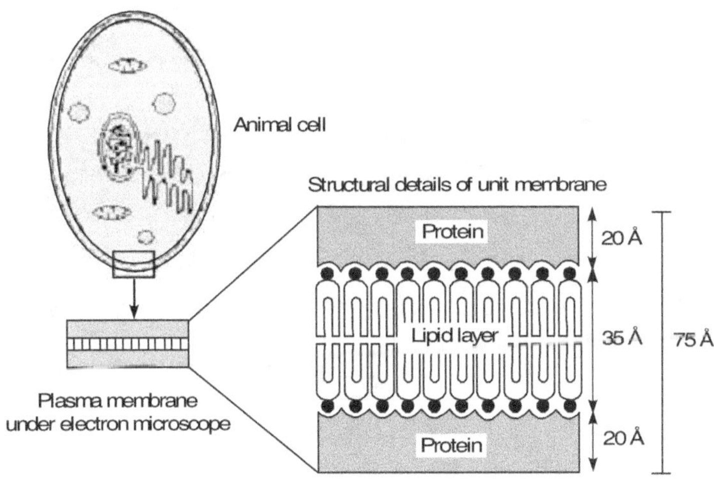

Figure 2.5 Robertson's unit membrane model for the plasma membrane

The membranes of cell organelles, plasma membrane, and endoplasmic reticulum are thought to have unit membrane structure. But later on it has been shown that the arrangement of lipids and proteins is variable in different membranes. Robertson's unit membrane concept was based on the study of the myelin sheath of a nerve fibre, which is a non-typical membrane and hence cannot be considered as a representative structure.

Danielli–Davson Model

In 1935, James Danielli and Hugh Davson studied the starfish egg membrane from the point of its surface tension. They proposed that the membrane is a bimolecular lipid layer, with the polar ends of the lipids facing outward and hydrophilic (water-loving) proteins coating the polar ends. Electron microscopic picture of plasma membranes revealed a three-layered structure—two 2.5 nm (1 nm = 10^{-9} m) thick dense lines separated by a clear 4-nm space (Figure 2.6).

This sandwich model was the first structural interpretation to account for the physical characteristics of the membrane and later on various modifications have been proposed for the model.

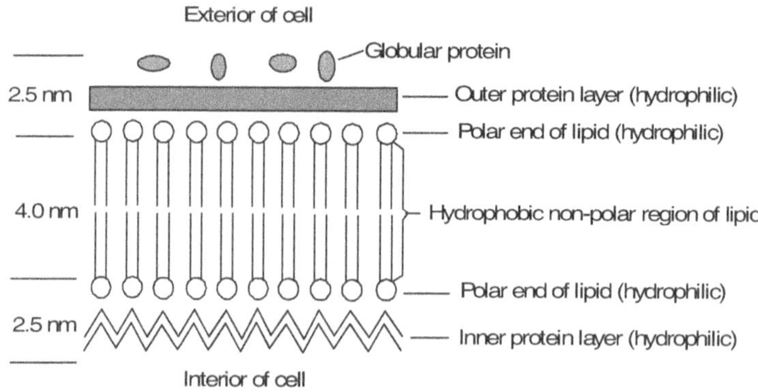

Figure 2.6   The Danielli–Davson model

The unit membrane model is similar to Danielli–Davson model except that the protein layers at the exterior and interior are different. The outer surface has mucoproteins while inner surface proteins are non-mucoid proteins.

The Fluid Mosaic Model

Experiments conducted in the late 1960s led to a different concept of membrane structure as detailed in fluid mosaic model proposed in 1972 by S. J. Singer and G. Nicolson. In this model, attention is focused on the physical state of the lipid, though the lipid bilayer is retained as the core of the membrane. Here the lipid molecules are present in a fluid state capable of rotating and moving laterally within the membrane. The structure and arrangement of membrane proteins are dramatically different from those of previous models. In this model, the proteins occur as a "mosaic" of discontinuous particles that penetrates deeply into the lipid sheet (Figure 2.7). In a way, these protein particles resemble floating icebergs in the sea of lipids. Thus the plasma membrane is a double layer of phospholipids with their polar heads away from one another and the globular molecules of proteins and sterols scattered in between. Integrated protein molecules within the lipid bilayer partially project out from both the extracellular and cytoplasmic sides of the membrane.

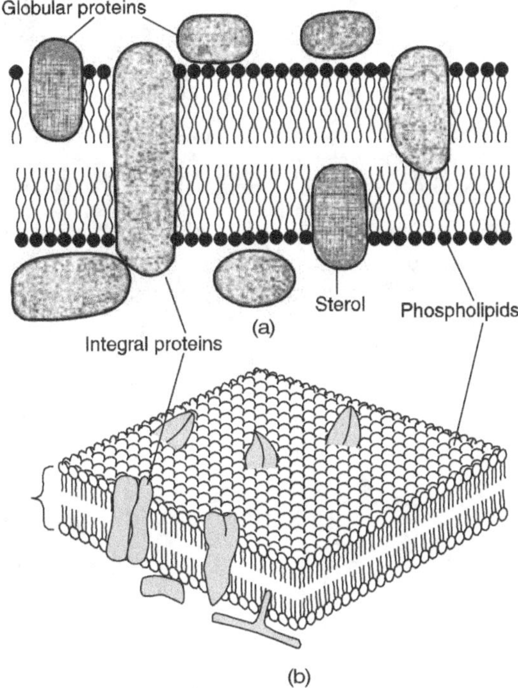

Figure 2.7   (a) Fluid mosaic model of eukaryotic plasma membrane, (b) A bacterial cell membrane

Micellar Model

Many cell biologists suggested that the membrane might consist of closely packed repeating units, which are similar in structure. In 1970, Green described this concept as "fused repeating units" and further pointed out that these repeating units are of two types, **monopartite** and **multipartite**. The membrane without projections is called monopartite; for example, the outer membrane of mitochondria is monopartite membrane. But in multipartite membrane, repeating units project from the membrane surface similar to the inner membrane of mitochondria (Figure 2.8a). Each repeating unit has a base, a stalk and a head (Figure 2.8b).

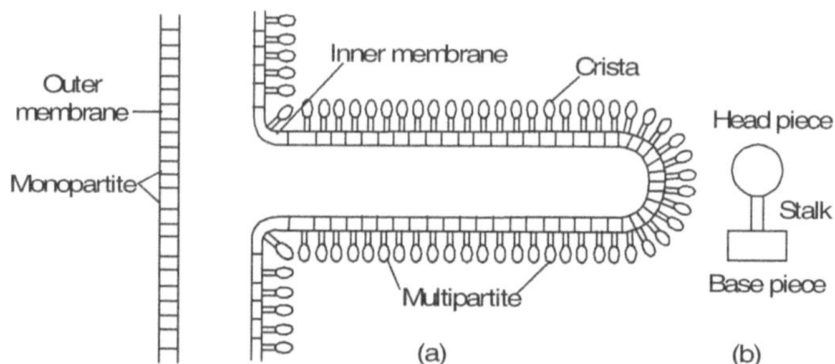

Figure 2.8   Green's model for monopartite and multipartite membranes

Lucy (1964) proposed another hypothetical membrane model known as micellar model, based on the concept that a micelle is a colloidal aggregation of molecules. In this model for natural membranes, the globular lipid micelle can be in a dynamic equilibrium with the bimolecular lipids. Electron microscopic study provides the basis for both double-layered structure as well as the micellar (subunit) structure. Hence it becomes difficult to generalize the basic structure of the membrane. Figure 2.9 shows that in nature there may be the transformation of bilayer structure into micellar structure under certain cases.

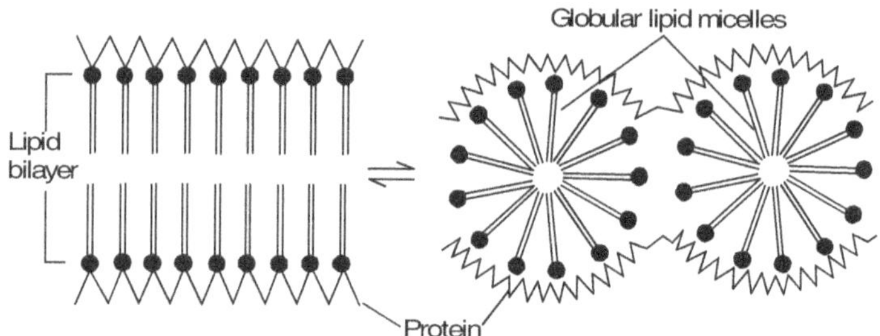

Figure 2.9   Transformation of bilayer structure into globular micellar form

FUNCTIONS OF BIOLOGICAL MEMBRANE

The important functions of membranes are as follows:

1. **Compartmentalization**   Membranes are living boundaries of the cytoplasm and cellular inclusions. They form continuous and unbroken sheets that divide the living matter into self-sustaining units to perform

specialized functions. Plasma membrane encloses the entire cellular contents, while nuclear and cytoplasmic membranes enclose various internal cellular spaces where they regulate different activities.

2. **Acting as a selectively permeable barrier**　In a living cell, there are several substances that enter and leave the cell. However, if the entry and exit of the matter is not regulated, it will be disastrous for the cell. Therefore plasma membrane regulates the entry of appropriate substances into cytoplasm and ensures that inappropriate substances are kept out. In this way it acts as a selectively permeable barrier.

3. **Transporting machinery for metabolic essentials**　Plasma membrane contains the machinery for physically transporting substances, for example, gases, solutes, and other biochemicals from one side of the membrane to the other. There are varieties of processes like passive diffusion, facilitated diffusion, active transport, pinocytosis, etc. involved in transport of metabolic essentials across the membrane. Plasma membrane is also able to establish ionic gradient across itself, as in the case of nerve and muscle cell, and plays a vital role in allowing the cell to give response to a stimulus.

4. **Responding to stimuli**　All biological membranes have on their surfaces specific protein molecules acting as receptors, which unite with complementary substances or ligands providing external stimuli to the cell. The complex of receptor and ligand initiates the cellular response through a cascade of reactions. The well-studied ligands are hormones, growth factors and neurotransmitters.

5. **Interaction between cells**　The plasma membrane which is the living boundary of the cell, mediates the interaction between the cells of multicellular organisms. To achieve harmony in the system, the plasma membrane allows the cells to coordinate with one another and to exchange materials and information.

6. **Site for biochemical activities**　Cells have to perform different biochemical activities in the presence of enzymes, and membranes provide the structural framework for the location of various enzymes. As a result, there is an effective coordination of biochemical reactions. The presence of an orderly enzymatic system in the mitochondrial membrane to achieve electron transport is one of the best examples to understand the significance of membranes.

7. **Association with energy transformation**　Membranes are closely associated with conversion of one form of energy to another. For example, in green plants, membrane-bound pigments transform light energy into chemical energy in the form of carbohydrates. Membranes also have the ability to convert carbohydrates into ATP—the energy-rich compound. The

machinery for the capture of energy and its conversion is present in the membranes of chloroplasts and mitochondria.

8.   **Other functions**   In addition to the above functions, membranes provide mechanical strength, assist in the movement of the cell, act as electrical insulators, etc.

MOVEMENT OF SUBSTANCES ACROSS CELL MEMBRANE

As mentioned earlier, in order to survive, the cell must interact with its environment. Metabolic essentials such as oxygen, salts, and nutrients must get into the cell, and waste products such as carbon dioxide and other metabolites must pass out of the cell. Since the contents of a cell are completely surrounded by its plasma membrane, all communication between cell and its extracellular environment are mediated by the cell membrane. Thus, plasma membrane has a dual responsibility. On one hand, it must retain the dissolved materials of the cell so that they do not simply leak out into the environment, while on the other hand, it must allow necessary exchange of materials into and out of the cell. The plasma membrane is a **selectively permeable barrier**, i.e., it is not equally permeable to all types of solutes. This selectivity allows the concentration of substances inside the cell to be strikingly different from that on the outside, a condition necessary for the sustainability of the life of the cell.

There are variety of processes by which the essential substances move across the cell membrane. Some of these are discussed in the following section.

Passive Diffusion

It is a spontaneous process in which water, oxygen, or carbon dioxide move through the cell membrane from a region of high concentration to a region of low concentration, eventually eliminating the difference in the concentration between the two regions. The rate and speed of diffusion is generally proportional to the difference in concentration of substance between the two areas. When the concentration in both regions becomes equal, the net diffusion ceases.

The movement of water molecules takes place much more rapidly through a cell membrane than do dissolved ions or small polar solutes that are commonly present in the cell and are essentially non-permeable. Because of this difference in the penetrability of water verses solutes, membranes are said to be **semipermeable**. Water moves readily through a semipermeable membrane from a region of a lower solute concentration to a region of higher solute concentration, the phenomenon being known as **osmosis**. Thus, osmosis is the diffusion of water across the semipermeable membrane that can be demonstrated by placing a cell into a solution containing a solute concentration different that from present within the cell itself (Figure 2.10). When the cell is placed in **hypotonic** solution, there is movement of water molecules from lower solute concentration (outside the cell) to higher solute concentration (within the cell) through the semipermeable membrane. As a result, the cell rapidly gains water and swells, when it is called **endosmosis**. Conversely, when the cell is placed in a **hypertonic** solution (a solution having higher solute concentration than that present within the cell), the cell loses water by **exosmosis** and as a result it shrinks. The solution is said to be **isotonic**, when there is similar solute concentration outside and inside the cell, and hence the cell placed in such solution shows no net loss or gain of water molecules and there is no change in its size.

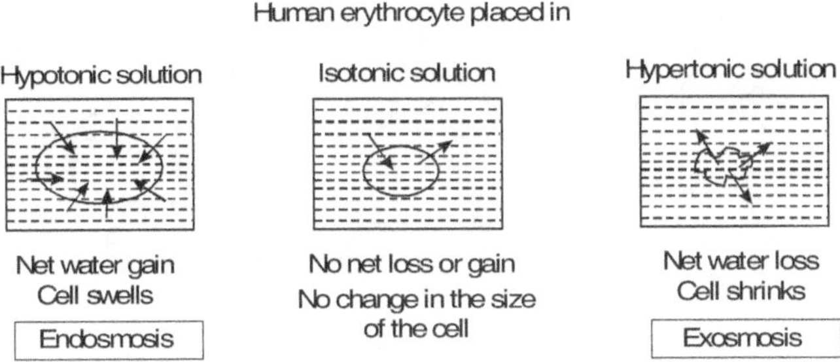

Figure 2.10   Osmosis. The effect of differences in solute concentration on either sides of the plasma membrane. Arrow indicates the movement of water molecule across the cell membrane.

Facilitated Diffusion

Essential substances like glucose and amino acids move across the membrane as a function of a concentration differential. However, these substances do so only up to a certain concentration level. When the concentration of glucose outside the cell increases to a certain level, the rate of diffusion no longer accelerates. This is due to saturation of certain specific reactive sites with the glucose molecules on the cell surface. This theory proposes that there are specific membrane-associated carrier molecules that form a complex with glucose and facilitate the diffusion of glucose into the cell through the membrane. Since glucose is the body's primary source of direct energy, most cells contain a membrane protein (glucose transporter) that facilitates the diffusion of glucose from the bloodstream into the cell (Figure 2.11).

Active Transport

Sometimes a substance(s) is required to be transported through the membrane against the concentration gradient and hence there is expenditure of energy.

This is known as active transport. Like facilitated diffusion, active transport depends on integral membrane proteins that are capable of selectively binding a particular substance and moving that substance across the membrane, driven by changes in protein's conformation. Unlike facilitated diffusion, however, movement of substance against a gradient requires the expenditure of energy (Figure 2.11).

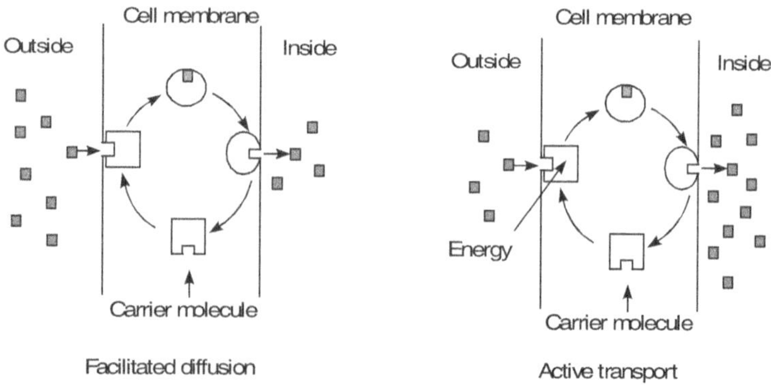

Figure 2.11   Movement of substance across the cell membrane. Facilitated diffusion uses carrier molecule and works with a concentration gradient; however, it may cease before equilibrium is reached. Active transport works against a concentration gradient and requires a carrier molecule and expenditure of energy. The number of dark squares indicates the differences in concentration. In both cases, there is change in configuration of carrier molecule that allows rapid substrate diffusion.

Active transport often involves charged ions, such as Na^+ and K^+, and electrolytes. These substances do not necessarily move across the cell membrane in response to a concentration differential. In fact, active transport is a mechanism that promotes the movement against a concentration gradient and results in build-up of a marked concentration differential. In the case of Na^+ and K^+, the cell maintains a higher concentration of Na^+ outside and vice versa for K^+. The Na^+ concentration is about 150 mM on the outside of the cell and 10 to 20 mM inside the plasma membrane, while the K^+ concentration inside a mammalian cell is about 100 mM and only about 5 mM outside the cell. The concentration difference for Ca^{2+} is even greater; in the typical cell, the cytoplasmic concentration of Ca^{2+} is 10^{-7} M, which is 1,000 to 10,000 times less than that outside the cell. The ability of a cell to generate such a steep concentration gradient across its plasma membrane cannot be achieved either by simple or by facilitated diffusion. It is possible only by active transport.

Since there is movement of substances across the plasma membrane against their concentration gradient, the membrane protein involved in active transport requires a "driver" to provide energy, and this is achieved by the hydrolysis of ATP. In 1957, Jens Skou, a Danish physiologist, discovered an ATP-hydrolysing enzyme in the nerve cells of crab that was active only in the presence of both sodium and potassium ions (as well as Mg^{2+}, which acts as a cofactor). Since the enzyme was responsible for ATP hydrolysis and active in transporting Na^+ and K^+, it was called the Na^+–K^+ ATPase, or the sodium–potassium pump. It is found only in animal cells. It is thought that sodium–potassium pump evolved in primitive animals as the primary means to maintain cell volume and as the mechanism to generate the steep Na^+ and K^+ gradient that play significant roles in the formation of impulses in nerve and muscle cells.

In plant cells, there is H^+-transporting, a P-type pump ("P" stands for "phosphorylation" of carrier protein present in plasma membrane) that plays a key role in the transport of solutes, in the control of cytoplasmic pH, and possibly in the control of cell growth by means of acidification of plant cell wall. Another well-studied P-type pump is the Ca^{2+}-ATPase present in both plasma membrane and the membranes of endoplasmic reticulum (ER). This Ca^{2+}-pump plays an important role in transport of calcium ions out of the cytoplasm into either the extracellular space or the lumen of the ER.

The carrier molecule called transporter mediates both facilitated diffusion and active transport. The transporters are made up of proteins that are encoded by the genes. Therefore, there are certain human genetic disorders, which are due to defective protein carrier molecules. Cystic fibrosis is one of such diseases in which the patient produces a thickened, sticky mucus that is very hard to propel out of the respiratory tract. As a result, these individuals typically suffer from chronic lung infections, which progressively destroy pulmonary function. The condition is due to a defect in the gene that codes the ATP-binding cassette (ABC) transporter, which is present in the epithelial cell membrane of lung. The defective ABC transporters cannot regulate the movement of water and chloride ions across the cell membrane. There is decrease in chloride efflux from the cells, which leads to an increase in the concentration and hence viscosity of bodily secretions.

Bulk Transport

Sometimes the entry of macromolecules such as proteins inside the cytoplasm of the cell is very much essential. But because of the large size and ionic charge, the passage of proteins across the cell membrane is very much limited. They do not transverse the lipid bilayer directly but do so indirectly via membrane-bound vesicles. In endocytosis, the cell membrane invaginates and fuses around an extracellular macromolecule (ligand), internalizing it into the cytoplasm of the cell. Material taken up by endocytosis is usually delivered to a network of tubules and vesicles that are collectively called endosomes. These are further classified as the early endosomes, which are located near the peripheral region of the cell and the late endosomes, which are typically located closer to the nucleus. Materials taken up at the cell surface by endocytosis are first transported to the early endosomes, where the material is sorted out. Some materials that enter the early endosomes are sent to the late endosomes, others are sent back to the plasma membrane, and some of the substances are conveyed towards the Golgi complex. The late endosomes can be considered as "prelysosomal" compartments since they have acidic pH and low concentration of lysosomal enzymes that had been synthesized in rough endoplasmic reticulum and processed in *trans* Golgi network. The endosomal pathway is shown in Figure 2.12

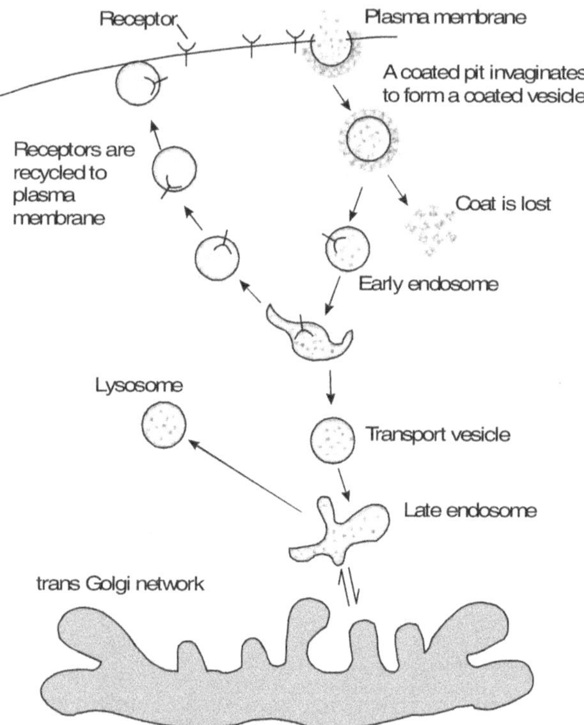

Figure 2.12   Diagrammatic representation of endosomal pathway

Conversely in exocytosis ("cell vomiting"), an intracellular vesicle containing a substance targeted for extracellular release fuses with the cell membrane and then releases its contents to the outer medium. This process of membrane fusion and content discharge is called exocytosis. A well-known example of this process is the release of neurotransmitter molecules by exocytosis in the presence of Ca^{2+} ions. The arrival of nerve impulse at the terminal knob of a neuron leads to an increase in the influx of Ca^{2+} ions and as a result, secretory vesicles containing neurotransmitters discharge their contents into the synapse.

When the cell has to take up extracellular particles greater than 0.2 μm in diameter, the endocytotic vesicle forms tightly around the particles, excluding most of the extracellular fluid. This process is **phagocytosis** ("cell eating"). Phagocytosis is carried out extensively by a few types of cells specialized for the uptake of particulate material from the environment and its delivery to lysosomes. In single-celled organisms such as amoebae and ciliates, ingestion of food occurs almost entirely by this mechanism. They trap the food particles and smaller organisms and enclosing them within folds of the plasma membrane. The folds fuse to form a vacuole (**or phagosome**) that pinches off from the plasma membrane. The phagosome fuses with a lysosome, and the material is digested intracellularly in the presence of the lysosomal digestive enzymes. In mammals, **macrophages** phagocytose inert particles, bacteria, antigen–antibody complexes, and even dead cells. Macrophages are present in blood sinusoids of the liver where they clear unwanted material. In lungs, they phagocytose foreign particles that are present in inhaled air and maintain sterility in airways and alveoli. Neutrophils (leucocytes having polymorphic nucleus), travel through the bloodstream, perform phagocytosis and protect the body from infecting bacteria.

Pinocytosis ("cell drinking") which is the process of cellular uptake of smaller particles (e.g. proteins), also involves vesicles, and some extracellular fluid. It occurs in many cell types and in many organisms. In 1925, for the first time the formation of a small vacuole at the cell surface was noted in amoeba. Electron microscopy has revealed the process in many mammalian tissues, most notably the endothelium, which lines blood vessels, kidney tubule cells, and the placenta. Under the electron microscope, a cross-section of a small-calibre capillary shows a single layer of endothelium that separates the blood from the surrounding tissue. Numerous pinocytotic vesicles are present in the plasma membrane of endothelial cells, which show selectivity in the uptake of the material. For example, serum albumin, a protein indispensable for maintenance of blood osmotic pressure, is excluded, but albumin glycosylated with glucose does enter the cells. This is clinically significant because the high blood glucose levels in diabetes can lead to non-enzymatic glycosylation of proteins such as haemoglobin and albumin. If proteins are abnormally taken up by pinocytosis in the endothelial cells, the structural integrity of capillary could be damaged, a situation often seen in diabetic patients.

SUMMARY

- The cytoplasm in both plant and animal cells is bounded by plasma membrane or plasmalemma, while the cell membrane is the limiting membrane of the cell organelles. Both of them are collectively known as biological membranes.

- Chemically, plasma membrane is mainly composed of lipoproteins—a complex of lipids with proteins.

- Cell biologists proposed different models for plasma membrane, which are classified into two basic categories: bilayer models where the proteins and lipid are arranged in the form of layers and micellar or subunit models in which there is the presence of number of small and similar independent subunits.

- E. Gorter and F. Grendal proposed the Lipid Bilayer Model for the plasma membrane.

- Unit Membrane Model was proposed by Robertson who found that all biological membranes appear to be made up of a lipid layer having the thickness of 35 Å and a dense band of proteins (20 Å thick) present on either side.

- James Danielli and Hugh Davson proposed that the membrane is a bimolecular lipid layer, with the polar ends of the lipids facing outward and hydrophilic (water-loving) proteins coating the polar ends.

- The Fluid Mosaic Model was proposed by Singer and Nicolson, according to which the lipid molecules are present in a fluid state capable of rotating and moving laterally within the membrane. And the proteins occur as a "mosaic" of discontinuous particles that penetrates deeply into the lipid sheet.

- Green described the concept of micellar model, where the plasma membrane might consist of closely packed repeating units, which may be of monopartite and multipartite type.

- Plasma membrane is a remarkably thin, delicate structure, yet it plays a key role in many of the cell's most important functions. It separates the living cell from its environment; it provides a selective barrier that allows the exchange of certain substances, while preventing the passage of others.

- All biological membranes contain transport machinery that physically transport substances from one side of the membrane to the other; they also show response to specific stimuli via receptors which bind with specific ligands in external space and relay the information to the cell's internal compartments.

- Plasma membrane mediates interactions with other cells; it provides a framework in which components can be organized; it also acts as a site where energy is transduced from one type to another.

REVIEW QUESTIONS

1. What is plasma membrane? Explain briefly its chemical composition.

2. Describe the structure and properties of different models of plasma membrane proposed by various cell biologists.

3. Why is membrane fluidity important to a cell?

4. How can the two sides of a plasma membrane have different ionic charges?

5. List some of the significant functions of biological membranes in the life of a eukaryotic cell.

6. Explain the effects of putting a cell into a hypotonic, hypertonic, or isotonic medium.

7. How does the Na–K–ATPase complex illustrate the sidedness of the plasma membrane?

8. Write short notes on:

    i. Unit membrane model of plasma membrane

    ii. Singer and Nicolson's fluid mosaic model

    iii. Selective permeability of the plasma membrane

    iv. Facilitated diffusion

    v. Active transport

    vi. ATP-binding cassette (ABC) transporter

    vii. Endocytosis and exocytosis

    viii. Phagocytosis and pinocytosis

CYTOPLASM

Cytoplasm is the colloidal material present in the cell bounded by the plasma membrane. It includes several substances either in dissolved or suspended condition. Several theories are there to explain the structure of cytoplasm; some of them are Butschill's alveolar theory, Altman's granular theory, and Fisher and Flemming's fibrillar theory. The physical and biochemical properties of cytoplasm are as follows:

Physical Properties of Cytoplasm

The various physical properties of the cytoplasm are due to its colloidal nature. A colloidal system can be defined as a system of liquid medium containing suspended particles whose diameter ranges between 10^{-6} to 10^{-4} nm. The following are some physical properties.

1. **Viscosity** The suspended particles present in the cytoplasm are responsible for the viscous nature of the cytoplasm.

2. **Elasticity** As per the necessity, the cytoplasm can expand or contract to a certain extent.

3. **Cohesiveness** The suspended particles have mutual attraction; hence there is cohesiveness in the cytoplasm.

4. **Phase reversal** The cytoplasm can be differentiated into interchangeable sol and gel states where sol is the liquid state and gel is the semi-solid state.

5. **Brownian movement** The suspended particles of the cytoplasm are in the state of to and fro movements.

6. **Cyclosis** In certain cells, the cytoplasm performs streaming movements, which are known as cyclosis as in the case of *Paramecium* and in the leaf cells of *Elodea*.

Biological Properties

Some of the important biological activities of the cytoplasm are growth, reproduction, respiration, digestion, assimilation, metabolism, excretion, irritability, etc.

Chemical Nature of Cytoplasm

Cytoplasm contains about 90% of water and remaining 10% is constituted by organic and inorganic substances.

The organic substances include carbohydrates, proteins, amino acids, fatty acids, glycerol, hormones, vitamins, and nucleic acids.

The common inorganic constituents are oxygen (60%), carbon (18%), hydrogen (10%), and nitrogen (3%) while the less common inorganic constituents are calcium, phosphorus, sodium, magnesium, iodine, and chlorine.

Cytoplasmic Inclusions

There are a number of living and non-living inclusions present in the cytoplasm. The living inclusions are nucleus, mitochondria, plastids, endoplasmic reticulum, Golgi complex, lysosomes, centrioles, microbodies, etc.

NUCLEUS

It is the most prominent organelle within the cell that was first discovered by Robert Brown in 1831, in an orchid cell. All metabolically active eukaryotic cells have a nucleus except for the brief period when they are dividing. The nucleus is also absent in matured erythrocytes in mammals. Similarly the phloem cells that conduct sugar in plants are anucleated. Bacteria and blue-green algae do not have true nucleus.

POSITION OF NUCLEUS

In most of the cells, nucleus is usually present at the centre of the cell. Sometimes cell inclusions or cell secretions accumulate inside a cell and nucleus may be shifted to an eccentric position. In some cases, it is peripheral or it may occupy various positions as in *Acetabularia*.

SHAPE, SIZE AND NUMBER

Since the nucleus has immense importance in storage and utilization of genetic information, the nucleus of eukaryotic cells has a rather undistinguished morphology (Figure 2.13). Depending on the type of the cell, the shape of the nucleus is variable. Generally its shape is rounded, oval, spherical, ellipsoidal, or disc shaped. The nucleus has smooth appearance or its surface may be drawn out into lobes and shows various forms (polymorphic nucleus) as in leucocytes. It may be elongated and horse-shoe shaped as in *Vorticella*.

Depending on the cell, the size of the nucleus is also variable. Its size is directly proportional to the cytoplasm. In a particular cell, the size of the nucleus can be determined by applying Heywig's formula, which is as follows:

$$NP = \frac{V_n}{V_c - V_n}$$

where,

NP = Nucleoprotein index,

V_n = Volume of the nucleus and

V_c = Volume of the cell.

Most of the cells are uninucleated, while some protozoans have binucleated (e.g. *Paramecium*) or multinucleated (e.g. *Opalina*) condition and similar multinucleate cells have been reported in *Cladophora*, *Pithophora*, lactiferous cells in angiosperms, etc. In certain fungi like *Mucor*, where there are no septa dividing the plant body into cells, several nuclei occur making the condition coenocytic.

ULTRASTRUCTURE OF NUCLEUS

A typical nucleus in its ultrastructure shows the following components:

1. Nuclear envelope,
2. Nuclear sap,
3. Chromatin, and
4. Nucleolus

Nuclear Envelope

It is a limiting membrane of the nucleus in all eukaryotic cells. It is responsible for the effective communication between the nucleus and cytoplasm. The contents of nucleus are present as a viscous, amorphous mass of the material enclosed by a complex nuclear envelope. The separation of the genetic material of the cell from the surrounding cytoplasm is one of the most significant features that distinguish eukaryotes from prokaryotes. The appearance of nuclear envelope is a landmark in biological evolution.

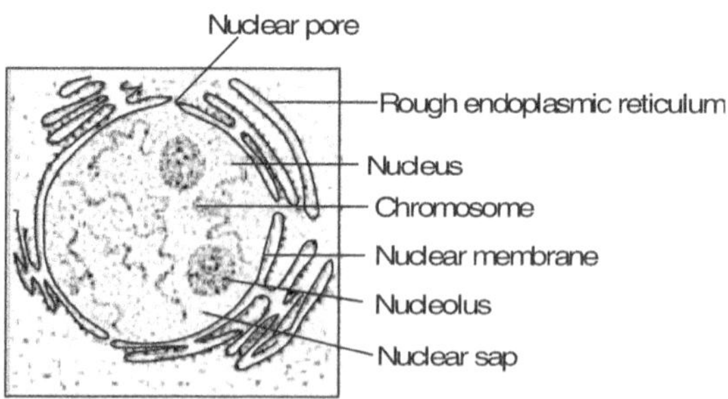

Figure 2.13 Diagrammatic representation of nucleus and its associated structures

The nuclear envelope or karyotheca is a double-membrane structure (Figure 2.13) separated by perinuclear or intermembrane space of 10 to 50 nm. The inner and outer membranes of the nuclear envelope have the thickness of about 70 to 80 Å and are mainly composed of proteins and lipids. On its cytoplasmic surface, the outer membrane is usually studded with ribosomes and it is often seen to be continuous with the membrane of the rough endoplasmic reticulum, whereas the inner membrane on its nucleoplasmic side is lined by a dense fibrillar network, known as **nuclear lamina** of varying thickness (about 100 nm), that which gives structural support to the nuclear envelope and serves as a site for the attachment of chromatin fibres at the nuclear periphery.

Though both membranes are separated by the perinuclear space for most of the area, they fuse with each other at irregular intervals leading to develop **nuclear pores** (Figure 2.14), which form a distinctive feature of the nuclear envelope. E. Hertwig first observed nuclear pores in 1876 using the light microscope. Electron microscopic study revealed that each nuclear pore has a diameter of about 60 to 90 nm and the number of pores varies between cell type and species from 1 to 100 per square micrometer of nuclear surface. The average mammalian cell consists of approximately 3,000 nuclear pores.

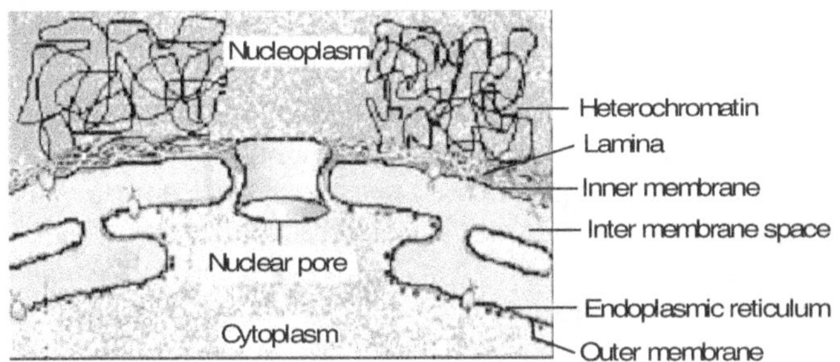

Figure 2.14 Schematic diagram showing the double membrane, nuclear pore, nuclear lamina, and continuity of outer membrane with rough endoplasmic reticulum

Nuclear Pore Complex

The nuclear envelope acts as barrier between the nucleus and the cytoplasm, while the nuclear pores are the gateways across that barrier. Afzelius (1955) first thought that a diaphragm across covered the nuclear pores. However a cross section passing through nuclear pore seen under the electron microscope shows that a sort of ring or hollow cylinder is fitted in the pore. This hollow cylinder is called annulus composed of electron-dense material. Nuclear pore contains a complex, basket-like apparatus called the **nuclear pore complex** as shown in the Figure 2.15. This complex fills the pore like a stopper, projecting into both cytoplasm and nucleoplasm.

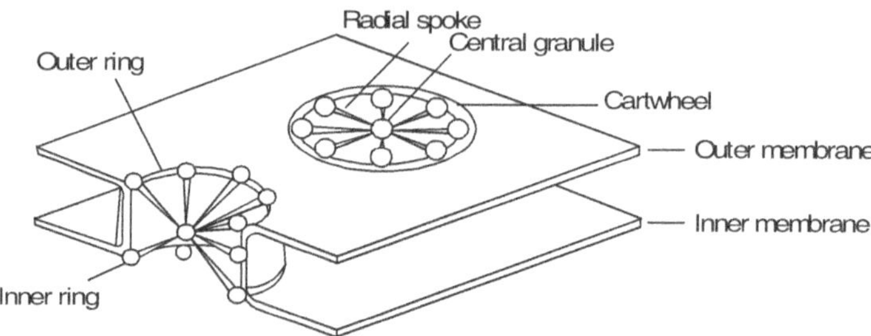

Figure 2.15   Diagrammatic representation of nuclear pore complex

W. Franke, *et al.* (1970) suggested that the nuclear pore complex includes two distinct annuli, an outer annulus and an inner annulus. Both annuli are made up of 8 to 9 granules having a diameter of about 100 to 250 Å. Both the rings fit over the outer and inner ends of the pore, one on the cytoplasmic side and the other on the nucleoplasmic side. In between the two rings, the side walls of the pore show finger-like projections, which are due to peripheral granules. Within the pore, a central granule with a diameter of about 40 to 300 Å is present that acts as a transporter. The central granule is suspended with annular granules by radiating fibrils producing a cartwheel-like structure.

The nuclear pore complex is a huge molecular complex showing octagonal symmetry that contains granules with an estimated 100 to 200 polypeptides. The cytoplasmic face of the nuclear envelope shows the peripheral cytoplasmic granules, while its nuclear face shows a basket-like appearance of the inner portion of the complex. The nuclear lamina is attached to the innermost side of the nuclear pore complex.

Functions of Nuclear Envelope

Since the nuclear envelope acts as a dynamic barrier between the cytoplasm and nucleus of a cell, it regulates the exchange of electrons, ions and molecules of various compounds. Besides this, the nuclear envelope also provides the attachment site for the chromatin material.

Exchange of materials   Many substances have to cross the nuclear envelope to reach the nucleus from the cytoplasm and vice versa. These substances are ions and small molecules, for example, nucleotides, macromolecules such as nucleoproteins and aggregates of ribonucleoprotein particles. The passage of these molecules particularly depends on their size. Molecules having a diameter of about 145 Å, i.e., smaller than the size of the pore can pass through. Small organic substances and ions easily diffuse through the nuclear pores. The transport of the molecules across the nuclear envelope occurs in two steps:

1. **Binding**   The transport molecule first binds to proteins at the nuclear pore complex. This event requires the nuclear signal sequence and this signal binding is a chemical affinity process.

2. **Translocation**   This step requires the continued presence of ATP, suggesting that some active transport and the nuclear-membrane-bound Mg^{2+}-ATPase may be involved.

Blebbing   In certain cases, the nucleocytoplasmic exchange could be directly through the nuclear bilayers, via the continuity with the cisternae of the endoplasmic reticulum or by pinching off or fusing with the nuclear envelope. Small sacs are developed by outpushings of the nuclear envelope, which are then cut off and form small vesicles. This process is called **blebbing**. These vesicles thus released from the annulate lamellae in turn form cisternae of the endoplasmic reticulum.

Transportation via endoplasmic reticulum   The intermembrane space present in the nuclear envelope is also called perinuclear cavity that is continuous with the cavity of cisternae of the endoplasmic reticulum. It is suggested that ribosomes pass through the inner membrane of the nuclear envelope into perinuclear space and are then distributed through the cisternae of the endoplasmic reticulum.

Other functions   The nuclear envelope provides the sites of attachment for the chromatin fibres particularly heterochromatin, which probably suggests that annuli are the centres for the synthesis of DNA.

In addition, the enzymes for the transport of electrons are reported to be present in the nuclear envelope. Since their concentration is similar to that in mitochondria, it is suggested that electron transport and phosphorylation occur in the nuclear envelope.

Nuclear Sap

It a semi-fluid substance present in the nucleus and variously known as **nuclear matrix, karyoplasm** or **nucleoplasm**. It is bounded by the nuclear envelope and contains nucleoproteins, nucleic acids, enzymes and minerals. In plant cells, the nucleoplasm probably contributes to formation of spindles. When the nucleus is subjected to a moderately high temperature (37°C), the nuclear envelope is solubilized with a detergent, chromatin is solubilized with a high salt concentration, and nucleic acids are digested by nucleases, the network of fibrils retaining the shape of the nucleus remains. This network of fibrils is known as **nuclear matrix**, which consists of three components: a residual envelop with pore complex, a residual nucleolus, and an internal matrix that is similar to the nucleoplasm of the intact nucleus.

The nuclear matrix has structural and functional aspects. In an intact condition, over 50% of nuclear DNA remains attached to a fibrillar component of the nuclear matrix. And this is often the DNA that is transcribed to form the proteins. Thus nuclear matrix is associated with transcription and it can be supported by the fact that most of the nascent RNA in the nucleus are found attached to the matrix where much of the post-transcriptional modifications of RNA take place. In addition, the initiation of replication of DNA occurs on a component of the nuclear matrix. Thus regulation of transcription and replication of DNA may also occur on the nuclear matrix.

SUMMARY

- The cytoplasm of the eukaryotic cell is the colloidal material that is bounded by the plasma membrane.

- Cytoplasm shows various physical properties including viscosity, elasticity, cohesiveness, phase reversal, Brownian movement and cyclosis, which are due to its colloidal nature.

- Nucleus is present in all metabolically active eukaryotic cells while it is absent in matured erythrocytes in mammals, and also in bacteria and blue-green algae.

- A typical nucleus consists of nuclear envelope, nuclear sap, chromatin, and nucleolus.

- Nuclear envelop is a limiting membrane of the nucleus in all eukaryotic cells. It acts as a dynamic barrier between cytoplasm and nucleus of a eukaryotic cell and also provides the attachment site for the chromatin material.

- Nuclear pore complex is a basket-like apparatus that fills the pore like a stopper and projects into both cytoplasm and nucleoplasm.

- Nucleocytoplasmic exchange could occur through blebbing to form small vesicles, which in turn form cisternae of endoplasmic reticulum.

- Karyoplasm or nucleoplasm is a semi-fluid substance present in the nucleus which contains nucleoproteins, nucleic acids, enzymes and minerals.

REVIEW QUESTIONS

1. What are the different physical and chemical properties of cytoplasm?

2. Is there any difference between nucleus of prokaryotic and that of eukaryotic cell?

3. Describe the structure of nuclear pore complex.

4. Explain the structure and functions of nuclear envelope.

5. Define karyoplasm. Describe the chemical composition of nucleoplasm.

6. Write short notes on:

    i.   Nucleoprotein index

    ii. Nuclear lamina

    iii. Cyclosis

    iv. Blebbing

    v.   Nuclear matrix

CHROMATIN

Enclosed in the nucleoplasm and visible in the interphase nucleus are found a number of fibrillar structures that constitute a network called **chromatin fibrils or chromonemata**. During cell division, the chromatin material condenses to form species-specific number of chromosomes. The term "**chromosome**" is used to describe the nucleic acid molecule that is responsible for genetic information in a virus, bacterium, eukaryotic cell, or cell organelle. It also refers to the densely coloured bodies seen in the eukaryotic cell stained with a nuclear stain like basic fuchsin.

In eukaryotes, individual chromosomes contain a single DNA molecule and because of the huge size (an average human diploid cell contains a nucleus of $10\,\mu$m diameter in which 6 billion base pairs of DNA having a length of about 2 metres divided in 46 chromosomes with about 33,000 genes are packed) it is folded in some orderly fashion. Even during the interphase of the cell cycle (the G1, S, and G2 phase), some folding is necessary simply to pack such long molecules into the nucleus of cell. There is slight difference in the pattern of folding in different regions of genome that play a significant role in determining which genes are active. The folding of DNA is attributed to the presence of specific proteins in the nucleus. **Chromatin** is a tight complex formed between eukaryotic DNA and nuclear protein. The key proteins in chromatin involved in orderly packaging of eukaryotic DNA are the **histones**—the basic proteins that contain a high proportion of positively charged amino acids (lysine and arginine). These proteins bind tightly to negatively charged DNA. There are five basic classes of histone molecules known as **H1, H2A, H2B, H3** and **H4**. They have a slightly different composition and molecular weight but occur in equal number, except for H1, which is present at one-half the level of each of the other (Table 2.1).

Table 2.1   Properties of histones associated with chromatin

Histone	Composition	Number of residues	Molecular weight	Relative molar concentration
H1	Rich in lysine	215	21,000	1
H2A	Slightly rich in lysine	129	14,500	2
H2B	Slightly rich in lysine	125	13,700	2
H3	Rich in arginine	135	15,300	2
H4	Rich in arginine	102	11,300	2

When interphase nucleus spread preparation is observed under the electron microscope, it appears like a "**Beads-on-a string**" configuration (Figure 2.16a). The chromatin has a beaded appearance built up of spherical, 11-nm particles that are connected by the thin fibre made up of linker DNA (Figure 2.16b). The DNA in a chromatin is very tightly associated with histones that package and order the DNA into repeating structural units called **nucleosome**. The amount of DNA tightly folded with each of nucleosome bead is approximately 200 base pairs (Figure 2.16c). The number of base pairs associated with each bead varies from about 150 to 250 and it mainly depends on the organism and the type of the tissue under investigation. If the linker DNA that connects adjacent nucleosomes is cut off from either side by the action of nuclease, it yields a highly protected DNA segment of 146 base pairs (Figure 2.16d).

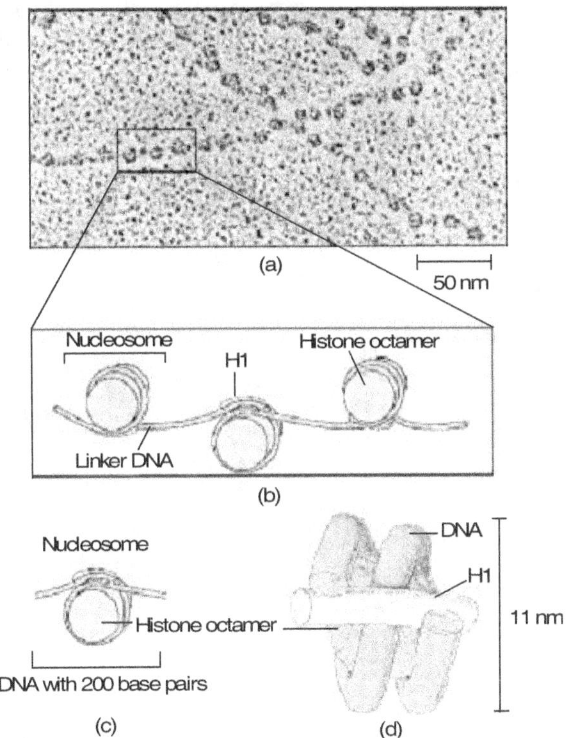

Figure 2.16   (a) An electron micrograph of chromatin showing "Beads-on-a-string" configuration. (b), (c), and (d) Diagrammatic representation of magnified view of nucleosome and associated structure.

Each nucleosome contains a **nucleosome core particle**, having 146 base pairs of supercoiled DNA wrapped almost twice around a wedge-shaped complex of eight histone molecules. The histone core complex is called **histone octamer**, since it encloses two molecules each of histones H2A, H2B, H3 and H4. The nucleosome core particles are about 11nm in diameter and the naked linker DNA is about 2 nm in diameter, appearing as a "beads-on-a-string" form under the electron microscope. One molecule of H1 histone is normally located just outside of each nucleosome core particle and is associated with both ends of DNA as it enters and leaves from the core particle as shown in Figure 2.16 (d). In addition to histones, the **non-histone chromosomal proteins** are also associated with chromatin. The non-histone proteins are acidic (negatively charged at neutral pH) and include a large number of widely diverse structural, enzymatic (like DNA and RNA polymerases) and regulatory proteins more or less firmly attached to the repeating nucleosome array. They are likely candidates for roles in regulation of expression of specific genes or sets of genes. Non-histone proteins do not take part in the basic chromatin structure; their abundance and identity can vary greatly from one cell type to other in the same organism.

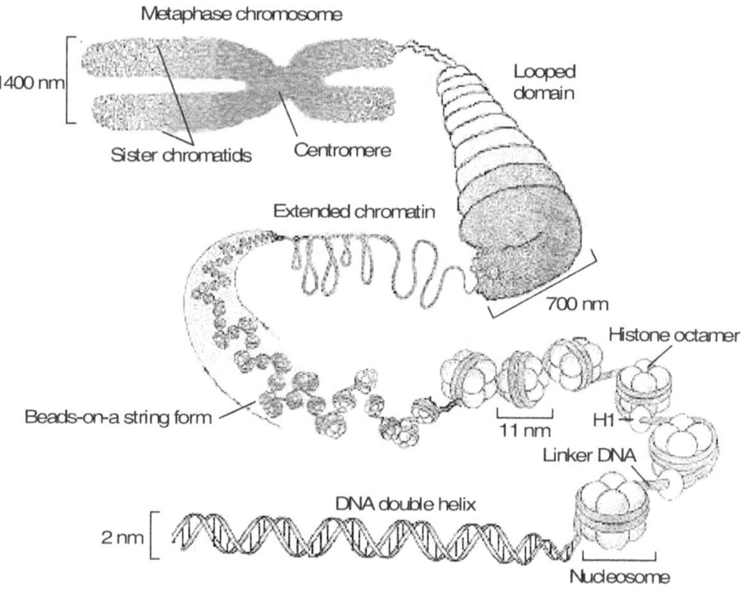

Figure 2.17   The levels of organization of chromatin

The DNA molecule wrapped around the nucleosome core particle forms the fundamental organizational unit of chromatin. In the next stage of its hierarchy, DNA molecules are further packed into 30-nm chromatin filaments and then into supercoiled loops. Each loop contains about 10 to 150 kilobases of DNA and is linked at its base by specific proteins. Finally supercoiled loops condensed to form metaphase chromosomes (Figure 2.17). The non-histone chromosomal proteins form a "scaffold" that is involved in condensation of the 30-nm chromatin fibre into tightly packed metaphase chromosomes. The mitotic chromosome represents the higher level of organization of chromatin and the ultimate in its compactness. A mitotic chromosome having a length of 1 μm typically contains approximately 1 cm of DNA.

HETEROCHROMATIN AND EUCHROMATIN

Once mitosis has been completed, most of the chromatin that composes the highly compacted chromosomes becomes dispersed and resumes its diffuse interphase condition. However in most of the cells, about 10% of chromatin material remains condensed throughout the interphase and visible in the light microscope, and such chromatin stains deeply. In 1929, E. Heitz used the term **heterochromatin** to describe such regions of chromatin. It is different from **euchromatin**, which is a less condensed region of chromatin and stains faintly. Heterochromatin is the metabolically inert region of the chromatin since it has little or no transcriptional activity, while euchromatin region is relatively more active in the synthesis of mRNA.

Depending on the permanence of compactness, heterochromatin is classified two types, viz. constitutive heterochromatin and facultative heterochromatin. Constitutive heterochromatin remains always condensed in all cells. It is present in mammalian mitotic cells near the centromeres or at the end of each chromosome. The base sequences in the DNA of constitutive heterochromatin are highly repetitive and they do not transcribe to form protein, whereas facultative heterochromatin is condensed for only part of the life of the cell and specifically inactivated during certain phases in the life of organism. Its best example is the mammalian **sex chromatin**. In mammalian male cell, there is a much larger X chromosome and a tiny Y chromosome. Both sex chromosomes have very few genes in common. In the female cell, there are two X chromosomes and only one of the pair is active and transcribes to form mRNA and then proteins. The other X chromosome remains condensed and inactive as heterochromatin as a darkly stained structure in the nucleus of female cell (Figure 2.18), called Barr body after the researcher who discovered it in 1949.

Inactivated euchromatin is essentially a type of heterochromatin in which the DNA sequences are heterochromatic for a period, then they are euchromatic,

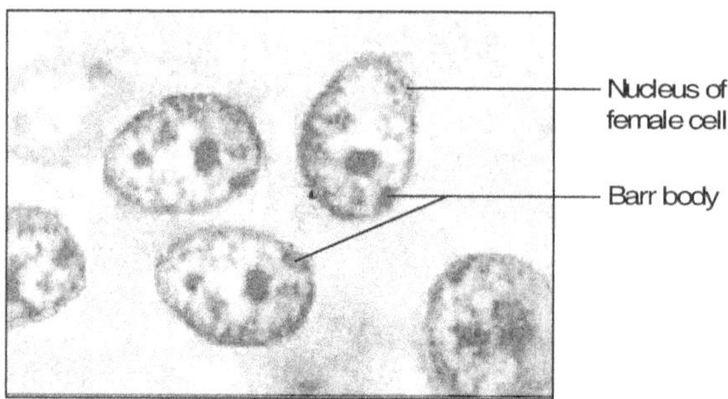

Figure 2.18 The inactive X chromosome in female nucleus is the example of facultative heterochromatin that is represented as Barr body

and then they can revert to heterochromatic again and so on. Thus there is a cycle of inactivation and reactivation of euchromatin throughout the lifetime of different tissues. Inactivated euchromatin is transcriptionally suppressed, although all the suppressed genes are not heterochromatic. This is true for the mature mammalian sperm cells where the entire nucleus is heterochromatic and their DNA sequences do not transcribe and translate to form RNA and proteins respectively. Similar is the case in terminally differentiated avian erythrocytes.

Eukaryotic genes are inactivated or silenced when the cytosine bases of certain nucleotides residing in G-C-rich regions are methylated. There are enzymes that can remove methyl groups as well as add methyl groups to cytosine residues in DNA. Methylation is a dynamic modification of genes, when a methyl group is added to a gene; it leads to inactivation, i.e., the gene does not transcribe, whereas the removal of methyl group is correlated with an increased level of transcription. Methylation is particularly evident in chromatin that has been rendered transcriptionally inactive by heterochromatization, such as the inactive X chromosome in the cells of female mammals as mentioned above.

NUCLEOLUS

Morphologically, the nucleolus appears as a spheroid body within the nucleus of eukaryotic cells. Its size can vary from approximately 1 to 5 μm. The number of nucleoli per nucleus is quite variable. In some cell types, the number of nucleoli is very stable and may be one or two per cell, but in other cells it may vary from one to many. For example, in cancer cells there is wide fluctuation in the number of nucleoli while normal amphibian eggs (oocytes) have hundreds of extra small nucleoli (Figure 2.19).

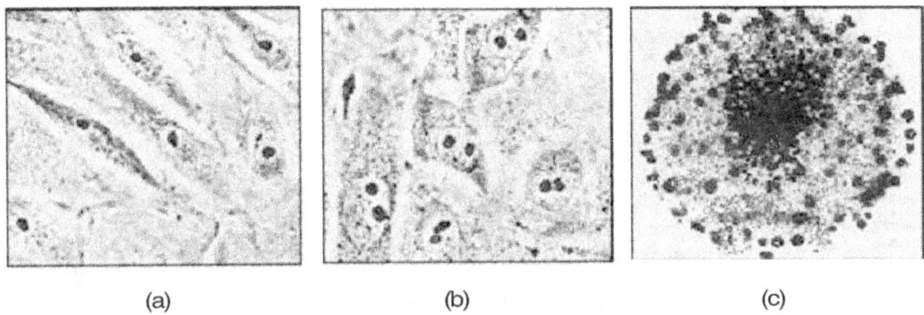

(a)   (b)   (c)

Figure 2.19   (a) Light micrograph of male rat kangaroo cells having single nucleolus cell, (b) Light micrograph of female rat kangaroo cells with two nucleoli per cell, and (c) Amphibian egg with many free-floating nucleoli.

The structural features of the nucleoli can alter rapidly in response to the environmental changes. For example, the nucleolus may become vacuolar due to cold shock and when it is treated with certain drugs, there may be separation of the dark and light components of the nucleolus. Thus the nucleolus can be considered as a good barometer of the general health of the cell since it shows alteration in its structure under certain conditions.

The electron microscopic structure of a typical nucleolus shows that there are two major components.

1. Dense granular components of about 150 Å in diameter, which may be either scattered throughout the nucleolus or localized in discrete regions around the periphery, and

2. A fibrillar component made up of fibrils having a length of about 50 to 80 Å that are scattered throughout the nucleolus. When the nucleolus is subjected to enzymatic digestion, it shows that the fibrillar and granular components contain RNA and ribonucleoproteins. Under the treatment of drugs like actinomycin D, granular and fibrillar components of nucleolus are segregated from each other, and this can be observed under the light microscope. When this drug is removed from the cells,

the segregation phenomenon is reversed, and the granular and fibrillar components of nucleolus become dispersed. Such experiments indicate that nucleolus demonstrates plasticity in its structural organization since it shows alteration under environmental stress.

NUCLEOLAR ORGANIZER

Extensive study has been carried out to understand the genetic basis of the nucleolus and many researchers have provided considerable information not only about nucleolar control but also about the general organization and control of genetic information. It was found that specific genes are associated with the nucleolus and they have the following properties.

1. The genes are very rich in specific nucleotide bases.

2. They occur in groups of multiple repeated sequences.

3. They are localized in one or few regions of the chromosomes.

4. Genes can be visualized and analysed under the electron microscope.

The nucleolar genes are responsible for the synthesis of ribosomal RNA (rRNA) and proteins that eventually become part of the ribosomes. Eukaryotic cells contain millions of ribosomes, each of which contains several molecules of rRNA together with dozens of ribosomal proteins. In fact, over 80% of the RNA of a cell is rRNA. To generate such a large number of transcripts in a cell, the DNA sequences encoding rRNA are normally repeated hundreds of times. The DNA responsible for the synthesis of rRNA is called **rDNA** and it is typically clustered in one or few regions on the chromosomes. Consequently, regions of a chromosome, which contain rDNA genes are called **nucleolar organizers**. In a non-dividing (interphase) cell, the clusters of rDNA genes are gathered together as a part of one or more nucleoli, which function as ribosome-producing organelles. Nucleoli disappear during mitosis and then reappear in the nuclei of daughter cells and subsequently control the synthesis of rRNA and protein.

Perhaps the most striking and direct knowledge about the rDNA genes is obtained from the study of amphibian oocytes, since they are typically large cells (up to 2 mm in diameter). During the development of amphibian oocytes, ribosomal genes are selectively amplified by increasing the number of nucleoli [Figure 2.19 (c)]. This DNA amplification is for the provision of large number of ribosomes that are necessary for protein synthesis, on behalf of which the fertilized egg will continue its embryonic development. Using special techniques, it is possible to gently disperse the fibrillar cores of oocyte nucleoli and to reveal the presence of a circular fibre. When these fibres are examined under the electron microscope, they are seen to resemble a chain of "Christmas trees" as shown in Figure 2.20.

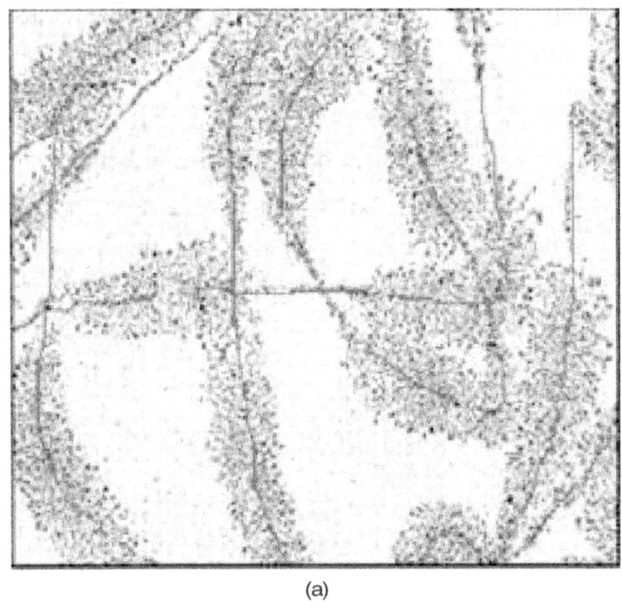

(a)

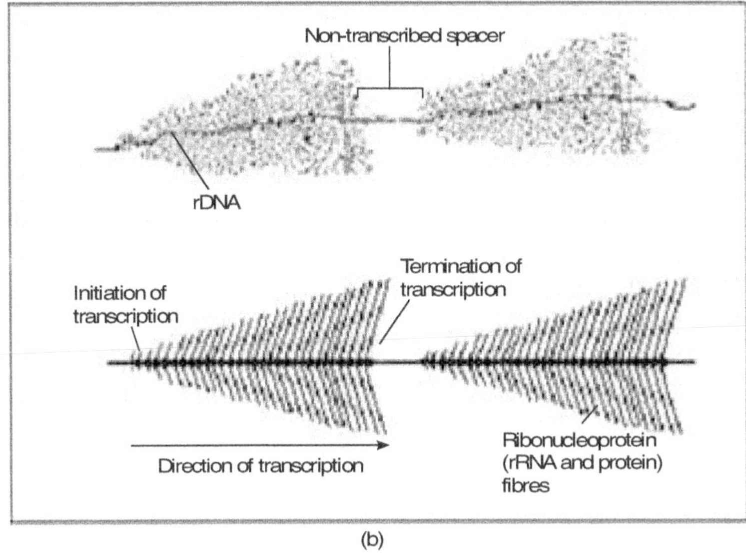

(b)

Figure 2.20 (a) Electron micrograph of rDNA isolated from amphibian oocyte, and (b) Tandem repeats of transcriptional units of ribosomal genes separated by non-transcribed spacer

Electron microscopic observation of the nucleolar fibre at higher magnification reveals the following aspects of nucleolar activity and rRNA synthesis:

1. There is tandem arrangement of repeated transcriptional units of ribosomal genes that resemble the "Christmas tree".

2. Non-transcribed spacer DNA separates each transcriptional unit.

3. Each transcriptional unit has a point of initiation and point of termination.

4. The dark line in the microphotograph is DNA, from which 100 or so fibrils emerging from the DNA as a branch of a "Christmas tree" are nascent ribonucleoprotein (rRNA and protein) fibres of increasing length.

5. The dark granule at the base of each fibril is the molecule of the enzyme RNA polymerase I, which is responsible for the synthesis of the nascent transcript.

RIBOSOMES

These are the small cytoplasmic particles or granules acting as the sites for protein synthesis. G. Granier (1899) noted the presence of some particulate basophilic units in exocrine gland cells and called them as the **ergastoplasm**. Palade, in 1953, studied these electron-opaque particles found attached to the outer surfaces of the endoplasmic reticulum and its vesicles in animal cells (Figure 2.21a). In some metabolically active cells, such particles may also be free within ground substance. The number of such particles which are free and attached varies; it is less in inactive cells or in cells that have been starved for a time, while the number of such particles is especially large in cells carrying on active synthesis. The cancerous cells may have large number of particles.

Caspersson, Brachet and others studied the biochemical nature of these granules and showed that their basophilic nature is due to the presence of ribonucleic acid (RNA), which is found attached to proteins, forming ribonucleoproteins (RNP). R. Roberts (1958) called them ribosomes. Electron microscopic study of ribosomes revealed that each **ribosome** is composed of two non-identical subunits (Figure 2.21b). In eukaryotic cells, the large subunit has the dimensions of $11.5 \times 14 \times 23$ nm, whereas the smaller and rounded subunit has a diameter of about 20 nm. The prokaryotic cells have somewhat smaller subunits, with dimensions of $15 \times 20 \times 20$ nm (large) and $6 \times 20 \times 22$ nm (small).

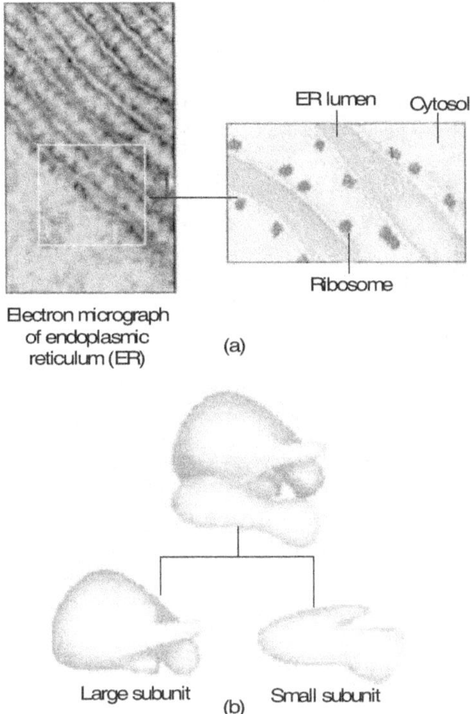

Figure 2.21 (a) Electron micrograph of exocrine pancreas showing endoplasmic reticulum and ribosomes attached to its outer surface. b) Each ribosome is composed of large and small subunits

Relatively pure preparations of ribosomes can be obtained by fractional centrifugation of a homogenized suspension of cells or a tissue in a sucrose medium or in non-polar solvents. The rate of movement of ribosomes under centrifugal force is given in terms of the sedimentation coefficient (ratio of the rate of acceleration), termed Svedberg unit (abbreviated as 'S') in honour of the investigator. The sedimentation coefficient is related with the size and molecular weight of the ribosomes.

Ribosomes are found in every type of cell, even in the smallest PPLO examined. They have been studied most in bacteria, but ribosomes of animal and plant cells have many properties common with bacteria. Mammalian RBCs do not have ribosomes. After extensive X-ray diffraction and fine-structure analysis, a three-dimensional model of *E. coli* ribosome was proposed that showed a large subunit having three protuberances sticking out of its upper side, as well as a flattened area on one of its surfaces on which the small subunit resides. The small subunit has a head and body separated by a

Table 2.2   Overall composition of prokaryotic and eukaryotic ribosomes

Prokaryotic ribosome (70S) MW ~ 2.7 × 10⁶		Eukaryotic ribosome (80S) MW ~ 4.2 × 10⁶	
Large subunit (50S) MW ~ 1.8 × 10⁶	Small subunit (30S) MW ~ 0.9 × 10⁶	Large subunit (60S) MW ~ 2.8 × 10⁶	Small subunit (40S) MW ~ 1.4 × 10⁶
5S ribosomal RNA (120 nucleotides)	16S ribosomal RNA (1,540 nucleotides)	5S ribosomal RNA (120 nucleotides)	18S ribosomal RNA (1,900 nucleotides)
23S ribosomal RNA (3,200 nucleiotides)	~ 21 specific ribosomal proteins	28S ribosomal RNA (4,700 nucleotides)	~ 33 specific ribosomal proteins
~ 35 specific ribosomal proteins		5.8S ribosomal RNA (160 nucleiotides)	
		~ 49 specific ribosomal proteins	
RNA–Protein ratio 2 : 1		RNA–Protein ratio 1 : 1	

constriction. The body occupies about two-thirds of the volume of this subunit. The assembled ribosome, therefore, has distinct structural features as shown in Figure 2.21b.

Each *E. coli* cell contains 20,000 or more ribosomes, making up almost a quarter of the dry weight of the cell. Bacterial ribosomes contain about 65% RNA and 35% protein: they have a diameter of about 18 nm and are composed of two unequal subunits with a sedimentation constant of 50S (large subunit) and 30S (small subunit) and a combined sedimentation constant of 70S. Both subunits contain dozens of ribosomal proteins and at least one large rRNA. The large subunit consists of 5S and 23S rRNA and approximately 35 ribosomal proteins, whereas the small subunit is composed of 16S rRNA and approximately 21 ribosomal proteins.

In some eukaryotic cells studied, the ribosomes appear to be 80S particles with a molecular weight of about 5 million. It is composed of two subunits; large is referred to as the 60S and the smaller, the 40S. When both subunits are biochemically analysed, it is found that they are composed entirely of RNA and ribosomal proteins that are strongly basic at cytoplasmic pH due to their high lysine content. Each subunit appears to contain a different ribosomal RNA molecule. The 60S subunit contains single copies of three different rRNAs, which sediment at 28S, 5.8S and 5S. The RNAs of the small subunit (40S) sediment at 18S. Both of these RNA molecules are derived from large RNA molecules produced in the nucleolus. Eukaryotic cells also contain 70S ribosomes in their mitochondria, and, in plants, there are 70S ribosomes in the plastids. The biochemical composition of prokaryotic and eukaryotic ribosomes is given in Table 2.2.

BIOGENESIS OF RIBOSOMES

In eukaryotes, the formation of ribosomes involves the synthesis of rRNA and proteins, their assembly into particles in the nucleolus, and transport of these particles to the cytoplasm where they provide the attachment sites for mRNA and tRNA for protein synthesis. The following three cellular RNA polymerases are involved in the transcription of genes that are responsible for biogenesis of ribosomes.

- ✢ RNA polymerase I catalyses the process of transcription of ribosomal genes (rDNA) present in nucleolar organizer for synthesis of 18S, 28S, and 5.8S rRNA.

- ✢ RNA polymerase II catalyses the transcription of rDNA genes to form mRNA that is responsible for synthesis of ribosomal proteins.

- ✢ RNA polymerase III catalyses the transcription of genes localized elsewhere on a different chromosome to form 5S rRNA.

Figure 2.22 shows a diagrammatic representation of the processing of the rDNA and other genes for the biogenesis of ribosomes.

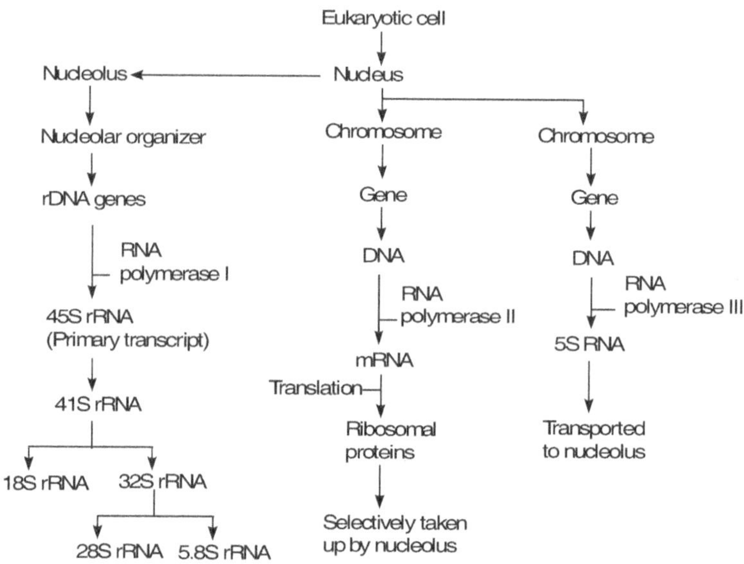

Figure 2.22   Scheme for the processing of mammalian rRNA and ribosomal proteins for biogenesis of ribosomes. The rDNA is sequentially transcribed to form 18S, 28S, and 5.8S rRNA. The 5S rRNA is made by RNA polymerase III from the genes located on different chromosome.

POLYRIBOSOMES

Following Zamecnik's discovery that ribosomes are complexes responsible for protein synthesis, and following the elucidation of the genetic code, the study of ribosomes accelerated. Masayasu Nomura and his colleagues in the late 1960s demonstrated that both ribosomal subunits can be broken down into their RNA and protein components, and then reconstituted *in vitro*. Under appropriate experimental conditions, RNA and proteins reassemble spontaneously to re-form the active 70S ribosome. Several ribosomes can assemble on the mRNA molecule to form **polyribosomes** or simply **polysomes** to synthesize multiple copies of protein in a metabolically active cell. The number of ribosomes associated in a polysome depends on the size of the protein being synthesized, for example, in rapidly growing *E. coli* cells, there may be an association of as many as 15,000 ribosomes per mRNA molecule, whereas in the reticulocytes (immature red blood cells which have lost the nucleus), only four to six ribosomes are attached to the messenger RNA that synthesizes only one protein (haemoglobin). In HeLa tumour cell, where many proteins are being synthesized, the number of ribosomes per mRNA varies from few to 30 or even more.

SUMMARY

- Chromatin fibrils or chromonemata are enclosed in the nucleoplasm and visible in the interphase nucleus as fibrillar structures.

- During cell division, the chromatin complex formed between eukaryotic DNA and nuclear protein condense to form species-specific number of chromosomes, which carries the genes, the units of heredity.

- There are five basic classes of histones (basic proteins), known as H1, H2A, H2B, H3 and H4 that contain high proportion of positively charged amino acids (lysine and arginine), which bind tightly to negatively charged DNA.

- Under the electron microscope, the interphase nucleus spread preparation appears like a "Beads-on-a string" configuration, where the beads are nucleosomes and the string is the double-stranded DNA.

- Each nucleosome contains a nucleosome core particle, having 146 base pairs of supercoiled DNA wrapped almost twice around a wedge-shaped complex of eight histone molecules called histone octamers.

- Heterochromatin is the deeply stained region of chromatin that remains condensed throughout the interphase whereas euchromatin is the less condensed region of chromatin and stains faintly.

- The base sequences in the DNA of constitutive heterochromatin are highly repetitive and they do not transcribe to form protein, while facultative heterochromatin is condensed for only part of the life of the cell and specifically inactivated during certain phases in the life of organism.

- Barr body is a darkly stained structure in the nucleus of the female cell; it is a heterochromatin present in one of the X chromosomes that remains inactive.

- Nucleolus appears as a spheroid body within the nucleus of eukaryotic cells. It shows plasticity in structural organization since it undergoes alteration under environmental stress.

- An electron microscopic study of a typical nucleolus revealed that there are two major components—dense granular components, and a fibrillar component.

SUMMARY

- The nucleolar genes synthesize ribosomal RNA (rRNA) and proteins that eventually become part of the ribosomes.

- Nucleolar organizers are regions of a chromosome, which contain rDNA genes responsible for the synthesis of rRNA.

- During the development of amphibian oocytes, ribosomal genes are selectively amplified by increasing the number of nucleoli. Electron microscope spread picture of fibrillar cores of oocyte nucleoli shows a tandem arrangement of repeated transcriptional units of ribosomal genes that resemble the "Christmas tree".

- Ribosomes are ribonucleoprotein (RNP) particles found in a variety of cells either in free state or found attached to the endoplasmic reticulum. Their number is especially large in cells carrying on active synthesis.

- Bacterial ribosomes are composed of two unequal subunits with a sedimentation constant of 50S (large subunit) and 30S (small subunit) and a combined sedimentation constant of 70S. Eukaryotic ribosomes are 80S particles with two subunits; the large is the 60S and the smaller, the 40S. Both subunits contain ribosomal proteins and rRNA.

- Three cellular RNA polymerases (I, II, and III) are involved in the transcription of genes that are responsible for biogenesis of ribosomes.

- Polyribosome or polysome is the assembly of ribosomes on mRNA molecules to synthesize multiple copies of protein in a metabolically active cell. The number of ribosomes associated in a polysome depends on the size of the protein to be synthesized.

REVIEW QUESTIONS

1. What are the differences between prokaryotic genome and eukaryotic genome?

2. Describe the ultrastructure of chromatin found in eukaryotic cell.

3. How can you distinguish between euchromatin and heterochromatin?

4. Explain the different classes and the role of basic proteins associated with DNA in eukaryotes.

5. Define Barr body. Explain its position, structure and significance.

6. Describe the ultrastructure of nucleolus with its function.

7. What is nucleolar organizer? Explain its functional aspect.

8. Explain the transcriptional units of ribosomal genes.

9. Describe the different classes of ribosomes with their structural features.

10. Explain the different types of RNA polymerases involved in the biogenesis of ribosomes.

11. Write short notes on:

    i. Nucleosome

    ii. Histone octamer

    iii. Ribonucleoproteins

    iv. Sex chromatin

    v. Polyribosomes

    vi. Svedberg unit

    vii. Bacterial ribosomes

    viii. Actinomycin D and nucleolus

MITOCHONDRIA

Mitochondria are filamentous or granular cytoplasmic organelles present in all eukaryotic cells. They have a characteristic structure and unique physiological function in relation to the production of usable chemical energy, hence they are popularly known as "powerhouses" of cells.

Mitochondria were first seen in 1888 by A. Kolliker who teased them out of insect flight muscle tissue. Later on they were observed by Altmann (1894) who called them **bioblasts**. From their thread-like appearance during spermatogenesis, C.Benda (1898) coined the term mitochondria [from the Greek *mitos* (thread) and *chondros* (granule)]. L. Michaelis (1900) discovered that mitochondria in living cells could specifically be stained by the dye Janus Green B. In 1934, Bensley and Hoerr isolated mitochondria from liver cells. Later on in the late 1940s, methods were developed for the large-scale isolation of mitochondria from tissues by differential centrifugation. Hogeboom *et al.* in 1948 found that mitochondria are the centres for cellular respiration.

SHAPE AND SIZE

Mitochondria are dynamic organelles, which move about the cell and alter their shape. Lewis and Lewis (1915) concluded that mitochondria are extremely variable bodies that are continuously moving and changing shape in the cytoplasm. These changes may have physiological importance. If the morphological features of mitochondria are monitored by light scattering and electron microscopy during respiration, changes occur depending on the metabolic state of the cell. Under certain environmental and physiological conditions, a long mitochondrion may swell at one end to become club-shaped or tennis-racket shaped (De Robertis *et al.*, 1970). Their shape may be vesicular or rod shaped.

The size of the mitochondria is also variable and the swelling of mitochondria is promoted by the osmotic pressure and pH of the medium. An average mitochondrion is about 3–4 μ long and about 0.5 to 1 μ in diameter. Mitochondria in muscle cells and pancreatic cells have somewhat large size that ranges in between 8 to 12 μ.

NUMBER AND DISTRIBUTION

Mitochondria are present in cells that perform aerobic respiration while they are absent in bacteria in which the body wall, i.e., plasma membrane contains respiratory enzymes. The number of the mitochondria per cell varies considerably, usually depending on the metabolic activity of the cell, so that active cells have more mitochondria than less active ones. Frog oocyte has

30,000 mitochondria, liver hepatocyte has 1,500 mitochondria, the cells of renal tubules have 300–400 mitochondria, while spermatocyte contains 100 mitochondria. In *Chaos chaos*, the giant multinucleated amoeba, about 500,000 mitochondria are present whereas some algal cells may have only 1 mitochondrion. But for a given type of cell, their number is more or less constant.

In most of the cells, mitochondria appear to be randomly distributed in the cytoplasm. But in the cells of some tissues, they show non-random distribution. Usually, mitochondria are distributed within the cell as per the demand of the energy. In the sperm middle piece, they are present next to the tail where its energy is utilized. Similar arrangements occur in muscles (mitochondria near the myofibrils), epithelial cells (near the ciliated border), and the kidney tubules (near the plasma membrane). Thus, mitochondria are present near the location where they are needed, presumably to decrease the time required for diffusion of ATP to its site of hydrolysis. Mitochondria are also prominent in many plant cells where they are the primary suppliers of ATP in non-photosynthetic tissues, as well as a source of ATP in photosynthetic leaf cells during periods of darkness.

STRUCTURE OF MITOCHONDRIA

A typical mitochondrion is a sausage-shaped body having a thickness of about 0.5 μ. Electron microscopic study reveals that it is enveloped by two lipoproteinous membranes, viz. inner membrane and outer membrane, lying parallel to one another (Figure 2.23a and b). Both mitochondrial membranes, under the electron microscope, show opaque lines that represent a double layer of lipid molecules. The lipid bilayer is covered on either surface by a layer of proteins like that of plasma membrane. The arrangement of the layers is similar to that in a unit membrane and must be changing as per the need of permeability. The membrane of mitochondrion is capable of active transport, accumulating ions against concentration gradient. Accumulation of ions is regulated by the inner membrane, the outer membrane being freely permeable to inorganic ions. Accumulation of ions in mitochondrion affects its structure and results in swelling. Isolated mitochondria, placed in hypotonic solutions, may swell to five times their normal volume yet remain intact.

The outer membrane shows uniform thickness of about 60 Å but the thickness of the inner membrane is variable. The space enclosed by the inner membrane is called inner chamber while the space between the outer membrane and inner membrane is called outer chamber or the intermembrane space. This space is 40–70 Å in width and is filled by a watery fluid. The inner chamber or matrix is usually filled up by proteinaceous material. The matrix may be homogeneous or may contain a granular substance. Within the mitochondrial matrix are dense granules, small ribosomes and a double-stranded circular DNA. The granules consist of insoluble inorganic salts and are believed to be the binding sites of divalent ions like Mg^{++} and Ca^{++}. The ribosomes present in the mitochondrion are of 70S type and similar to those of bacteria, and take part in the protein synthesis.

The non-chromosomal DNA found in mitochondria is significant because it encodes a small number of mitochondrial polypeptides (13 in human) that integrate into the inner mitochondria together with polypeptides encoded by genes present within the nucleus and are imported from their site of synthesis in the cytoplasm. The mitochondrial DNA may be circular as in most of the animal cells or linear as in plant cells. But it does not appear to have histones associated with it as in the nuclear DNA of higher organisms. In this respect, mitochondrial DNA resembles the bacterial DNA. There may be one or more DNA molecules in a single mitochondrion, which can replicate itself as it also contains DNA polymerase. This provides the mitochondrion the ability to replicate itself.

The inner membrane is thrown into a series of irregular folds, which form incomplete partition-like structures and are called **cristae** or **cristae mitochondrials**. These cristae are delicate and finger shaped and may be branched or tubular. The outer surface of the outer membrane and the inner surface of the inner membrane are covered by thousands of small particles. Those on the outer surface are called the **subunits of Parson** and are without stalks. Those on the inner surface are called the **subunits of Fernandez–Moran**. These are stalked particles and are called elementary particles or F1 particles (Figure 2.23c).

Each elementary particle is about 70–100 Å in diameter and the particles are placed apart a distance of 100 Å from one another. Each F1 particle is attached to the inner membrane by a short stalk of 35–50 Å long, while the head and base are about 80–100 Å (Figure 2.23d). It was supposed that each F1 particle contained all enzymes for electron transport and therefore these particles were called the **electron transport particles** or ETP. According to current thinking, the electron transfer chain lies in the basal piece and contains a part of respiratory enzymes. The stalk and head piece take no part in electron transport. The head piece contains ATP synthetase and is concerned with phosphorylation. The stalk contains oligomycin-sensitive protein.

The F1 particles are the seats of two significant steps in cellular respiration, viz., oxidative phosphorylation and electron transport system. The mitochondria are aptly called the powerhouses of the cell, since ATP which is the energy currency, is produced here through oxidative phosphorylation. The enzymes necessary for this are present in the matrix and over the inner membrane in a definite sequence. The energy and reactants are therefore passed from one enzyme to the other in a fixed sequence. In addition, the enzymes of Krebs' cycle are found in the matrix.

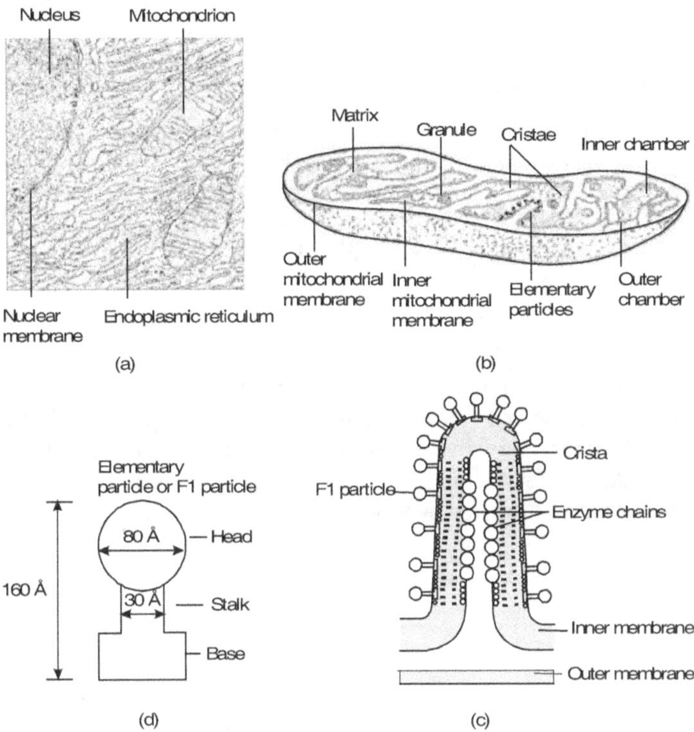

Figure 2.23 (a) Electron micrograph of a portion of a cell showing endoplasmic reticulum and mitochondria. (b) Diagrammatic representation of a section through a mitochondrion. (c) Mitochondrial crista showing sequential arrangement of elementary particles and internal organization. (d) Enlarged portion of F1 particle.

RESPIRATORY CHAIN

The respiratory chain is located in the inner membrane of mitochondria. It can be broken into four complexes and two mobile carriers.

Complex I (NADH–ubiquinone oxidoreductase) This consists of NADH dehydrogenase with 9 distinct iron–sulphur centres in addition to flavoprotein.

Complex II (Succinate–ubiquinone oxidoreductase) It is composed of succinate dehydrogenase with FAD-containing enzyme.

Complex III (UQ H_2–Cytochrome c oxidoreductase) This group consists of cytochrome b and cytochrome c.

Complex IV (Cytochrome c–O_2 oxidoreductase or Cytochrome c–oxidase)   It forms the final step in electron transport in a mitochondrion, which transfers the electrons from cytochrome c to oxygen and also translocates protons across the inner membrane.

Between complex I and III and also between complex II and III, there is a mobile carrier the coenzyme Q (CoQ) or the ubiquinone (UQ). Between complexes III and IV the mobile carrier is cytochrome c.

ORIGIN OF MITOCHONDRIA

Each mitochondrion has an average lifespan of about 5–10 days. There are the following three views in relation to the origin of mitochondria.

From folding of membranes   The infolding of plasma membrane or budding from pre-existing cell forms a double-membrane vesicle. Further the cristae would appear as ingrowths of the inner membrane (Figure 2.24). The conversion of endoplasmic reticulum into mitochondria was observed in cray fish. Both the views were supported by Robertson (1964).

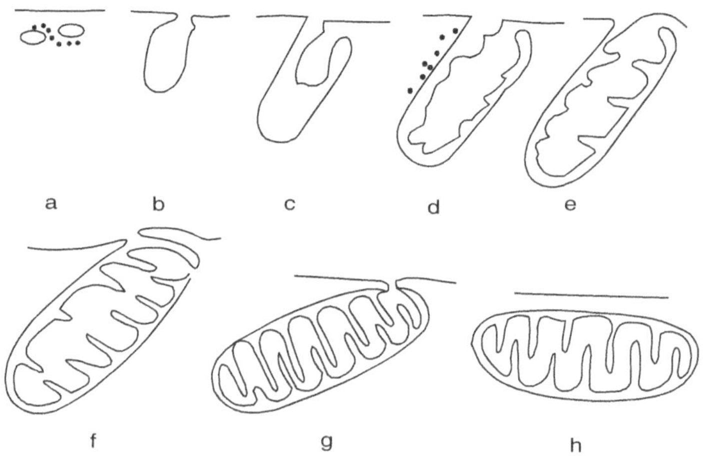

Figure 2.24 Possible mechanism in the formation of mitochondria from plasma membrane

de novo origin   This view was supported by Harvey (1951) who subjected sea urchin eggs to centrifugation and separated two halves, a nucleated half (without mitochondria) and a non-nucleated half (with mitochondria). Further it has been suggested that mitochondria released messenger RNA that enters the endoplasmic reticulum and with the help of ribosomes, membrane proteins have been synthesized that helped in the formation of mitochondria. Similar experimental results were obtained in *Guppies* and *Neurospora* that confirmed the view.

By division of mature mitochondria   According to Luck and Rich (1964), mitochondria have the capacity of self reproduction since they have their own DNA, RNA and ribosomes and enzymes helping in replication, transcription and translation. It is possible for mitochondria to grow and divide as autonomous, self-propagating particles.

Bacterial origin   The fact that mitochondria are similar in size to that of rod-shaped bacteria is more than coincidental, since a body of evidence suggests that these organelles have evolved from bacteria that lived symbiotically within other cells.

OXIDATIVE METABOLIC PATHWAYS IN CYTOPLASM AND MITOCHONDRIA

Metabolism is the collection of biochemical reactions that occur within the cell. It involves two processes, the breakdown of complex organic substances (the process is referred to as catabolism) and again to build up the complex molecules essential for the body from simple ones (the process is known as anabolism). Catabolism is usually an oxidative reaction while anabolism is accomplished by reduction reactions. Both processes require energy and 95% of this energy is provided by mitochondria formed in it through oxidative metabolism. It is the point to note that not only mitochondria form ATP but even the cytoplasm of a cell can form ATP. About 5% of the energy is generated in the cytoplasm by a process referred to as anaerobic respiration. Let us assume that the cell is going to break down one molecule of glucose and in the process extract all the possible energy from this molecule in the form of ATP. The overall reaction is as follows:

$$C_6H_{12}O_6 + 6O_2 \rightarrow 6CO_2 + 6H_2O + \text{Energy (38 ATP)}$$

Under aerobic conditions, glucose metabolism is carried out in four stages, glycolysis, pyruvic acid oxidation, Krebs' citric acid cycle and oxidative phosphorylation in hydrogen/electron transport chain.

Glycolysis

It is the stepwise enzymatic breakdown of one molecule of glucose into two smaller molecules of pyruvic acid and release of hydrogen atoms. The enzymes involved in the process are situated in the cytoplasm of a cell. Glycolysis can take place in the presence (aerobic) or absence (anaerobic) of oxygen.

The glycolytic pathway is composed of a series of ten reactions, which are summarized in Figure 2.25. Glycolysis is also known as Embden–Meyerhoff–Parnas (EMP) pathway after the German physiologists who first discovered the pathway. Sometimes the process may start with fructose or glycogen instead of glucose. Glycogen is broken down to glucose in the process known as glycogenolysis and then it enters the EMP pathway. Glycolysis begins with the activation of a molecule of glucose by phosphorylation at the expense of two molecules of ATP and through a series of steps, it ends in the formation of two molecules of pyruvic acid. Each step is enzymatically catalysed and is reversible.

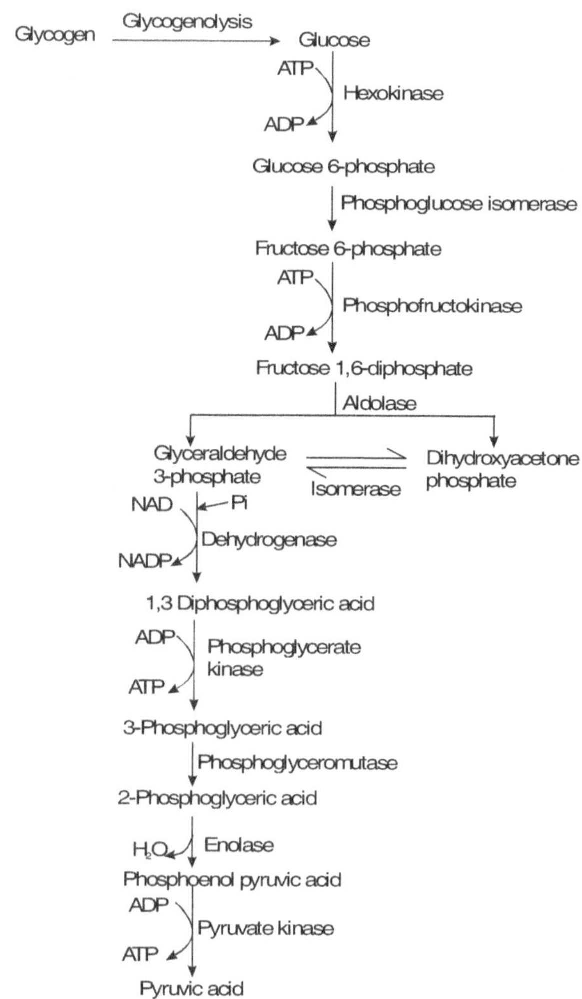

Figure 2.25   The steps of EMP pathway

The overall equation of glycolysis is as follows:

$$Glucose + 2ADP + 2Pi + 2NAD^+ \rightarrow 2 \text{ Pyruvic acid} + 2 \text{ ATP} + 2NADH + 2H^+ + 2H_2O$$

Pyruvic Acid Oxidation

The pyruvic acid molecule formed at the end of EMP pathway is then translocated through the inner membrane to the mitochondrial matrix, where it is first metabolized by the enzyme pyruvate dehydrogenase to form acetyl CoA. This is the aerobic pathway for oxidation of pyruvate. Pyruvate dehydrogenase is a giant multimeric protein complex that consists of 60 polypeptide chains constituting three different enzymes.

$$\text{Pyruvic acid} + CoA\text{-}SH + NAD \rightarrow \text{Acetyl CoA} + NADH + H^+ + CO_2$$

The above equation indicates the oxidative decarboxylation of pyruvate in which CO_2 is released, a pair of electrons is transferred to NAD^+ (oxidation), and an acetyl group is transferred to CoA (coenzyme A is a complex organic compound derived from vitamin pantothenic acid). Acetyl CoA is then fed into the citric acid cycle.

Fermentation or anaerobic oxidation of pyruvate Depending on the cell type, the pyruvic acid may break down further to form either ethyl alcohol and CO_2 or two molecules of lactic acid, the process known as anaerobic respiration or fermentation (Figure 2.26). Many bacteria, fungi like yeast, germinating seeds and certain types of skeletal muscle cells (which are required to contract repeatedly and oxygen supply may not be able to keep pace with the cell's metabolic demand) perform anaerobic respiration.

In alcoholic fermentation, two molecules of pyruvic acid formed at the end of glycolysis, instead of entering into Krebs' cycle, are decarboxylated to form acetaldehyde, and CO_2 is released. The acetaldehyde acts as the final acceptor of hydrogen. The $NADH_2$ reduces acetaldehyde to form ethyl alcohol (or ethanol) and NAD^+ is reformed. Thus ethanol is formed as the end product of alcoholic fermentation, the commercially important process that is carried out by the yeast cells.

Lactic acid fermentation is carried out by the bacteria like lactobacilli, clostridia, streptococci, etc, and also by the rapidly contracting skeletal muscle cells under diminished supply of oxygen. Pyruvic acid is acted upon by the enzyme lactate dehydrogenase to form lactic acid. The objective of anaerobic oxidation is not to produce lactate but to re-oxidize NADH and thus permit continued ATP production from glycolysis.

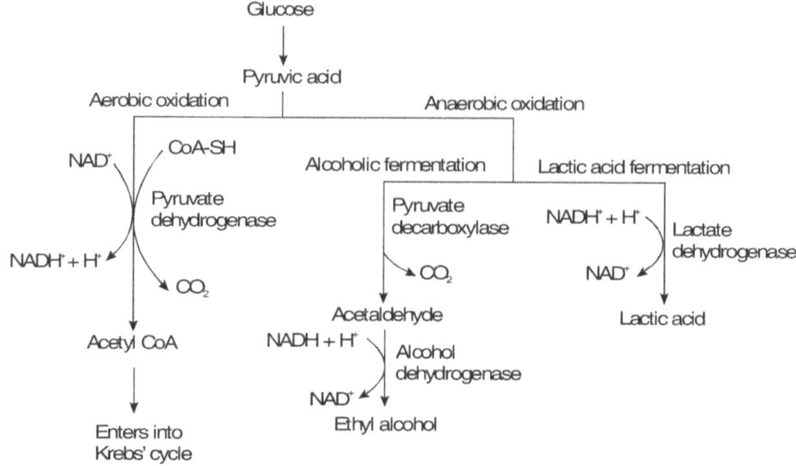

Figure 2.26 Fermentation: Most cells perform aerobic oxidation in the presence of oxygen. If oxygen supply is diminished, as in a strenuously contracting skeletal muscle cell or a yeast cell living under anaerobic condition, these cells are able to regenerate NAD^+ by fermentation. Muscle cells carry out fermentation and form lactate, whereas yeast cell forms ethanol.

Krebs' Citric Acid Cycle

Acetyl CoA formed as a result of oxidation of the pyruvic acid is fed into a cyclic pathway called **tricarboxylic acid (TCA) cycle** where the substrate is oxidized and its energy is conserved. The TCA cycle is also known as the **Krebs' cycle** after the British biochemist Hans Krebs, who worked out the pathway. The TCA cycle is the central pathway of the cell. If the position of the TCA cycle in the overall metabolism is considered, the metabolites of this cycle are found to be the same compounds generated by most of the cell's catabolic pathways. Not only carbohydrates like glucose but also amino acids (AAs) formed after breakdown of proteins, and fatty acids produced after lipolysis undergo a series of reactions to produce metabolites of the TCA cycle. The oxidation of these metabolites takes place through the Krebs citric acid cycle. Thus, the mitochondria become the focus for the final energy-conserving steps in metabolism regardless of the nature of the starting material. With the exception of succinate dehydrogenase that is bound to the inner membrane, all the enzymes of TCA cycle are present in the soluble phase of the mitochondrial matrix.

The first step in the TCA cycle is the condensation of the two-carbon acetyl group with a four-carbon oxaloacetic acid to form six-carbon citric acid. Since the citric acid contains three carboxylic (–COOH) groups, the cycle is also called the tricarboxylic acid cycle as mentioned earlier. The citric acid undergoes a series of degradation steps to reform oxaloacetic acid, which again enters the TCA (Figure 2.27).

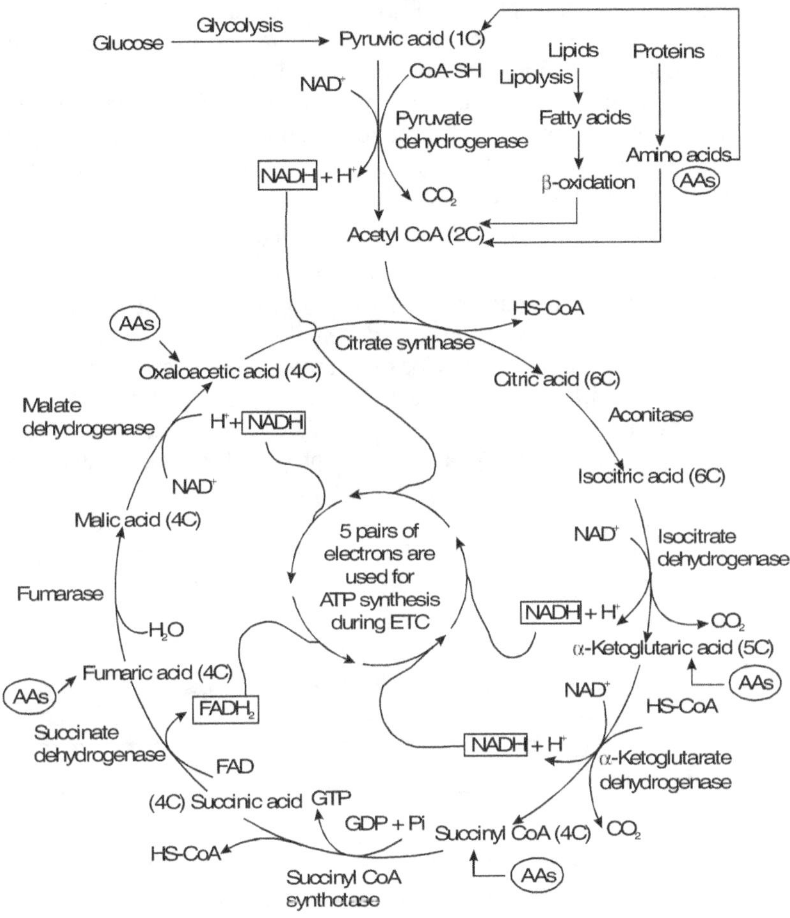

Figure 2.27 The tricarboxylic acid cycle, also called the Krebs' cycle after the person who formulated it, or the citric acid cycle after the first compound formed in it. The five pairs of electrons removed from substrate molecules by pyruvate dehydrogenase and the enzymes of the TCA cycle and transferred to NAD⁺ and FAD are passed along electron-transport chain (ETC) and used in ATP synthesis. The names of enzymes catalysing each step are provided. The products formed during the breakdown of proteins [amino acids (AAs)] and lipids also provide inputs for TCA cycle. Number of carbon atoms present in each metabolite is mentioned in brackets.

Electron Transport Chain (ETC)

During glycolysis and Krebs' cycle, hydrogen atom is removed at various steps. In a living system, the removal of hydrogen from the substrate is the oxidation and addition of hydrogen indicates **reduction** process. The removal of hydrogen atom means the removal of a proton and an electron. When one substance releases hydrogen atoms, another substance accepts them. The substance that releases hydrogen is called the **hydrogen donor** and is said to be **oxidized** while the other substance that accepts the hydrogen is called the **hydrogen accepts** and is said to be **reduced**.

Hydrogen removed from the substrate during cellular respiration is picked up by NAD (nicotinamide adenine dinucleotide), FAD (flavin adenine dinucleotide) and other acceptors. This hydrogen is split into proton (H^+) and electron (e^-). The electrons are transported along the electron transport chain (or respiratory chain), which is located in the inner mitochondrial membrane- and protons are dragged behind them. ETC is composed of four types of membrane bound electron carriers—flavoproteins, cytochromes (a, a_3, b, c, and c_1), ubiquinone (or coenzyme Q), and iron–sulphur proteins (Fe-S). All of them are arranged in a sequence from higher energy content (redox potential) to lower energy content, i.e., in decreasing order of energy content. Hydrogen is passed from one acceptor to the next in ETC, step-by-step down an energy gradient, until it is finally combined with O_2 to form water; the process is called **terminal oxidation** (Figure 2.28).

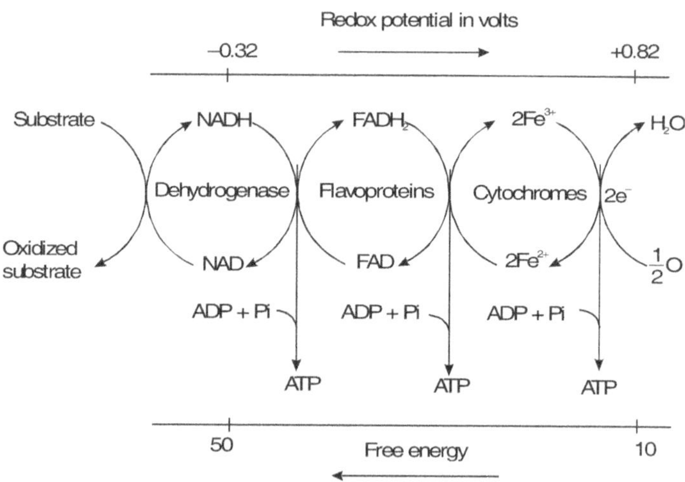

Figure 2.28   Electron transport chain

During electron transport, the electrons removed from oxidized substrate release the energy that is harnessed to synthesize ATP. This process is referred to as **oxidative phosphorylation** that is different from the substrate-level phosphorylation in which ATP is formed directly by transfer of a phosphate group from a substrate molecule to ADP. Each pair of electrons transferred from NADH to oxygen by means of the electron transport chain releases sufficient energy to synthesize approximately 3 molecules of ATP. Each pair donated by $FADH_2$ releases sufficient energy for the formation of approximately 2 molecules of ATP. If one adds up all the ATPs formed from one molecule of glucose completely catabolized in glycolysis and Krebs' cycle, the net gain is 36 ATPs (including GTP formed in the TCA cycle), as shown in Table 2.3.

Thus, in eukaryotic cellular respiration, one molecule of glucose generates 36 or 38 ATP molecules in addition to CO_2 and H_2O. In relation to oxygen used in the process, 6 O_2 are used in terminal oxidation of 12 molecules of reduced coenzyme, which are formed in aerobic respiration. Thus, in all, twelve 1/2 O_2 are used in terminal oxidation while 12 H_2O molecules are formed out of which $6H_2O$ molecules are utilized in various steps of TCA cycle. Therefore, there is net gain of 6 water molecules.

Table 2.3   Total number of ATP molecules formed from the complete oxidation of one molecule of glucose

Phase	$NADH_2 = 3$ ATP	$FADH_2 = 2$ ATP	GTP = ATP	Total ATP
Glycolysis		--	--	
ATP expended	–2 ATP			
Substrate phosphorylation	4 ATP			
Reduction of NAD (2 or 3 ATP)	4 or 6 ATP	--	--	6 or 8
Acetyl CoA formation (2 NAD reduced)	6 ATP	--	--	6
Krebs' cycle	$6 \times 3 = 18$ ATP	$2 \times 2 = 4$ ATP	$2 \times 1 = 2$ ATP	24

Net gain of ATP = 36 or 38

FUNCTIONS OF MITOCHONDRIA

Energy Transduction

The mitochondrion transduces chemical energy that is utilized by the cell in synthetic reactions and active transport. The sources of this chemical energy are three major foodstuffs—carbohydrates, lipids, and proteins. These compounds are broken down into intermediate metabolites that are translocated into the mitochondrial matrix, where, through the citric acid cycle (Krebs' cycle), they are oxidized to CO_2 (Figure 2.29). This oxidation is coupled with the reduction of coenzymes such as NAD and FAD, reducing energy to oxygen in electron transport chain, located in the inner membrane. The oxidation of components of this chain is coupled with the phosphorylation of ADP to ATP, the energy-rich compound, utilized by the cell to carry out the vital processes. Thus the flow of energy in the mitochondrion can be represented as:

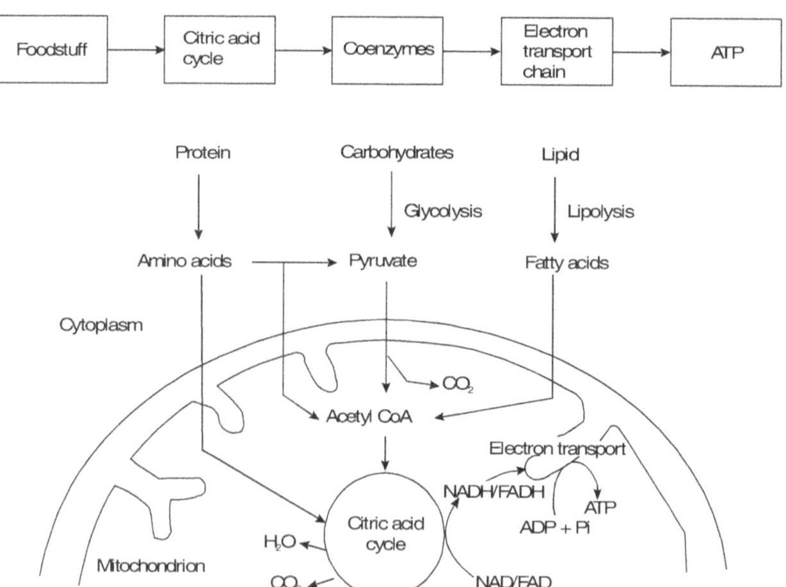

Figure 2.29 An overview of breakdown of carbohydrates, proteins and lipids into intermediate metabolites that enter into mitochondrion where they are oxidized through citric acid cycle, and generate NADH and $FADH_2$. High-energy electrons in these molecules are passed along the electron transport chain embedded in the inner mitochondrial membrane, to molecular O_2. The energy released during electron transport is used to form ATP.

Transport of ATP Molecules

Cellular respiration results in the formation of energy-rich ATP molecules, which are stored inside the mitochondrion. As per the necessity of energy, the mitochondrion moves to the place where energy requirement is high. The hydrostatic pressure inside the matrix increases, as a result the mitochondrial membranes contract and then water and ATP molecules are squeezed out.

Involvement in Haem Synthesis

Haem is a protoporphyrin complexed with iron, the prosthetic group of haemoglobin. The mitochondrial fraction of rat liver cells and avian blood corpuscles contains δ-aminolevulinate synthetase that plays significant role in the synthesis of δ-aminolevulinate. It is an important intermediate in protoporphyrin synthesis. Thus mitochondria are also involved in the biogenesis of haem.

Thermogenesis

The inner mitochondrial membrane contains approximately 20 electron carriers that make up the electron transport chain, which set up the proton gradient. In addition to ATP synthesis, this proton gradient is used to produce metabolic heat in newborn babies, hibernating animals, and cold-stressed animals. This process is known as thermogenesis, and it occurs in a specialized tissue, brown fat, and in its constituent cells, the adipocytes. Usually, animals generate heat by physical activity such as shivering or by blood circulatory mechanisms, or protect their body from cold with the help of their hairy or furry coat, but in newborns, the body is without much hair and other heat-generating mechanisms are underdeveloped. In such situation, the adipocytes play a significant role in thermogenesis.

A series of events occur in the adipocytes of brown fat to generate heat that protects the newborns. The hormone, norepinephrine, binds to the receptor present in the plasma membrane of adipocytes, and this leads to the formation of cyclic AMP inside the cell. The cyclic AMP acts as a second messenger and activates protein kinase that in turn activities lipase to hydrolyse triglycerides into fatty acids and glycerol. Fatty acids activate the thermogenic pump in the inner mitochondrial membrane. This pump upsets the proton gradient set up by the electron carriers. Instead of ATP synthesis, a third mitochondrial inner proton pump formed by the protein called **thermogenin** is activated and uses the free energy of the electrochemical gradient to produce heat to protect the newborn child which has not fully developed the other thermogenic mechanisms.

ABNORMAL MITOCHONDRIA AND HUMAN DISORDERS

There are number of human diseases that can be due to alterations in mitochondrial structure and function. In the **mitochondrial myopathies**, staining characteristics of the organelles are used as a sign in clinical diagnosis. Muscle biopsies taken from such patients when examined by light microscopy show "ragged-red fibres", which represent aggregates of abnormal mitochondria. When these abnormalities are observed under electron microscope, they showed crystals in the matrix, concentric cristae, etc. Patients with mitochondrial myopathies show a wide variety of muscular disorders, ranging from inadequate contraction to fatigue. **Leber's hereditary neuropathy** is another mitochondrial disease, which causes blindness in young adults, especially men. Most of the disorders are due to defects in any one of the four complexes of the electron transport chain, enzymes of citric acid cycle, use of certain substrates, and the uptake of metabolites.

SUMMARY

- Mitochondria are filamentous or granular cytoplasmic organelles acting as powerhouses in all eukaryotic cells.

- Mitochondria are dynamic organelles showing plasticity in their shape and size. And the changes related to these parameters may have physiological importance.

- Cells performing aerobic respiration have several mitochondria since they are the centres for oxidative metabolism, while bacteria do not have mitochondria. Their number usually varies as per the metabolic activity of the cell. Mitochondria are distributed within the cell as per the demand of energy.

- Electron microscopic study reveals that a mitochondrion is comprised of two lipoproteinous outer and inner membranes, enclosing the matrix that contains dense granules, small ribosomes, a double-stranded circular DNA and the enzymes of Krebs' cycle.

- Cristae are a series of irregular folds in the inner membrane of a mitochondrion that are covered by thousands of small particles called subunits of Fernandez–Moran or elementary particles or F1 particles.

- Each F1 particle is the seat of two significant steps in cellular respiration, viz., oxidative phosphorylation and electron transport system.

- The electron transport chain contains four different types of carriers: haem containing cytochromes, flavin nucleotide-containing flavoproteins, iron–sulphur proteins, and quinones.

SUMMARY

- Biogenesis of mitochondria can take place from folding of plasma membranes, by the growth and division of pre-existing mitochondria or possibly by de novo synthesis. Many of the mitochondrial properties can be accounted for their evolution from symbiotic bacteria.

- Mitochondria transduce chemical energy (ATP production) by oxidation of foodstuff via citric acid cycle (Krebs' cycle) and electron transport chain; the machinery required for these processes is located in the inner membrane.

- In addition to ATP synthesis, mitochondria play a significant role in thermogenesis that occur in a specialized tissue, brown fat, and in its constituent cells, the adipocytes by setting up a proton gradient in the electron transport chain. This specificity is used to produce metabolic heat in newborn babies, hibernating animals, and cold-stressed animals.

- Mitochondria are also involved in the biogenesis of haem, since the mitochondrial fraction is known to contain δ-aminolevulinate synthetase the enzyme that is essential for synthesis of δ-aminolevulinate, an important intermediate in protoporphyrin synthesis.

- Alterations in mitochondrial structure and function can lead to a number of human diseases known as mitochondrial myopathies that show a wide variety of symptoms including muscular disorders, ranging from inadequate contraction to fatigue or even blindness in young adults.

REVIEW QUESTIONS

1. Describe the ultrastructure of a mitochondrion and compare the properties of the inner and outer mitochondrial membranes.

2. Explain the steps in oxidative metabolism that occur in the mitochondrial matrix and F1 particle.

3. Why is the TCA cycle considered to be the central pathway in the energy metabolism of a cell?

4. Define terminal oxidation. Explain the process that leads to formation of water molecule.

5. What are the points of differences between aerobic and anaerobic respiration.

6.   Describe some of the major activities of the mitochondria.

7.   Write short notes on:

    i.   Mitochondrial DNA

    ii.   Number of mitochondria in a cell

    iii.   Origin of mitochondria

    iv.   EMP pathway

    v.   Fermentation

    vi.   Proton pump in ETC

    vii.   Thermogenesis by mitochondria

    viii.   Mitochondrial myopathies

    xi.   ATP generation in oxidative metabolism

PLASTIDS

Plastids are small inclusions found in most of the plant cells and some protozoans. They are generally absent in the animal cells, bacteria, and fungi. Plastids are of two types—leucoplasts (colourless plastids) and chromoplasts (coloured plastids).

LEUCOPLASTS

These plastids are without pigment or colouring substance and hence known as leucoplasts. They are usually present in the cells that do not receive direct light. They are also found in the plant gametes and embryonic cells. Leucoplasts mainly serve for the storage of food—**amyloplasts** are related with storage of starch and are found in tubers, cotyledons and endosperm, **aleuroplasts** store proteins while **elaioplasts** store oils.

CHROMOPLASTS

These are the commonest coloured plastids having various coloured pigments. Chromoplasts act as primary sites for trapping and converting solar energy and hence they are very essential for the existence of not only the green plants, but for the entire living world. On the basis of their coloured pigments, chromoplasts are of following types.

1. **Chloroplasts** These are green coloured plastids present in most of the plant cells. They contain the green pigment called **chlorophyll** that is of two types **chlorophyll *a*** and **chlorophyll *b*** that play an indispensable role in photosynthesis.

2. **Phaeoplasts** These are brown coloured plastids having the brown pigment called **fucoxanthin**. They are present in brown algae, diatoms and in some dinoflagellates. Phaeoplasts have the capacity to absorb light energy and to transfer it to chlorophyll to carry out photosynthesis.

3. **Rhodoplasts** These are chromoplasts containing red pigment called **phycoerythrin** and appear red, e.g. Rhodophyceae or the red algae.

4. **Blue-green chromoplasts** These are present in blue-green algae containing the pigments called **phycocyanin, chlorophyll *a*** and **carotenoids**.

5. The colours of the flowers and the fruits like tomato, carrot, etc. are due to chromoplasts other than chloroplasts. Such pigments do not take part in photosynthesis.

CHLOROPLASTS

These are the most abundant and common plastids found in all green plants. They show great variation in size, shape and number. A cell may have only one large chloroplast, measuring over 100 μm in diameter as in polyploid plants, algae like *Chlamydomonas* and shade-loving plants, or two as in *Zygnema*, or over 100 smaller ones as in higher plants. Chloroplasts may have oval, spheroid or disc-shaped structure, or it may be band-shaped or ribbon-shaped as in *Spirogyra*. Indeed in the algae, the morphology of the chloroplasts is so diverse and species-specific that it is used as a criterion for taxonomic classification.

Chloroplasts are usually lens-shaped in higher plants, 4 to 6 μm in length, and 1 μm thick. There are about half a million chloroplasts per square millimetre of the leaf surface and they are in constant motion due to cytoplasmic streaming or cyclosis and hence are not found at a fixed location. In some cases, it is reported that chloroplasts move actively and show amoeboid movements. The concentration of chloroplasts depends on the light intensity. They are most abundant in the tissues of green plant such as leaves that perform photosynthesis and less in number or absent in the root. Chloroplasts multiply by division and disintegrate if their number exceeds beyond a required level.

Ultrastructure of Chloroplast

In 1947, S. Granick and K. Porter published the first electron micrograph of the isolated chloroplasts and showed that the grana appeared as stacks of internal discs. The outer envelope of a chloroplast is made up of a double membrane with a structure very similar to that of plasma membrane. Each membrane is 6 to 8 nm thick and the space between them is 10 to 20 nm. The outer membrane is rather rigid structurally. The ground substance enclosed by the walls is called the **stroma** or matrix, which consists of lipoproteins, enzymes to carry out photosynthesis, several kinds of discrete bodies including plastoglobuli (lipid deposits about 0.1 μm in diameter), starch grains, ribosomes and 2.5 nm diameter fibrils that represent the chloroplast DNA (Figure 2.30).

The electron micrograph of a thin section of chloroplast also shows a complicated lamellar system that is composed of small cylindrical structures called the grana. The number of grana per chloroplast is variable. There may be 10 to 1000 grana in various chloroplasts. Each granum is a small cylindrical structure about 0.5 μm in diameter and made up of small disc-shaped sacs one over the other to form a pile. In 1962, W. Menke coined the term thylakoid ("sac-like") to describe these membranous structures. In each granum there may be 2 to 100 stacked thylakoids. Adjacent grana are connected together by small tubules called intergrana or the stroma lamellae. The extensive folding of lamellar system to form the grana plays a significant role in increasing the surface area of the chloroplast. A. Wellburn has calculated that a typical leaf has 600 times its surface area of photosynthetic membranes for each cell layer containing chloroplasts.

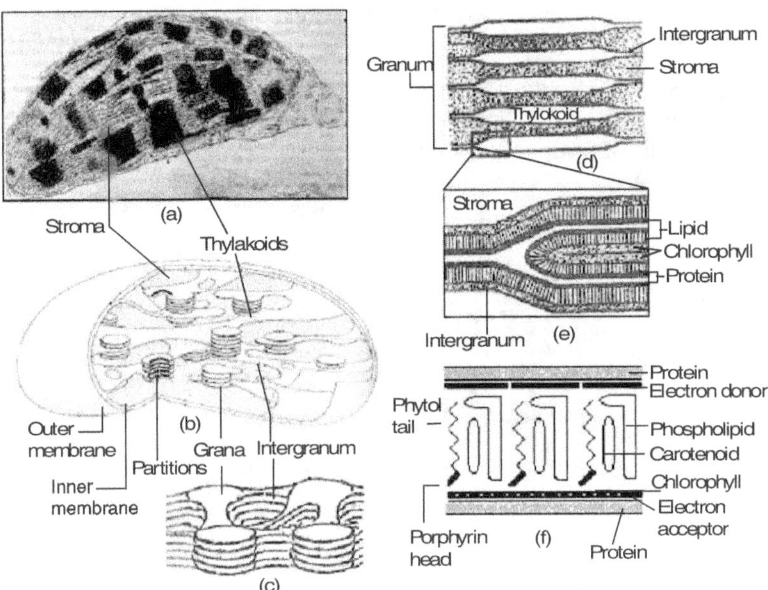

Figure 2.30 Chloroplast structure. (a) Electron micrograph of cross section of chloroplast, (b) Diagram of structures within chloroplast, (c) Grana and intergrana, (d) Cross section of single granum showing stack of thylakoids, (e) Magnified view of lamellar structure, and (f) Arrangement of molecules in the membrane of a granum.

The structure of lamellae and granum has been determined from studies with polarizing and electron microscopes. The latter shows that the various layers making a single membrane of a granum differ in electronic opacity. The lamellae and outer membrane of chloroplast are composed of lipoprotein subunits, each subunit having a protein core surrounded by a lipid layer. The membranes of chloroplast have a single layer of subunit, while the thylakoid membrane is double due to juxtaposition of the membranes. The thylakoid membranes enclose the space, called **fret channel** and the space between two thylakoids is called **loculus**. The small discs or flattened spheroids are known as **quantosomes**, and the photosynthetic units are present inside the fret channel. Each quantosome is about 155–158 Å in diameter and about 100 Å in thickness, and consists of about 250 molecules of chlorophyll along with carotenoids and other chemical compounds. The chlorophyll molecules are sandwiched between the layers of electron donors and electron acceptors.

Chemical Composition of Chloroplast

The average chemical composition of chloroplast on a dry weight basis consists of 40–50% protein, 23–25% phospholipids, 5–10% chlorophyll, 1–2% carotenoids, 3% RNA and about 2–3% DNA. Mg is present in about 2–3% of total ash content. The other elements such as Fe and Cu are found in traces.

Chlorophylls   The most distinctive lipid components of the chloroplast are the pigments. Among the chlorophylls, chlorophyll *a* and *b* are well known and are most widely distributed amongst higher plants. In bacteria, the photosynthetic pigment is bacteriochlorophylls (**B Chl**). The sulphur bacteria have B Chl and small amount of chlorobium. The chlorophyll molecule is a complex structure basically made up of a "head" and "tail" like a tennis racket.

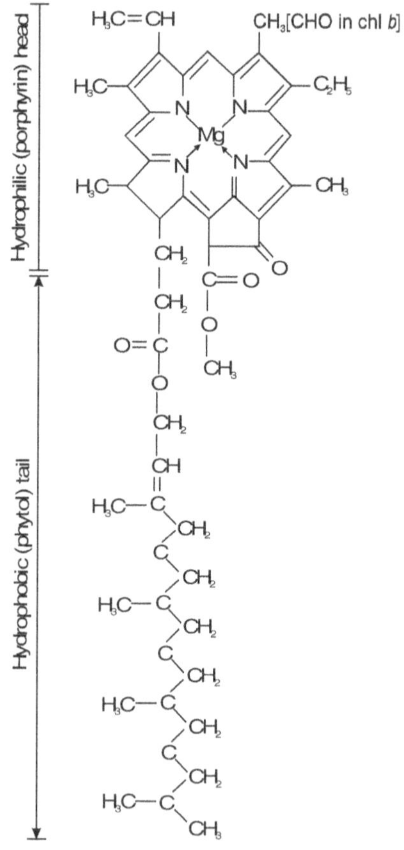

Figure 2.31   Structure of chlorophyll *a* and chlorophyll *b*

The head is a hydrophilic porphyrin structure made up of four pyrrole rings attached to each other (forming a tetrapyrrole structure) at the centre by an isocyclic ring containing Mg atom at the centre. Extending from one of the pyrrole rings is the hydrophobic long tail (phytol). The phytol chain is esterified on the C atom of one of the pyrrole rings and has only one double bond (Figure 2.31). The phytol is an alcohol chain with the formula $C_{20}H_{39}$. It may be considered as hydrogenated carotene.

The structure of chlorophyll molecule is similar to that of haemoglobin except that in the centre of the tetrapyrrole head, an iron atom is present instead of an atom of magnesium. The arrangement of chlorophyll molecule in the membrane of thylakoid is very similar to that of phospholipids, with the phytol chain (tail) perpendicular to the plane of the lamella and the porphyrins ring (head) associated with the membrane proteins (Figure 2.30f).

Chlorophyll a is a blue-green crystalline structure having the formula $C_{55}H_{72}O_5N_4Mg$. Chlorophyll b is a green-black crystalline structure with the formula $C_{55}H_{70}O_6N_4Mg$ and it differs from **chl a** in that at third carbon atom, methyl group (CH_3) is substituted by aldehyde group (CHO). In addition to this point of difference, both these photosynthetic pigments have different absorption spectra. The peaks of the absorption of chlorophyll a are 429, 410, and 660 nm, whereas the peaks of absorption spectra for chlorophyll b are 430, 453, and 442 nm.

Carotenoid pigments   In addition to chlorophyll pigments, carotenoids act as accessory pigments. Carotenoid pigments belong to a large group of compounds called terpenoids. They are of two types, hydrocarbon carotenes and their oxygen derivatives or xanthophylls. These lipid compounds have a colour ranging from yellow to purple. Carotenoids absorb radiant energy in the middle of the visible light spectrum in between 449 and 490 nm.

Carotenoids show wide distribution in both plants and animals. The major carotenoids of higher plants are β-carotene, lutein, violaxanthin and neoxanthin. In some plants, α-carotene is also present. Carotenoids are the derivatives of a red pigment called **lycopene** found in many plants. They are also present in microorganisms including red algae, cyanobacteria, photosynthetic bacteria, fungi, etc. Like chlorophyll, the carotenoids occur in their natural states as protein complexes. According to Goodwin (1960), the chlorophyll and carotenoids may be attached to the same portion forming a complex called **photosynthin**. Carotenoids play a significant role in the prevention of photooxidation of chlorophyll and they absorb solar energy and transfer it to chlorophyll a.

Phycobilins   These are also accessory pigments that are involved in the transfer of light energy to chlorophyll. Phycobilins are of two types—the phycoerythrin (red) and phycocyanin (blue)—and are found only in the algal divisions, Rhodophyta, Cyanophyta, and Cryptophyta. The isolation of these pigments in the pure state is difficult due to their strong association with proteins.

Origin of Chloroplasts

The photosynthetic organelles, the choloroplasts arise from the pre-existing **proplastids**, which have a small spheroid body surrounded by double-membrane wall. In the presence of sunlight, proplastid grows to a size of about 1μ in diameter and invaginates its membranous walls to form a vesicle. These vesicles later on are inter-connected by the formation of intergrana tubules. Prelamellar body or primary granum may form in the absence of the sunlight. In due course of development, the primary granum gives rise to a crystalline lattice inside this photosynthetic organelle.

PHOTOSYNTHESIS

Photosynthesis is the single most important physico-chemical process of the world that makes possible the existence of life on earth. The major function of chloroplast is to convert light energy into chemical energy. Green plants carry out photosynthesis in the presence of sunlight and manufacture sugars from water and carbon dioxide. During photosynthesis, water is oxidized to oxygen while carbon dioxide is reduced to sugar. The overall equation of photosynthesis is as follows.

$$6CO_2 + 12H_2O \xrightarrow[\text{Chlorophyll}]{\text{Light energy}} \underset{\text{(Glucose)}}{C_6H_{12}O_6} + 6H_2O + 6O_2 \uparrow$$

Park and Beggins (1964) coined the term quantosomes for the photosynthetic units with distinct morphological features, which are arranged in the form of monolayer in the thylakoids of chloroplast. In the middle of each quantosome there is a **photosynthetic reaction centre** . Thus the thylakoid is the location of "light" reactions of photosynthesis, where radiant energy is converted into chemical energy in the form of ATP and NADPH. The two photosytems involved in light reaction of photosynthesis are called **photosystems I and II**. Each system has its own type of chlorophyll. **Photosystem I (PS I)** absorbs far red light while **photosystem II** (PS II) absorbs shorter wavelengths of red light. Both the photosystems contain chlorophyll *a*, chlorophyll *b* and carotenoids. But PSI consists of more carotenes and less xanthophylls than PSII. Both the photosystems work in cooperation to absorb radiant energy. Along with the two photosystems, an electron transport system, and the oxidative phosphorylation mechanism are located in the lamellae. This chemical energy is used to fix carbon dioxide into sugars in the Calvin cycle, located in the stroma. In addition, the stroma plays an indispensable role in the anabolism of fatty acids and amino acids. Since chloroplasts also have genetic material, they replicate their own genome and synthesize some of its proteins.

Light Reaction

Photosynthesis involves two reactions, the first reaction is the Hill reaction, which is carried out in the presence of light, hence also called the **light reaction**. The second reaction in the process of photosynthesis is independent of light hence known as **Blackman reaction** or **dark reaction**. Light reaction is the primary process of energy transduction that takes place in the thylakoid membrane as a photochemical reaction. The light reaction involves phosphorylation and photolysis of water. The overall equation of light reaction can be represented as:

$$NADP + ADP + Pi + H_2O \xrightarrow{\text{Light}} ATP + NADPH_2 + O_2 \uparrow$$

Photolysis of Water

In 1937, Robert Hill demonstrated that isolated chloroplasts evolved oxygen when they were illuminated in the presence of a suitable electron acceptor, such as ferricyanide. The ferricyanide is reduced to ferrocyanide by the photolysis of water, i.e., splitting of water molecule in the presence of light. The reaction is called **Hill reaction** that explained the fact that oxygen evolved during light reaction is from water and not from CO_2. Using a heavy isotope of oxygen (O^{18}), Ruben, Randall and Kamen (1941) experimentally provided the direct proof that oxygen evolved during photosynthesis comes from water and not from carbon dioxide.

$$6CO_2 + 12H_2O \xrightarrow[\text{Chlorophyll}]{\text{Light energy}} \underset{\text{(Glucose)}}{C_6H_{12}O_6} + 6H_2O + 6O_2 \uparrow$$

Photophosphorylation

The input of energy in the process of photosynthesis is the visible light, which has the characteristics of both a particle and a wave. The energy of the particle, the photon, can be expressed in terms of its ability to do electric work, as in a solar cell. Light at a shorter wavelength (e.g. blue light) has photons with more energy than light at a longer wavelength (e.g. red light). The photosynthetic pigments present in thylakoids absorb the light energy in different wavelengths. The actual photochemical reaction occurs in the 1% of the photosynthetic pigments termed the reaction centres that are composed of chlorophyll *a*. The remaining molecules of chlorophyll *a* and all other pigments absorb light and transfer the energy to the reaction centre. With the absorption of light energy, the chlorophyll molecule becomes "excited". Generally the excited molecule quickly returns to its original energy level with the expulsion of electron. The loss of electron makes the chlorophyll molecule oxidized, the process that is ultimately coupled with the reduction of NADP. The high-energy electron is transported through the electron transport system (ETS) present in the thylakoid membrane. While doing so, the electron is de-energized, i.e., the energy released during ETS is used to build high-energy phosphate bonds in ATP. Since the synthesis of ATP takes place in the presence of light, the process is called **photophosphorylation** and can be represented as follows:

$$ADP + Pi + Energy \xrightarrow[\text{chlorophyll}]{\text{Light}} ATP$$

Arnon *et al.* (1954) were the first to show that isolated chloroplasts could produce ATP when exposed to light. They called this kind of phosphorylation as photophosphorylation since it is different from oxidative phosphorylation occuring in mitochondria in that it requires light as a source of energy and depends on the presence of oxygen. So far, the following two kinds of photophosphorylation have been recognized:

Non-cyclic photophosphorylation As the name indicates, in this type, the electron emitted by chlorophyll *a* molecule does not return. It involves two coupled photosystems PS II and PS I that are activated by two wavelengths 680 nm and 700 nm respectively. Robert Hill and F. Bendell (1960) proposed the **Z-scheme** for the electron transport in non-cyclic photophosphorylation. The Z-scheme (Figure 2.32) describes the flow of electrons along with their redox potential from water, through the two photosystems, to NADP. The electron carriers, plastoquinones, cytochrome *b*, cytochrome *f* and plastocyanins, are arranged in the series to form the electron transport chain, which transfer the electron emitted by the PS II excited by the photons of light (680 nm).

In photosystem II, the primary reaction is the oxidation of chlorophyll molecule due to light absorption of P680 reaction centre. This reduces the pigment phaeophytin that in turn reduces plastoquinone, the carrier to the cytochrome *b/f* complex. The oxidized P680 extracts an electron from the electron donor, called carrier Z, which in turn extracts an electron from donor M. The latter, a mangenese-containing component, has the ability to extract an electron from water, thereby producing oxygen. The cytochrome *b/f* mediates the transfer of electrons between the two photosystems via iron–sulphur complex.

A second photochemical reaction occurs in photosystem I. The PS I is excited by the photons of 700 nm. The subsequent loss of an electron causes the P700 to reoxidize plastocyanine. Reduced ferrodoxin is the ultimate product of this system that involves several intermediate electron carriers like chlorophyll *a*, phylloquinone (vitamin K), and X (an iron–sulphur centre). Finally the electron transferred through ferrodoxin and flavoprotein is used for reduction of NADP.

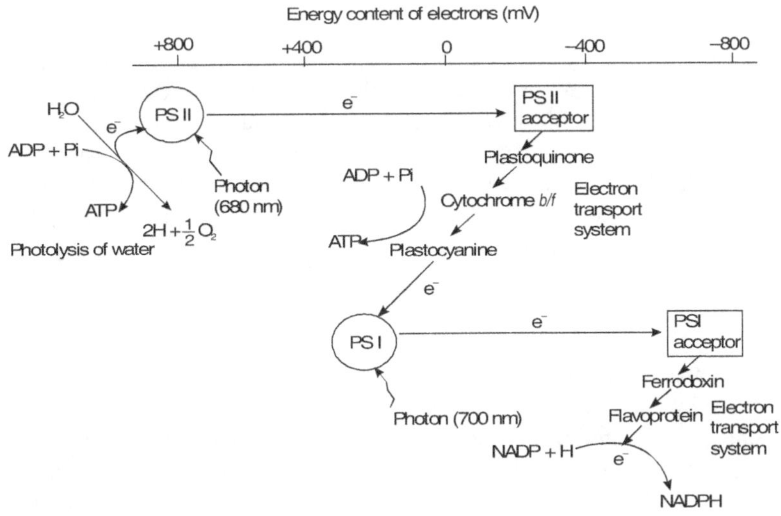

Figure 2.32   The Z-scheme showing the flow of electrons and non-cyclic photophosphorylation

The most significant feature of non-cyclic photophosphorylation is the release of oxygen, in addition to formation of ATP and NADPH that are further utilized in the dark reaction.

Cyclic photophosphorylation   In this type only one photosystem (PS I) is involved and the electron lost by the chlorophyll molecule returns to PS I hence the name cyclic photophosphorylation (Figure 2.33). The electron emitted by the chlorophyll molecule that is excited due to absorption of photon (700 nm) is accepted by electron acceptor, ferrodoxin. Then the electron passes through the electron transport chain consisting of electron carriers like plastoquinone, cytochrome *b/f* complex and finally plastocyanine. The cyclic photophosphorylation leads to formation of ATP and it does not need water as electron donor and there is no reduction of NADP.

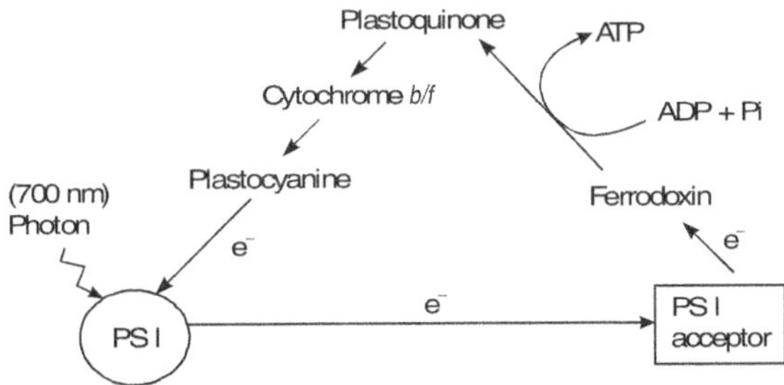

Figure 2.33   Diagrammatic representation of cyclic photophosphorylation

Dark Reaction

It is the second reaction in the process of photosynthesis, which is independent of light, hence known as **dark reaction**. The presence of dark reaction was first discovered by Blackman (1905). It is the thermochemical reaction in which NADPH and ATP are used to reduce atmospheric CO_2 to form carbohydrates. Thus the dark reaction is the carbon assimilation phase that occurs inside the stroma of chloroplast. The reaction can be represented as follows:

$$CO_2 + ATP + NADPH_2 \longrightarrow (CH_2O)_n + NADP + ADP + Pi$$

Calvin Cycle or C3 Pathway

The present knowledge about CO_2 fixation is the experimental outcome of efforts made by Calvin and his co-workers, who traced the path of carbon using its radioactive isotope (^{14}C). They observed that carbon reduction in the unicellular green algae is a cyclic process and the first stable product is 3-phosphoglyceric acid (3-carbon compound), hence such plants are called C3 plants. The entire process of photosynthesis runs in a cyclic manner and includes the formation of hexose sugars or starch and regeneration of ribulose diphosphate (RuDP), the process known as Calvin cycle which is summarized in Figure 2.34.

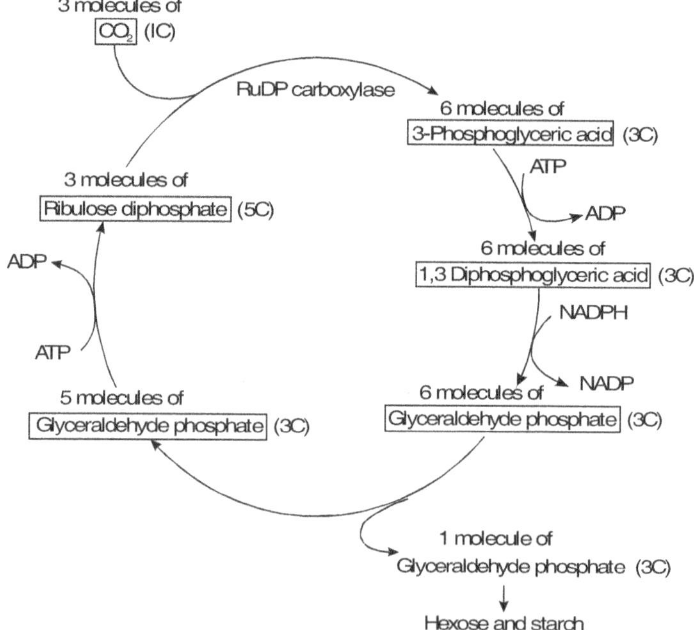

Figure 2.34   Schematic representation of Calvin cycle (C3 pathway)

HSK Pathway for C4 Plants

In spite of the significance of the Calvin cycle, its universal presence was questioned (Karpilov and Tarchevseky, 1963). During the same period, Kortschak *et al.* reported the presence of malic acid and aspartic acid in the leaves of sugar cane, which was later confirmed by Hatch and Slack in 1966. Plants living in warm tropical climate with bright sunshine, e.g. monocots like rice, maize, sugar cane, etc. form oxaloacetic acid (OAA) as the first stable 4-carbon compound. Therefore these plants are called C4 plants as they follow typical C4 photosynthetic pathway that is also known as HSK pathway or β-carboxylation pathway (Figure 2.34).

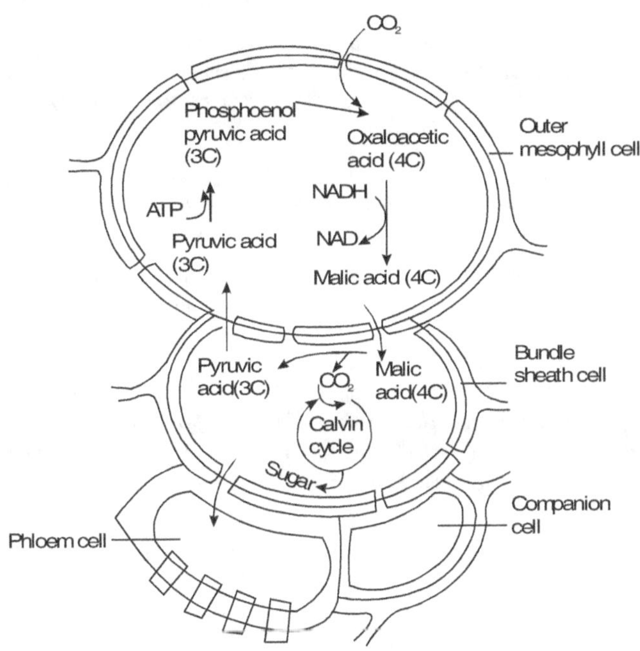

Figure 2.35 Schematic representation of HSK pathway in C4 plants

The C4 plants show the most distinguishing anatomical feature in their leaves called, **Kranz** ("wreath") anatomy. In these leaves, mesophyll is not differentiated into palisade and spongy parenchyma. Radially arranged chlorenchymatous bundle sheath cells surround the vascular bundles. The C4 plants have in their leaves two types of chloroplasts (dimorphic chloroplasts). The bundle sheath cells do not have grana while mesophyll cells possess well-developed grana. In the outer mesophyll cells, phosphoenolpyruvate (PEP) acts as CO_2 acceptor in the presence of the enzyme PEP carboxylase to form the 4-carbon compound, oxaloacetic acid (OAA), as the first stable compound. It is further reduced to malic and/or aspartic acid. These compounds then move towards the site of C3 photosynthesis, i.e., bundle sheath cell where these acids are converted enzymatically into pyruvic acid and CO_2. This carbon dioxide is then used in the Calvin cycle.

⊙ Plastids are small inclusions found in most of the plant cells and some protozoans and are generally absent in the animal cells, bacteria, and fungi. The plastids are of two types, viz. leucoplasts (colourless plastids), and chromoplasts (coloured plastids).

⊙ Leucoplasts are usually present in the cells that do not receive direct light and also in the plant gametes and embryonic cells. They mainly serve for the storage of food— amyloplasts store starch and are found in tubers, cotyledons and endosperm, aleuroplasts store proteins while elaioplasts store oils.

⊙ Chromoplasts are the coloured plastids having coloured pigments and they act as primary sites for trapping and converting solar energy which plays a significant role in the existence of not only the green plants, but also the entire living world.

⊙ On the basis of coloured pigments, chromoplasts are classified as a) chloroplasts (green coloured plastids) present in most of the plant cells and playing an indispensable role in photosynthesis, b) phaeoplasts (brown coloured plastids) having the brown pigment called fucoxanthin found in brown algae, diatoms and in some dinoflagellates, c) rhodoplasts (red coloured plastids) having phycoerythrin found in red algae, d) blue-green chromoplasts present in blue-green algae containing the pigments called phycocyanin, chlorophyll a and carotenoids. The colours of the flowers and the fruits are due to chromoplasts other than chloroplasts, which do not take part in photosynthesis.

⊙ Electron micrograph of chloroplast also shows a complicated lamellar system composed of small cylindrical structures called the grana. Each granum consists of stacked thylakoids. Adjacent grana are connected together by small tubules called intergrana or the stroma lamellae. The stroma or matrix is the ground substance enclosed by the walls of the chloroplast.

⊙ The thylakoid membranes enclose the space called fret channel, which consists of small discs or flattened spheroids known as quantosomes as the photosynthetic units. Each quantosome comprises of about 250 molecules of chlorophyll along with carotenoids.

⊙ Each chlorophyll molecule is made up of a hydrophilic "head" and hydrophobic long "tail". The "head" is the porphyrin structure having four pyrrole rings attached to each other forming tetrapyrrole structure and Mg atom at the centre. A long phytol "tail" extends from one of the pyrrole rings.

SUMMARY

⊙ *In addition to chlorophyll pigments, hydrocarbon carotenes and xanthophylls act as accessory pigments. Two types of phycobilins, the phycoerythrin and phycocyanin are also accessory pigments in algae.*

⊙ *The prime function of chloroplast in green plants is to perform photosynthesis in the presence of sunlight and manufacture sugars from water and carbon dioxide. During photosynthesis, water is oxidized to oxygen while carbon dioxide is reduced to sugar.*

⊙ *Photosynthesis involves two reactions, the Hill reaction, which is carried out in presence of light, hence also called the light reaction, the Blackman reaction, and the second reaction is independent of light hence known as dark reaction.*

⊙ *Light reaction takes place in the thylakoid membrane as a photochemical reaction. The light reaction involves phosphorylation and photolysis of water that leads to the synthesis of ATP and release of oxygen respectively.*

⊙ *Dark reaction is the thermochemical reaction in which NADPH and ATP are used to reduce atmospheric CO_2 to form carbohydrate. Thus dark reaction is the carbon assimilation phase that occurs inside the stroma of chloroplast.*

⊙ *Calvin traced the path of carbon using its radioactive isotope (^{14}C) and observed that carbon reduction in the unicellular green algae is a cyclic process and the first stable product is 3-PGA (3-carbon compound), hence such plants are called C3 plants.*

⊙ *Monocots like rice, maize, sugar cane, etc. thriving in warm tropical climate form oxalo-acetic acid (OAA) as the first stable 4-carbon compound. Therefore these plants are called C4 plants as they follow the typical C4 photosynthetic pathway, also known as HSK pathway. C4 plants have the most distinguishing anatomical feature in their leaves, called Kranz anatomy.*

REVIEW QUESTIONS

1. What are plastids? Explain the different types of plastids with their functions.

2. Describe the various classes of chromoplasts with examples and functions.

3. Explain the ultrastructure of chloroplast.

4. Describe the types of photosynthetic pigments with their structures and absorption spectra.

5. Define photophosphorylation. Distinguish between cyclic and non-cyclic photophosphorylation.

6. Describe the basic plan of the Calvin cycle with its significance.

7. What are C4 plants? Describe the HSK pathway with its importance.

8. What is the relationship between light and dark reactions?

9. What are PS I and PS II ? Explain the transfer of electrons in Z-scheme.

10. Write short notes on:

    i.   Leucoplasts

    ii. Quantosome

    iii. Photolysis of water

    iv. Kranz anatomy

    v.   Reaction centres

    vi. Chloroplast DNA

ENDOPLASMIC RETICULUM

Most of the plant and animal cells have a network of delicate strands and vesicles in their cytoplasm, which was previously referred to as **ergastoplasm** by Garnier (1897). Later on, Porter and Kaallman (1952) called them endoplasmic reticulum (ER). The ER is connected to the outer nuclear membrane with the ER space opening into the perinuclear space between the two membranes and it is known to be connected with the plasma membrane and opens to the outside (as shown in animal cell, Figure 1.2). Similar to mitochondria and plastids, ER provides a membrane surface to perform biochemical reactions and an internal compartment to segregate certain functions. Its membrane is used for the synthesis of membrane and secretory proteins, as well as many lipids. Within the ER lumen, proteins are modified (by the process like glycosylation) and targeted inside or outside the cell. In addition, the lumen serves as a storage depot for Ca^{2+}, especially in muscle cell.

STRUCTURE OF ER

The endoplasmic reticulum is found in both animal and plant cells and is absent in prokaryotic cells. Its structure varies from cell to cell. However, in general, morphologically it consists of three types of structures, viz. **cisternae, vesicles** and **tubules**.

The cisternae are flat, unbranched, and elongated membrane-bound spaces arranged parallel to one another representing a lamellated structure. They are 40–50 μm in diameter and generally are found in cells active in synthesis.

The **vesicles** are abundant in pancreatic cells and have sac-like structures or are in the form of oval bodies attached to the membrane with a diameter of about 25–500 μm.

The **tubules** show variation in shape and make their presence in the cells engaged in the synthesis of steroids.

TYPES OF ER

The endoplasmic reticulum is divided into two broad categories: the **rough endoplasmic reticulum** (RER) and the **smooth endoplasmic reticulum** (SER). Both types of ER are composed of a system of membranes that enclose the membrane-bound spaces. The fluid content of the cytoplasm is accordingly divided by ER into two compartments, that is, the space enclosed within its membranes is called **cisternal space** (usually 30 nm in diameter), and the region outside the membranes is referred to as **cytosolic space**. The cisternal space can increase considerably in certain situations, such as induction by drugs. The huge surface area of ER makes it the largest membrane in the cell.

For example, in the liver hepatocyte, ER occupies 15% of the cell volume, and its surface area of 63,000 μm^2 is 37 times that of the plasma membrane.

The outstanding morphological difference between RER and SER is the presence of ribosomes on the surface of RER facing the cytosolic space whereas ribosomes are absent in SER. The RER appears as an extensive membranous organelle composed primarily of cisternae separated by cytosolic space as shown in Figure 2.36a. The RER is commonly a network of parallel, interconnected sheets, surrounding the nucleus and radiating out towards the periphery of the cell. Numerous RER are present in the cells that are active in synthesis of proteins, especially membrane and secretory proteins. RER proliferates during hormone-induced cell differentiation. Although electron micrograph of RER shows separate compartments of cisternal spaces, it is thought that all cisternae of RER are continuous with one another.

In the electron micrograph of SER (Figure 2.36b), the membranous elements are typically tubular and form an interconnecting system of pipelines curving through the cytoplasm in which they are present. In addition to lacking ribosomes, SER often appears more vesicular, with swollen cisterane. It may be found along with RER in some cell types, like that in liver hepatocytes, but it is more commonly proliferated in specialized cells, which do not have RER. For example, SER is found in the cells of endocrine glands (e.g. testicular cells) that are actively synthesizing steroid hormones and in flower petal cells, SER may be correlated with the synthesis of floral pigments. When an animal is treated with a certain drug, liver SER proliferates in association with the metabolism of the chemical. In striated muscle cells, an extensive SER surrounds the muscle fibres; the complex structure is known as **sarcoplasmic reticulum** playing a significant role in muscle contraction. The SER is also present in the pigmented cells of retina in the form of tightly packed tubules and vesicles called myeloid bodies.

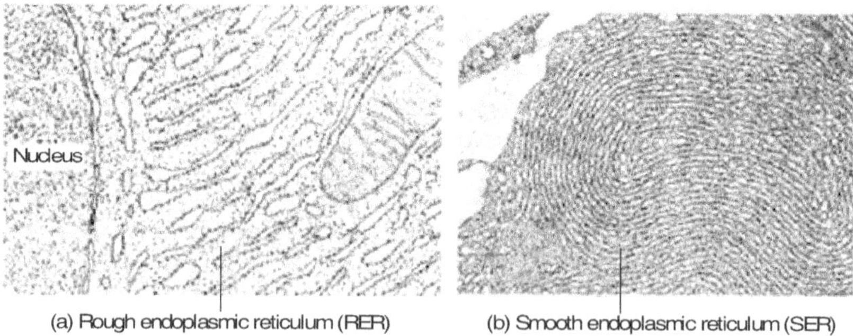

(a) Rough endoplasmic reticulum (RER)   (b) Smooth endoplasmic reticulum (SER)

Figure 2.36   Electron micrographs of rough and smooth endoplasmic reticulum

SARCOPLASMIC RETICULUM

It is a modified smooth endoplasmic reticulum complexed with the striated muscles. It was first reported by Veratti (1902) as a complicated network of vesicles, tubules and cisternae that are closely associated with muscle fibres. Electron microscopy showed that the sarcoplasmic reticulum (SR) forms a membranous sleeve around the myofibrils (Figure 2.37). Longitudinal tubules of ER merge with one another to form the outer vesicles. Two adjacent outer vesicles have a **transverse tubule** in between them. Both the outer vesicles and a transverse tubule together form a **triad** or **T-system** over the Z-line of myofibril. The adjacent triads are interconnected through sarcoplasmic tubules to form **central cisternae** that lie along the H-zone of the anisotopic (A-) band of the myofibril. Such a T-system is quite prominent in fibres of active muscles, but less so in cardiac and slow muscles. Because the T-system plays an indispensable role in excitation and relaxation of the muscle, it is more than a telegraphic connection signalling contraction. Approximately 80% of the integral protein of SR membrane consists of a Ca^{2+}–ATPase whose function is to transport Ca^{2+} ions out of the cytoplasm and into the lumen of SR where they are stored until needed.

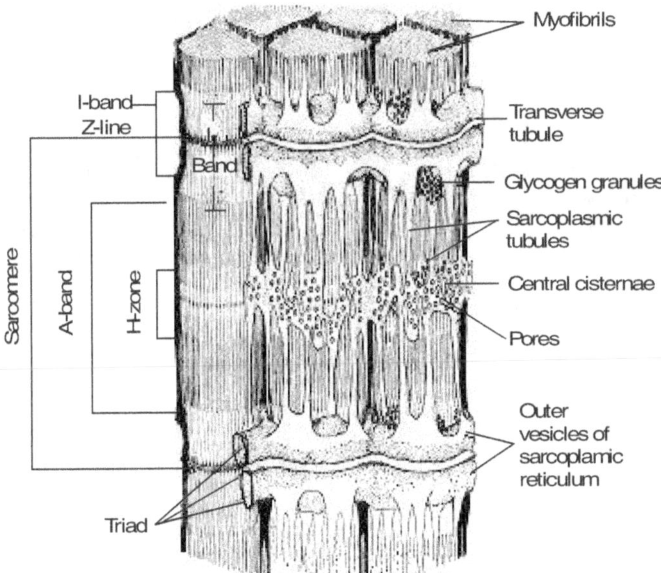

Figure 2.37 Functional anatomy of a sarcoplasmic reticulum

A muscle contracts in response to the stimulus it gets via the innervating motor nerve fibre at the myoneural junction. The excitation of the sarcolemma (plasma membrane of the muscle cell) at the arrival of nerve impulse induces the depolarization of T-system, which in turn, in some way, results in the liberation of calcium ions from the outer sarcoplasmic reticulum vesicles of the triad. The elevated calcium levels trigger the contraction in a muscle fibre in the presence of ATP.

MICROSOMES DERIVED FROM ER DURING CELL FRACTIONATION

With the advent of electron microscopy and autoradiography, it becomes possible to study the structural and functional aspects of cellular organelles. But these

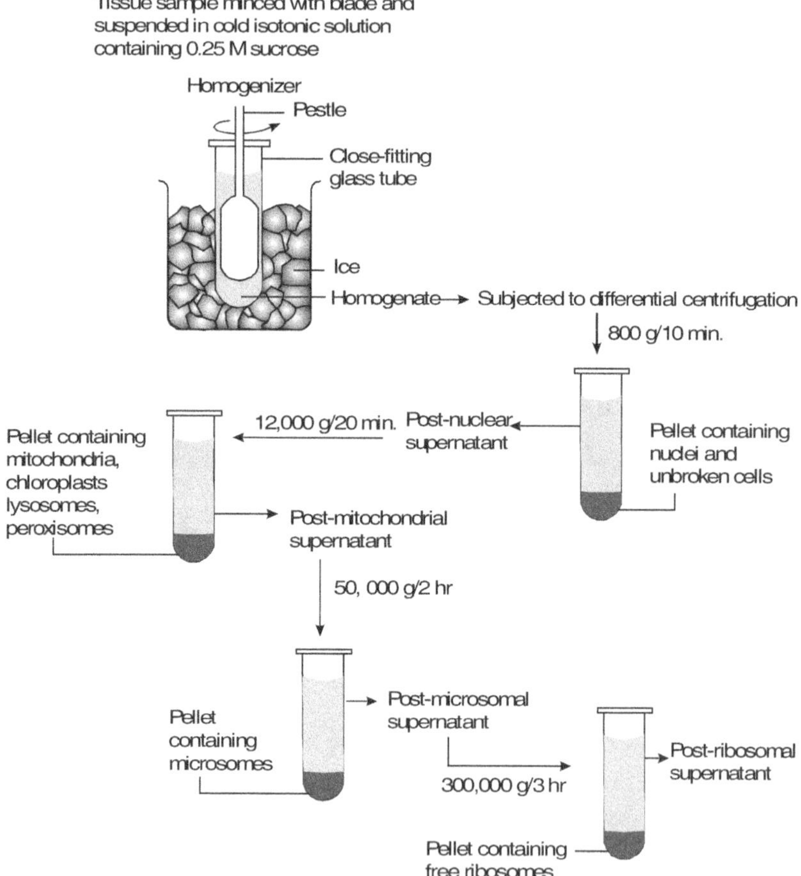

Figure 2.38   Process of cell fractionation. The step-by-step procedure for isolation of cell organelles by differential centrifugation.

techniques are unable to provide an insight into the biochemical composition of different cell organelles. In the 1950s and 1960s, A. Claude and C. De Duve pioneered in the development of the techniques that break up (**homogenize**) cells and help in isolation of the particular types of organelles. The isolation of a particular cell organelle in bulk quantity is generally accomplished by the technique called **differential centrifugation** (Figure 2.38). It is based on the principle of the density of a particular organelle. As long as they are more dense than the surrounding medium, particles of different size and shape travel towards the bottom of a centrifuge tube and form a pellet at different rates when subjected to centrifugation.

In a homogenizer, the cell is ruptured by mechanical disruption in an isotonic buffered solution (often contains sucrose). The homogenate is then subjected to sequential centrifugations, initially to low centrifugal force for a short period to separate the large cellular organelles, the nuclei that are sedimented into a pellet. By increasing successive centrifugal forces, larger cytoplasmic cell organelles like mitochondria, chloroplasts, lysosomes, and peroxisomes can be separated out from the suspension. Further the post-mitochondrial supernatant subjected to higher centrifugal force for a longer duration yields the **microsomes**, which are spherical vesicles having a diameter less than 100 nm derived from the fragments of endoplasmic reticulum. Finally the post-microsomal supernatant is subjected to ultracentrifugation to obtain free ribosomes.

Chemical Composition of Microsomes

When the cell disruption destroys the fragile interconnections of ER, the ends of the membrane fragments fuse to form vesicles called microsomes. In relation to the chemical composition, it can be stated that the microsomal membrane is approximately 40% lipids and 60% protein by weight. The lipids present in the microsomal fraction are mostly phospholipids that consist of phosphatidylcholine, phosphatidylethanolamine, phosphatidylinositol and phosphatidylglycerol. **Ribophorins** (or **ribosome receptor proteins**) are the membrane proteins of RER that help ribosomes to bind to the microsomal membrane while several proteins are permanent residents of the cisterna. These are called **reticuloplasmins** that include several enzymes that modify proteins after they are made (e.g. protein disulphide isomerase that catalyses disulphide bond formation), proteins involved in stabilization of protein structure (e.g. the binding protein), and the Ca^{2+}-binding proteins. The enzymes present in the liver ER membrane are ATPase, cytochrome c reductase, cytochrome P450, NADH-cytochrome b_5, NADH-cytochrome c reductase, nucleoside pyrophosphatase, and nucleotidase-5′. The RNA content determines the density of microsomes, with rough microsomes being denser than smooth ones. This permits a separation of RER from SER.

FUNCTIONS OF ER

Synthesis of Secretory Proteins and their Modification

Jamieson and Palade demonstrated the importance of ER in the synthesis of proteins and processing of integral membrane proteins in exocrine pancreas. The proteins are made on RER with a hydrophobic signal, by which they can embed in the lipid bilayer. In addition, the ER is the site of synthesis of a group of proteins targeted for storage within another membrane-bound organelle in the cell (such as lysosomes in animal cell, storage proteins in seeds and membrane proteins) or for export outside of the cell (secretion). This forms one class of proteins that are synthesized on the ribosomes attached to the outer (cytosolic) surface of the RER membranes. This is in contrast to another class of proteins, such as enzymes of glycolysis, proteins of cytoskeleton, and proteins normally found in microbodies, chloroplasts, and mitochondria, that are synthesized on the "free" ribosomes and subsequently released into the cytoplasm.

Gunter Blobel and David Sabatini proposed the **signal hypothesis**, which explained that the synthesis of a protein on a membrane-bound or free ribosome was determined by information contained in the N-terminal portion of the polypeptide that emerges first from the ribosome during protein synthesis. They suggested that secretory proteins contained a special **signal sequence** (or **signal peptide**) at their N-terminus that triggered the attachment of the ribosome to an ER membrane and the movement of nascent polypeptide into the cisternal space of the ER. The presence of signal sequence is a common feature of virtually all proteins that enter the endoplasmic reticulum. Sometimes during the subsequent passage of the growing polypeptide chain through the membrane, the signal sequence is clipped off by a special protease called **signal peptidase**. Once inside the lumen of the ER, the protein is further modified and then delivered to the final destination.

As mentioned earlier, the signal sequence is translated on a free ribosome, and then it binds the synthesizing complex to the ER. As the intermediary between translation and binding, there is a small RNA-protein complex referred to as **signal recognition particle** (SRP). This is composed of a molecule of RNA with about 300 nucleotides and six different tightly bound polypeptide chains. An interesting aspect of its mode of action is SRP's ability to bind to ribosome and stop protein synthesis. Generally, about 70 amino acids are translated before SRP stops translation. The SRP does not arrest translation of all mRNAs, but specially acts on those mRNA which have coded information for synthesis of secretory proteins. The translation block is reversed only when the SRP makes contact with the **docking protein**, an integral part of the ER membrane. Once the ribosome with its protein-synthesizing components binds to ER membrane surface via docking protein, SRP diffuses away to initiate transport of another polypeptide chain (Figure 2.39). During translocation, the attachment of ribosomes to ER membrane is primarily assisted by ribophorins as mentioned earlier. Soon after the signal sequence has traversed the ER membrane, and while the rest of the secretory or storage protein is still being translated, a signal peptidase associated with the ER membrane cleaves off-signal sequence. Cleavage of the signal peptide appears to be an essential event for the functioning of the protein translated after it.

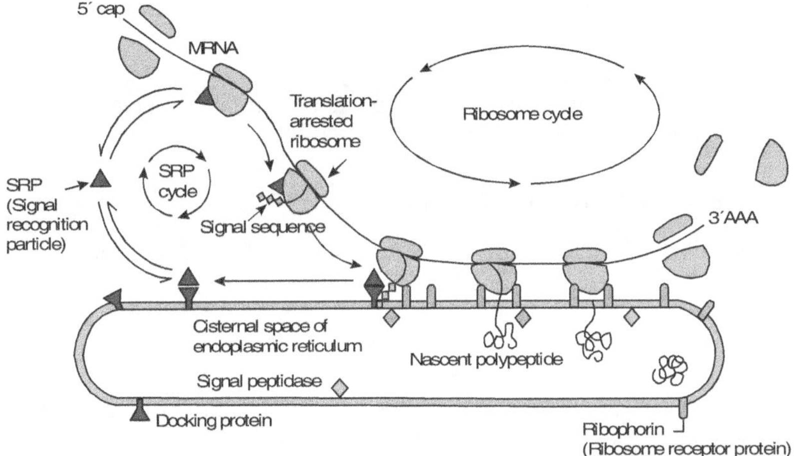

Figure 2.39 Role of SRP and docking protein in the translocation of proteins across the endoplasmic reticulum

After the signal sequence has been cleaved, but still during protein synthesis, four possible modifications of protein can take place. Enzymes present in inner ER membrane or cisternal space catalyse the modifying reactions, which change the physical properties of protein. These reactions are as follows:

Glycosylation of protein   It is the addition of pre-synthesized oligosaccharide chains attached to the lipid carrier, called dolichol phosphate which is a hydrophobic molecule built from more than 20 isoprene units and is embedded in the lipid bilayer of the ER membrane. The oligosaccharide chain consists of *N*-acetyl glucosamine, mannose, and glucose and its transfer to protein is catalysed by a group of membrane-bound enzymes called **glycosyltransferases**. Its attachment to protein ensures that the target protein will be in the cisternal space in a proper three-dimensional configuration. Nearly all of the proteins produced on membrane-bound ribosomes—whether integral components of a membrane, lysosomal or vacuolar enzymes, or parts of extracellular matrix—become glycoproteins as a result of glycosylation.

Formation of disulphide bonds   These bonds are formed between cysteine residues present in the polypeptide chain. As a result, different regions of the protein are covalently linked, which is essential for establishing the three-dimensional structure of protein. The reaction is catalysed by protein disulphide isomerase.

Hydroxylation of proline and lysine residue in a polypeptide chain   This causes the interchain linkages of the fibrous proteins like collagen and elastin to becomes stabilized. The reaction is catalysed by peptidylproline hydroxylase.

Carboxylation of glutamate residue   This produces the γ-glutamyl side chain providing the modified protein the ability to bind Ca^{2+} ions. The reaction is significant in the blood-clotting protein, prothrombin, where it binds tightly to clotting factors on cell surfaces.

Role of ER in the Formation of Transport Vesicle

Typically large, flattened cavities of the cisterane of the RER can act as transport channels for membrane and vacuolar proteins to be moved from their site of synthesis to the apical tips of the RER. The cisternae of RER having smooth surface at the apical tip are without ribosomes and are called transitional elements that cut off from ER to form transport vesicles in the biosynthetic pathway, which are targeted to Golgi complex for further processing to form secretory granules (Figure 2.40).

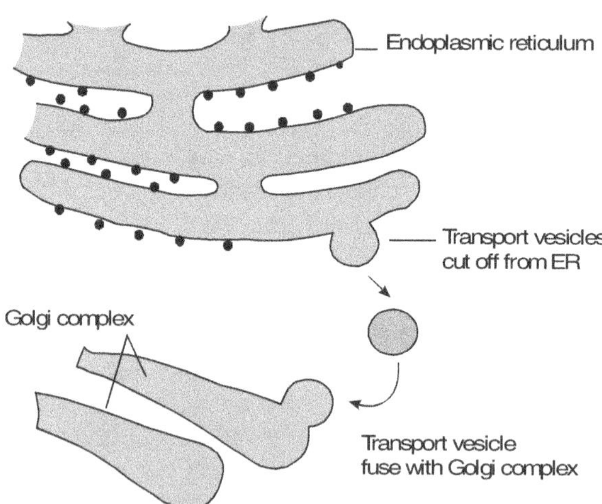

Figure 2.40   Transport of proteins from endoplasmic reticulum to Golgi complex

ER and Carbohydrate Metabolism

In plant cells SER is involved in the synthesis of cell wall polysaccharide, while in mammalian liver cells, glycogen is deposited on the surface of the SER. In a starving animal, the hormone glucagon stimulates a series of events to convert stored glycogen to the final product (glucose), which on oxidation provides energy to vital organs. This is especially true in fasting animals in which, when resumption of feeding occurs, accumulation of polysaccharides takes place. The sequence of biochemical reactions in the conversion of stored glycogen to glucose is as follows:

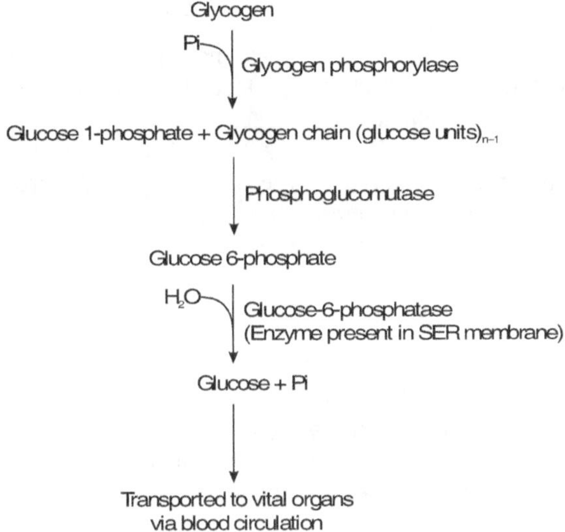

ER and Lipid Metabolism

The ER plays an indispensable role in lipid anabolism, as it contains a number of enzymes that are involved in the process. The significant anabolic pathways in lipid metabolism include incorporation of long chain fatty acids to lipids, acylation with coenzyme A, synthesis of fatty acids and desaturation, synthesis of glycolipids and lipoproteins, etc., which are carried out by the enzymes present in ER.

Specialized lipids like terpene oils and sticky fluid on the stigma of flowers are often made in the SER of plant tissues, whereas the animal organs like testes, ovary, adrenal gland, etc. have large amount of SER to synthesize steroid hormones. In addition, the ER in many tissues synthesizes prostaglandins from fatty acids.

Sarcoplasmic Reticulum (SR) and Muscle Contraction

As mentioned earlier, the SR is the filamentous network running between muscle fibres. The T-system in SR depolarizes at the arrival of a nerve impulse, which in turn depolarizes the sarcolemma. Through a large membrane-spanning protein, the signal is transduced to the SR that causes to release Ca^{2+}. At the muscle fibre, Ca^{2+} binds to troponin and allows the contractile proteins, actin and myosin, in sarcomere (functional unit of the muscle) to slide past one another for contraction.

ER and Formation of Nuclear Envelope

The ER is connected to the outer nuclear membrane with the ER space opening into the perinuclear space between two membranes. It is suggested that the ER originates from the nuclear envelope, which it resembles physically, and chemically to some extent; however, at the end of mitosis, the nuclear envelope re-forms from the ER, making this seem unlikely (DeRobertis *et al.*, 1970). In addition to the nuclear membrane, ER is also involved in the formation of most of the cell organelles like Golgi complex, mitochondria, lysosomes, cell plate, etc.

ER, Electron Transport and Detoxification

The cytochrome P450 is the electron acceptor between NADPH and O_2 that is known to be located in the SER outer surface. There are many forms of P450 carrying a variety of functions including electron transport from diverse substrates like bile acid precursors, cholesterol, fatty acids, geraniol, vitamin D_3, etc. and the detoxification or activation of xenobiotics such as drugs and pesticides.

In the body of vertebrates, microsomal P450 is found in the lungs, skin, gastrointestinal tract, placenta and other locations of the entry of xenobiotics. In liver cells, abundant SER with P450 carry out biotransformation of xenobiotics, which enter the body through the digestive system. In most cases, the potentially toxic substances like benzpyrene (component of cigarette smoke), codeine (narcotic analgesic), amphetamine (stimulant drug), pentobarbitol (anaesthetic), etc. are inactivated by P450 oxidation. There are few notable exceptions: P450 converts aflatoxins (metabolites of certain fungi that contaminate food) and nitrosamines (by-products of nitrites used to preserve meat product) into potent carcinogens that can transform a normal cell into malignant cell.

SUMMARY

- Endoplasmic reticulum (ER) is a network of delicate strands and vesicles in the cytoplasm of most of the plant and animal cells and is absent in prokaryotic cells.

- Morphologically, ER consists of three types of structures, viz. cisternae, vesicles and tubules. The cisternae are flat, unbranched, and elongated membrane-bound spaces arranged parallel to one another and representing a lamellated structure. The vesicles are sac-like structures or are in the form of oval bodies attached to the membrane, and the tubules show variation in shape and make their presence in cells engaged in the synthesis of steroids.

- ER is of two types—the rough endoplasmic reticulum (RER) and the smooth endoplasmic reticulum (SER). The striking difference between RER and SER is the presence of ribosomes on the surface of RER whereas SER are devoid of ribosomes.

- Sarcoplasmic reticulum is the complex structure formed in striated muscle cells, where an extensive SER surrounds the muscle fibres playing significant role in muscle contraction. SER includes two outer vesicles and a transverse tubule, which together form a triad or T-system over the Z-line of myofibril.

- During cell fractionation, the post-mitochondrial supernatant subjected to higher centrifugal force for a longer duration yields the microsomes, which are spherical vesicles derived from the fragments of endoplasmic reticulum.

- Chemically, the microsomal membrane is lipoproteinous. The lipids are mostly phospholipids, and ribophorins (or ribosome receptor proteins) are the membrane proteins of RER that help ribosomes. Reticuloplasmins are proteins found in the cisterna that includes several enzymes to modify proteins after they are made.

- ER is the site of synthesis of a group of proteins targeted for storage within another membrane-bound organelle in the cell such as lysosomes in animal cell, storage proteins in seeds and membrane proteins or secreted outside of the cell.

- Enzymes present in cisternal space of ER catalyse the modifying reactions, which change the physical properties of protein. The modifying reactions are glycosylation of proteins, formation of disulphide bonds between polypeptide chain, hydroxylation and carboxylation of some of the amino acids in a polypeptide chain.

- In addition to synthesis and modification of proteins, ER plays an indispensable role in the synthesis of glycogen, lipids, cholesterol and steroid hormones, and helps in formation of nuclear envelope and other cell organelles, detoxification of certain drugs and muscle contraction.

REVIEW QUESTIONS

1. Describe the ultrastructure of rough and smooth endoplasmic reticulum.

2. Explain how proteins are secreted through the endoplasmic reticulum.

3. What are the major morphological differences between RER and SER? Give an account of the major differences in their functions.

4. Describe the structure and function of sarcoplasmic reticulum.

5. Explain the steps that occur between the time a ribosome attaches to a messenger RNA encoding secretory protein and the time the protein leaves the RER.

6. Define differential centrifugation. Describe the steps to isolate the microsomal fraction from a tissue homogenate.

7. Explain the chemical composition of microsomes. Add a note on the function of ribophorins and reticuloplasmins.

8. What are the different methods of modification of protein that are to be secreted from RER.

9. Describe the various functional aspects of endoplasmic reticulum.

10. Write short notes on:
    i. T-system
    ii. Signal peptidase
    iii. Docking proteins
    iv. Signal recognition particles
    v. Transport vesicles
    vi. Biotransformation of xenobiotics in SER
    vii. Cell fractionation

GOLGI COMPLEX

This is a membrane-bound structure found in the cytoplasm that performs specific biochemical functions. In 1898, Camillo Golgi, an Italian Biologist, discovered a dark yellow network located near the nucleus in the nerve cells of barn owl. He termed this the **apparate reticulare interno** that came to be known as the Golgi complex. Later, C. Golgi won the Nobel Prize for this discovery in 1906. Golgi complex remained a controversy over the next half century between those who believed that the organelle actually existed in the living cells and those who believed it was an artifact (an artificial structure formed during preparation for microscopy). Only the development of phase contrast, especially electron microscopy, helped in the identification of the Golgi complex as a cell organelle in unfixed, freeze-fractured specimen. Baker (1953) described this complex as lipochondria as they have high lipid content.

SHAPE, SIZE AND POSITION

The Golgi complex, with rare exceptions (e.g. red blood cells), is present in all eukaryotic cells, but is perhaps more prominent in vertebrate cells than in invertebrate or plant cells, where it has been collectively called the **dictyosome**. The Golgi complex is organized in basically different ways in different types of cell, but its organization for any one kind of cell is usually the same. Thus, it is small in muscle fibres and many other types of cells, but it is large and well developed in secretory cells and in nerves, where it shows a clear reticulate structure. It occupies different positions in different kinds of cells. In higher plants, the Golgi complex is scattered all over the cytoplasm apparently without any location, whereas in animal cells, it is localized into an organelle.

STRUCTURE OF GOLGI COMPLEX

Its shape and form may vary depending on the cell type. Typically, however, the Golgi complex appears as a complex array of interconnecting **cisternae**, **tubules**, and **vesicles** (Figure 2.41). The flattened and disc-like cisternae having dilated rims are associated with vesicles and tubules to form the Golgi complex. Each Golgi body consists of 4–8 cisternae, each having a diameter of 0.5 to 1.0 μm, which are arranged in an orderly stack. Each cisterna is slightly curved, enclosed by the trilaminar membrane, 6 nm in thickness and separated from the other by a 100–150 Å space. In some cases, the intercisternal space is filled with fibres. The membranes of cisternae are fenestrated (porous) and the pores may be localized or present all over the surface. The tubules arise from the peripheral region of the cisternae and form an anastomosing network. The vesicles are goblet-like structures attached to the tubules and are pinched off from the dilated rims of the cisternal tubes. The vesicles are of two types: the

smooth vesicles often called secretory granules and the rough surfaced **non-clathrin-coated vesicles**, whose vesicle coats contain a complex of seven distinct, but related, coat proteins (COPs).

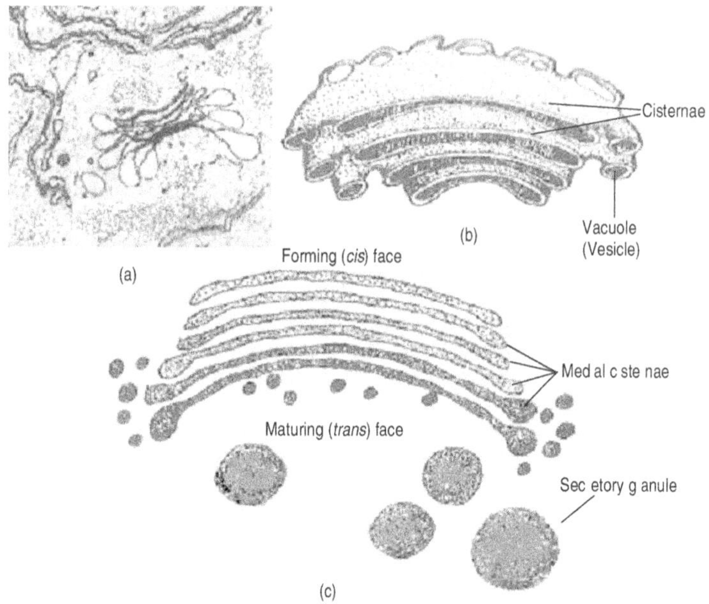

Figure 2.41   (a) Electron micrograph of Golgi complex, (b) Stereoscopic view and (c) Section view of Golgi complex

The cisternae of the Golgi complex show polarity, the cisterna closest to the ER is said to be at the *cis* face, while the cisternae at the opposite end of the stack present towards the plasma membrane is said to be at *trans* face. In between *cis* and *trans* cisternae the stack of cisternae are called **medial** cisternae. Differences in the composition of the membrane compartments from the *cis* to the *trans* face through the medial cisternae reflect a polarity in the function of Golgi complex. Newly synthesized membrane, secretory, and lysosomal proteins leave ER through the vesicles that bud off from dilated rims of ER and enter the Golgi complex through *cis* or "forming" face and then passing through **medial** cisternae and finally reach the *trans* or "mature" face. This overall polarity of Golgi complex has been observed in many cell types, which further supports the concept of a flow of membrane components and synthetic products from ER to the Golgi. On the mature face of the Golgi complex, there are numerous small vesicles that appear to be budding off. These vesicles apparently contain a synthetic product that has been modified and accumulated.

CHEMICAL COMPOSITION

The electron microscope cytochemistry study of Golgi complex reveals that different cisternae appear to have different chemical composition. Some of the enzymes are more concentrated in *cis* or *trans* cisternae of the Golgi complex indicating functional differences across the organelle. The enzymes present in the cisternae are galactosyl transferase, inosine diphosphatase, acid phosphatase, thiamine pyrophosphatase, etc. Some of these enzymes are not present elsewhere in the cell and hence considered as reliable marker enzymes of the Golgi complex.

About 60% lipids and 40% proteins by weight are present in the Golgi membrane. The chemical composition of Golgi membranes is somewhat intermediate between ER and plasma membrane. The lipids present in the membrane consist of cholesterol, sphingomyelin, phospholipids, fatty acids, etc. The complete biochemical analysis of *cis* to *trans* cisternae of Golgi complex suggests that the *cis* side of Golgi is similar in its chemical composition to that of ER, and the vesicles derived from *trans* side resemble the plasma membrane with which they may fuse. The proteins present in the Golgi membrane also show similar patterns of distribution among the *cis* and *trans* cisternae. In addition to proteins and lipids, some of the primary sugars like sialic acid, glucosamine, and hexoses such as mannose and galactose also mark their presence in the Golgi membrane.

FUNCTIONS OF GOLGI COMPLEX

Many functions have been attributed to the Golgi. Some of them are as follows:

Secretion of proteins   Secretory cells have well developed, enlarged Golgi, whereas non-secretory cells have a small, poorly developed Golgi. Autoradiography and electron microscopy studies have demonstrated that secretory proteins produced in the ER are stored in Golgi and subsequently are contained in the vesicles budded off from the *trans* Golgi network (TNG). These vesicles containing secretory proteins are called zymogen granules, which eventually release their content from the cell by the process called exocytosis [Figure 2.42(a)]. These secretory proteins may be enzymes, peptide hormones or lysosomal proteins. For several peptide hormones such as insulin and parathyroid hormone, post-ER proteolytic processing takes place in cisternae of Golgi complex prior to secretion. In the case of insulin, the initial protein synthesized in β-cells of islets of Langerhans in pancreas is **preproinsulin** having 103 amino acids in its polypeptide chain. In RER, the terminal 23 amino acid residues are removed; as a result, hormonally inactive **proinsulin** is formed. Later on, an internal 30-residue segment is excised from this chain in Golgi and secretory granules, leaving two polypeptide chains—A-chain with 21-amino acid residues and B-chain with 30-amino acid residues as the active from of the hormone insulin that is to be released from the cell by exocytosis.

Synthesis of carbohydrates   The Golgi complex is also the site of synthesis and concentration of carbohydrate-rich products, such as glycoproteins and mucopolysaccharides. The role of Golgi complex in glycosylation was first shown by autoradiography in goblet cells of intestinal epithelia, where mucin (a protein with very high sugar content) is formed, stored, and released.

In addition, the Golgi complex is also involved in the synthesis of most of a cell's complex polysaccharides, including, glycosaminoglycans found in the extracellular matrix of animal cells or the pectins and hemicellulose found in the cell wall of plants.

Cell division and Golgi complex   In plant cells, the Golgi complex plays an indispensable role in the formation of cell plate during mitosis.

Formation of lysosomes   The RER synthesizes hydrolytic enzymes with the help of ribosomes attached to its outer surface, which are then sent to *cis* cisternae of the Golgi complex through transport vesicles. Later on, the secretory vesicles referred to as primary vesicles containing hydrolytic enzymes are budded off from *trans* cisternae of Golgi complex. The primary vesicles may in turn fuse with endocytotic vesicles to form secondary lysosomes [Figure 2.42(b)].

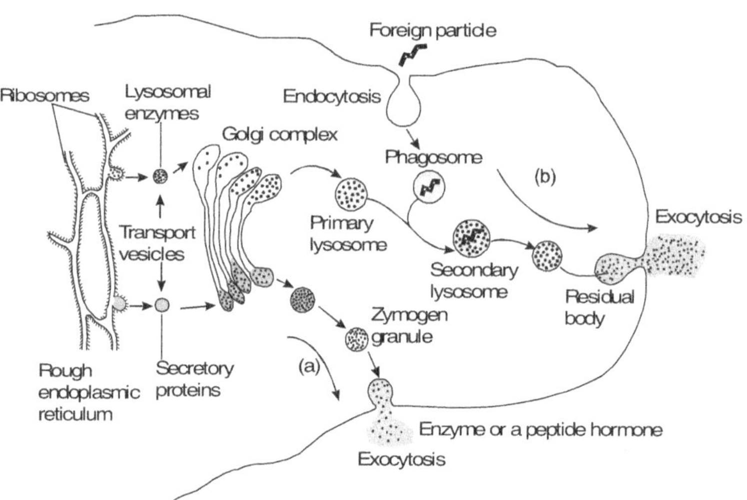

Figure 2.42   Role of Golgi complex in the formation of zymogen granule filled with digestive enzyme-secretory protein that is released from the cell by exocytosis as shown in (a) and, in the formation of primary lysosome, which fuses with endocytotic vesicle and forms secondary lysosome enclosing foreign particle to be digested by lysosomal enzymes as shown in (b). The residual body encloses the waste material formed after destruction of foreign particle (by hydrolytic lysosomal enzymes) that is thrown out from the cell by exocytosis.

Formation of acrosome   During spermiogenesis (development of motile spermatozoa from non-motile spermatids), the Golgi complex is seen to contribute to the formation of proacrosomic granules from its cisternae. The granules then fuse to form the acrosome of the spermatozoa.

Formation of melanin granules   In animals, the Golgi is involved in the segregation of hormone-secretion products and segregation of pigments. Pigment in the form of melanin granules, is found in the Golgi complex following its synthesis in ER of cancer cells. The oocyte of the salamander and retinal cells in the chick embryo show the Golgi complex associated with formation of melanin.

SUMMARY

- *Camillo Golgi discovered a dark yellow network known as the Golgi complex located near the nucleus in the nerve cells of barn owl.*

- *The Golgi complex is present in all eukaryotic cells, with rare exceptions (e.g. red blood cells). It is perhaps more prominent in vertebrate cells than in invertebrate or plant cells, where it has been collectively called the dictyosome.*

- *Morphologically, Golgi complex appears as a complex array of interconnecting cisternae, tubules, and vesicles.*

- *Cisternae of Golgi complex are differentiated as cis or "forming" face, medial cisternae and the trans or "mature" face.*

- *The biochemical analysis of cis to trans cisternae of Golgi complex suggests that the cis side of Golgi is similar in its chemical composition to that of ER, and the vesicles derived from trans side resemble the plasma membrane with which they may fuse.*

- *Secretory proteins produced in the ER are stored in Golgi and then vesicles filled with secretory proteins bud off from the trans Golgi network (TNG).*

- *Primary vesicles containing hydrolytic enzymes bud off from trans cisternae of Golgi complex, which in turn fuse with endocytotic vesicles to form secondary lysosomes.*

- *Golgi complex is the site of synthesis and concentration of glycoproteins and mucopolysaccharides. In plant cells, it plays an indispensable role in the formation of cell plate during mitosis.*

- *Golgi complex contributes to the formation of proacrosomic granules that fuse to form the acrosome of the spermatozoa. In addition, it is associated with the formation of melanin.*

REVIEW QUESTIONS

1. What is the contribution of Camillo Golgi as a cell biologist? Explain how the discovery of Golgi complex remain a controversy over a decade.

2. Describe the ultrastructure of the Golgi complex. Add a note on its *cis* and *trans* faces of cisternae.

3. Explain why the chemical composition of cis side of Golgi is similar to that of ER and *trans* side resemble the plasma membrane?

4. Describe the steps in the formation of primary lysosomes from TGN.

5. Explain the process of modification of insulin in the Golgi complex before its secretion.

6. Enumerate the functions of Golgi complex.

7. Write short notes on:

    i.   Dictyosome

    ii. *Trans* Golgi network

    iii. Glycosylation in Golgi complex

    iv. Zymogen granules

    v.   Proinsulin

    vi. Secondary lysosome

LYSOSOMES

These are dense rounded particles showing centrifugal properties between those of the mitochondria and the ribosomes. They were termed as **pericanalicular dense bodies**. It was C.de Duve (1949) who located the presence of hydrolytic enzymes in these bodies and renamed them as *lysosomes (lysis* = digestion, *soma* = body). As stated earlier and shown in the Figure 2.42, the **primary lysosomes** are pinched off from the *trans* cisternae of Golgi complex. The endocytotic vesicle formed as a result of invagination of cell membrane enclosing foreign particle may fuse with the primary lysosome to form **secondary lysosome**. The products of hydrolysis of foreign particle or macromolecule are released from secondary lysosome to the cytoplasm for reuse in biosynthesis or energy metabolism. The undigested and unabsorbed material present in **residual body** is expelled from the cell by the process called exocytosis.

OCCURRENCE

Lysosomes have been demonstrated in protozoa, insects, amphibians, mammals, and in few plant cells. The hepatocytes, pancreatic cells, kidney cells, etc. have large number of lysosomes. Numerous lysosomes are present in cells that take up macromolecules or large substances from their environment. For example, the intestinal epithelial cells and phagocytic reticuloendothelial cells can have up to several hundred lysosomes. On the other hand, secretory cells of the pancreas, which do not have absorptive role, contain very few lysosomes. Slime moulds also have lysosomes. Generally lysosomes are absent in the prokaryotic cells.

ULTRASTRUCTURE

The appearance of lysosomes under electron microscope is neither distinctive nor uniform. But in general, the diameter of these rounded or oval sacs ranges between 0.25 to 0.8 μm. The variation in lysosomal morphology is most evident in microphages, cells that are specialized for ingesting foreign particles by a process of phagocytosis.

Lysosomes are enclosed within a single membrane composed of lipoproteins. The lysosomal membrane is very delicate; it prevents the hydrolytic enzymes contained inside the lysosomes from reacting with the cytoplasmic substrates, and could be disrupted by the blender or could break down on refrigeration. The internal material enclosed by lysosomal membrane is quite heterogeneous, ranging from dense matrix to granular to flaky, and these variations in the structure of lysosomes are correlated with its functional state (Figure 2.43). Homogeneous and dense matrix is present in primary lysosomes whereas secondary lysosomes and residual bodies are more heterogeneous.

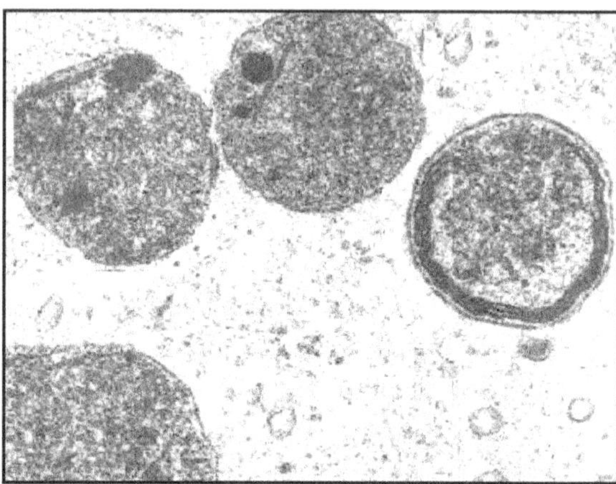

Figure 2.43 Lysosomes containing granular, and filamentous materials surrounded by a single membrane

CHEMICAL COMPOSITION

The lysosomes are mainly characterized by the presence of lytic or **hydrolytic enzymes**. A typical lysosome consists of a collection of approximately 60 different hydrolytic enzymes (Table 2.4), which are produced in the RER and targeted to lysosomes. The lysosomal enzymes are capable of hydrolysing virtually every type of biological macromolecule into low-molecular-weight products that can be transported to cytoplasm where they can be utilized in the anabolic pathway or production of energy. In general, the hydrolytic enzymes catalyse the reaction.

$$A - B + H_2O \longrightarrow AH + BOH$$

The enzymes are stored inside in a semi-crystalline or crystalline form and hence the lysosomes appear densely granular under the electron microscope. The lysosomal enzymes have their optimal activity at an acidic pH; hence they are collectively to referred to as **acid hydrolases**. The intralysosomal pH is approximately 4.6, whereas the surrounding cytoplasm has a pH of 6.5 to 7.5. The maintenance of lower pH in the lysosomal compartment is a property of the lysosomal membrane. A lower pH means higher proton concentration inside the lysosomes that is achieved by a membrane-bound, **ATP-driven proton transporter**.

The chemical composition of lysosomal membrane is rather similar to that of plasma membrane, especially in its protein : lipid ratio and high cholesterol content. The lysosomal membrane proteins analysed by gel electrophoresis indicate the presence of at least 40 polypeptides. Major species of closely related conjugated acidic proteins are lysosomal glycoproteins (lgp), which are heavily glycosylated and protect the membrane from attack by proteases present inside the lysosome.

Table 2.4 Lysosomal enzymes

Enzyme	Substrate	Product
Proteases	Cathepsin	Amino acids
	Collagen	Amino acids
Nucleases		
Acid ribonuclease	RNA	Ribose sugar, phosphate, and nitrogen base
Acid deoxyribonuclease	DNA	Deoxyribose sugar, phosphate and nitrogen base
Acid phosphatase	Phosphomonoesters	Monophosphates
Glycosidase	Glycogen	Monosaccharides
Acid lipase	Triglycerides	Fatty acids and alcohol
Phospholipase	Phospholipids	Fatty acids, alcohol, and phosphates
Heparin sulphamidase	Heparin	Monosaccharides, nitrogen, sulphates

FUNCTIONS OF LYSOSOMES

The lysosomes play a role in two major cell functions, phagocytosis and autophagy.

Phagocytosis

In 1893, E. Metchnikoff confirmed the role of lysosomes in the intracellular digestion of extracellular macromolecules that may include food or any other substances. The plasma membrane forms a small inpushing that is pinched off from the cell membrane and is set free to migrate in the cytoplasm in the form of endocytotic vacuole. If the substance engulfed is a liquid then the process is called **pinocytosis**, whereas in the process of **phagocytosis**, the cell engulfs solid substances. In either case, an endocytotic vacuole forms, and gradually becomes acidified. The endocytotic vacuoles enclosing foreign content are called **phagosomes**. The phagosome then fuses with **primary lysosome**, which sometimes may be developed from ER or sometimes indirectly from the Golgi complex as described earlier.

The fusion of phagosome with primary lysosome results in the formation of organelle called **secondary lysosome** in which the hydrolysis of macromolecules take place in the presence of hydrolytic enzymes. The products of such intracellular digestion are then released into the cytoplasm either for reuse in biosynthesis of anabolic substances or for energy production. The vacuole called **residual body** may still contain undigested and unabsorbed material that is eliminated from the cell by exocytosis. The process is shown diagrammatically in Figure 2.42(b).

Protozoa such as amoebae and ciliates exemplify the phenomenon of phagocytosis specifically for the digestion of food particles entrapped in food vacuole, which fuses with lysosome and food is digested intracellularly with the help of lysosomal hydrolytic enzymes. In most of the higher animals, phagocytosis is performed for the protection from foreign invaders, rather than as a mode of feeding. Inside the mammalian body, invading bacteria, antigen–antibody complexes, and even the dead cells are engulfed by macrophages and neutrophils, which play an important role in immunological defence. The uptake and digestion of haemoglobin by mammalian hepatocytes is one of the significant physiological processes. In humans, about 9 million aging red blood cells are destroyed and some of its components are recycled.

Lysosomes in some cells also perform reverse **phagocytosis**, i.e., release of its hydrolytic enzymes in extracellular space. For example, the lysosomes in epithelial cells of prostate gland fuse with plasma membrane and by exocytosis release acid phosphatase that becomes a component of prostatic fluid. The anterior tip of the sperm cell (acrosome) contains Golgi vesicles containing lysosomal enzymes, which when released on contact with the vitelline membrane (outer coat) of the egg, digest the coat and allow the entry of sperm nucleus into ooplasm (cytoplasm of the oocyte) to achieve fertilization.

Autophagy (Self-feeding)

It is an intracellular degradation of organelles or macromolecules. At the time of starvation, some cell organelles like mitochondria, ER, etc. can be broken down and digested by lysosomal enzymes to release energy without damaging or killing a cell. Since the lysosomal enzymes have the ability to break down or digest the cell, they are sometimes called **suicidal bags**.

Autophagy also provides the healthy cell with a continuous pool of free amino acids by degrading proteins present in the cytoplasm before they become non-functional. This process involves invagination of lysosomal membrane around the adjacent area of cytoplasm to enclose the old age proteins that must be degraded and resynthesized to perform their specific functions.

There are several other examples that can explain the process of autophagy; some of them are degradation of Mullerian ducts (which would form oviducts) in developing male, the involution of mammary gland when lactating mother stops lactation, and resorption of tail by a tadpole as it metamorphoses into frog. Hormonal secretions coordinate most of these events and it appears that the increased level of autophagy is due to increased synthesis of lysosomal enzymes.

Thus, lysosomes are not only involved solely in the destruction of materials that enter the cell from external environment but also play an indispensable role in organelle turnover, i.e., the destruction of organelles and their replacement. Hence, several **autophagic vacuoles** are seen in electron micrograph which are formed as a result of fusion of lysosomes with mitochondria or other organelles that are to be destructed. It is estimated that in a mammalian liver cell, one mitochondrion is destructed every 10 minutes or so by an autophagic vacuole. The starving cells have increased number of autophagic vacuoles indicating that their number depends on the physiological state of the cell.

HUMAN DISORDERS AND LYSOSOMAL FUNCTION

There are a number of **inborn errors of human metabolism** in which the activity of one or more of hydrolytic enzymes present in lysosome is absent or very low. Some of the genetic diseases due to lysosomal enzyme deficiencies leading to lysosomal storage disorders. Some of the lysosomal storage disorders with their symptoms are listed in Table 2.5. In all the cases, due to absence of an enzyme in the lysosome, its substrate accumulates in the cell, which can be detected in vacuoles by microscopy. Although diseases are caused by common mechanism, they show diverse clinical symptoms. Most of the diseases lead to neurological disorders and many to infantile death.

It becomes quite clear that disturbances in lysosomal function could have profound effects on human health. In addition to the above-mentioned human diseases, other disorders resulting from defects in lysosomal functions are silicosis, asbestosis, etc. **Silicosis** is a miner's disease that results from the uptake of silica particles by phagocytic cells in the lungs. The particles become enclosed within lysosomes but cannot be digested; instead, the chemical interaction between silicic acid and components of the lysosomal membrane cause it to leak. This results in release of hydrolytic enzymes into the cell, damaging the tissues of the lungs. The fibroblasts are stimulated to deposit collagen thereby decreasing the elasticity of the lung and further worsens the functionality of the lungs leading to death. A similar result occurs when asbestos dust is taken up by lung macrophages causing rupture of lysosome membrane. The cells die, releasing more dust particles, and this leads to the disease

asbestosis, in which tissues of the lung are severely affected and the condition even leads to death. Certain types of inflammatory diseases, such as **rheumatoid arthritis,** result, in part, from the release of lysosomal enzymes from immune cells into the extracellular space, causing damage to materials in the joints. Under certain circumstances, more uric acid is manufactured in the body that cannot be eliminated by the kidneys. It becomes deposited in the joints in the form of sodium urate crystals, which are engulfed by phagocytes and results in rupture of lysosomal enzymes causing swelling and inflammation of joints; the disease is referred to as **gout**.

Table 2.5 Some of the lysosomal storage disorders in human

Human disease	Enzyme deficiency	Substrate accumulated	Symptoms
Gaucher's disease	β-glucosidase	Glucocerebroside	Infantile mental retardation, liver and spleen enlargement, and erosion of long bones
Tay-Sach's disease	Hexominidase A	Ganglioside G_{M2}	Skeletal, cardiac and respiratory abnormalities, mental retardation, infantile death
Gangliosidosis G_{M1}	β-galactosidase G_{M1}	Ganglioside G_{M1}	Mental retardation, enlargement of liver, death by age 2
Fabry's disease	β-galactosidase A	Ceramide hexoside	Skin rash, kidney failure, pain in lower extremities
Farber's lipogranulomatosis	Ceramidase	Ceramide	Painful and deformed joints, skin nodules, infantile death
Krabbe's disease	Galactocerebrosidase	Galactocerebroside	Degeneration of myelin, mental retardation, infantile death
Pompe's disease	α-glucosidase	Glycogen	Liver and spleen enlargement, infantile death

It is experimentally proved that the lysosomes contain **deoxyribonuclease** or **DNase** and this enzyme can cause double-strand breaks in DNA of chromosomes, which are more difficult to repair than a single-strand break. In addition, the cause of the malignancy also lies in the fact that cancer cells contain abnormal chromosomes. In human beings, the partial deletion of chromosome 21 leads to blood cancer or **myeloid leukemia** and it is suspected that abnormal release of lysosomal enzymes may result in such chromosomal deficiency, but it has yet to be proved.

MICROBODIES

These are membrane-bound vesicles observed during electron microscopic study of mammalian cells like that of kidney tubule cells. Microbodies have approximately 0.5 to 1.0 μm diameter and a dense granular appearance. They contain different enzymes, which catalyse specific type of oxidative reactions. On the basis of their enzymatic composition, microbodies are categorized into three major types, liver peroxisomes, leaf peroxisomes and seed glyoxysomes.

LIVER PEROXISOMES

A typical hepatocyte contains approximately 1000 peroxisomes in the form of oval or spherical vesicles bound by a single membrane. Their number increases during foetal and postnatal growth and during liver regeneration. Drugs such as aspirin and clofibrate, a hypolipidemic agent can induce peroxisome proliferation. Morphologically, peroxisomes often contain a dense crystalline core rich in wide varieties of enzymes, in addition to flavin oxidase–catalase systems; other enzymes are glycolate oxidase, fatty acid CoA oxidase, amino acid oxidase, NADP-isocitrate dehydrogenase, acyl transferase, fatty acid β-oxidation enzymes, aminotransferase, and NADH-cytochrome c reductase. Peroxisome proteins (enzymes) are synthesized in the cytoplasm and once their translation is completed, they are imported into the organelle.

Peroxisome enzymes carry out the two-step reduction of molecular oxygen to water. In the first step, the enzyme oxidase removes electrons from a variety of substrates (RH_2), such as uric acid or amino acids and hydrogen peroxide (H_2O_2) is produced. In the second step, the enzyme catalase degrades hydrogen peroxide into water and oxygen.

$$O_2 \xrightarrow{\text{Oxidase}} H_2O_2$$
$$RH_2 \quad R_2$$

$$2H_2O_2 \xrightarrow{\text{Catalase}} 2H_2O$$
$$O_2$$

Since hydrogen peroxide is a highly reactive and toxic oxidizing agent, it would oxidatively destroy many cellular components if it had access to them. Because the generation of H_2O_2 is limited to the interior of peroxisomes, where catalase almost instantly breaks it down, the cellular level of H_2O_2 never reaches dangerous levels.

Functions   The peroxisome β-oxidation system acts on very long chain fatty acids, which are not oxidized by mitochondria until shortened. It also acts on some phenols, and a part of ingested alcohol. The syntheses of liver cholesterol and bile acids occur in peroxisomes. It is also evident that the enzyme luciferase that generates light emitted by firefly is also a peroxisome enzyme. In addition, enzymes present in peroxisomes mediate the initial steps in synthesizing plasmologens, a significant class of phospholipids in the brain.

In most animals, peroxisomal oxidase converts uric acid (a product of purine nucleotide digestion) into allantoin. But this enzyme is absent in primates and as a result, uric acid accumulates in the body fluid and forms crystals. The situation leads to gout that can be thought of as a genetic disease of primates and is due to loss of a peroxisomal protein. Acatalasemia is another example of a genetic disorder affecting a single peroxisomal enzyme. Patients with this disease show very low catalase activity, and many of them develop ulcers in the mouth, especially around the teeth. The condition seems to occur because of their inability to metabolize H_2O_2 produced by bacteria living in the buccal cavity.

It is also evident that peroxisomes can also be associated with cancer. There is a diverse set of drugs like herbicides, analgesics, and plasticizers used to decrease blood lipids, and the prolonged use of these chemicals can cause proliferation of liver peroxisomes and the condition can lead to liver cancer. The ability of catalase to metabolize the peroxide cannot keep pace with the over-production of H_2O_2. As a result, large amount of free radicals are produced inside the cell that act as carcinogens and damage DNA, thereby uncontrolled cell division leads to tumours in the liver.

LEAF PEROXISOMES

In certain green plant cells, especially in palisade cells of C3 leaves and bundle sheath cells of C4 leaves, peroxisomes help chloroplasts to perform photorespiration. The arrangement of closely apposed peroxisomes, chloroplasts and mitochondria in a leaf cell reflects the biochemical interdependence of these three cell organelles, where the product of one organelle is utilized as a substrate in another organelle (Figure 2.44). In the chloroplast, photorespiration starts with the reaction of ribulose 1,5-diphosphate (RuBP) with oxygen to form phosphoglycolate (2-carbon compound), which is later on converted to glycolate. It is then sent to the peroxisomes, where the enzyme glycolate oxidase converts glycolate into glyoxylate and H_2O_2 is liberated as by-product. Glyoxylate is then transaminated to form glycine, which is then transported to mitochondria where glycine is converted to serine. Peroxisomes utilize this serine from mitochondria and convert it to glycerate that can be transported to the chloroplast where it is utilized for synthesis of carbohydrates via formation of phosphoglyceric acid (3-PGA).

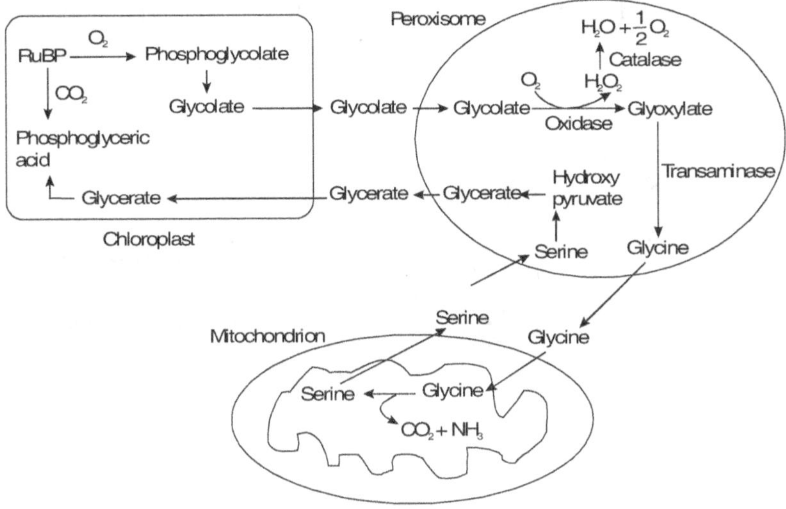

Figure 2.44 Interdependence of leaf peroxisome, mitochondria and chloroplast in photorespiration

GLYOXYSOMES

These are another type of microbodies most prominently found in plant seedlings of certain beans and nuts, where they store fats as energy reserve. Glyoxysomes are absent in mature and dry seeds. In addition to oxidase and catalase, they contain a number of other enzymes that are not found in animal cells. The unique enzymes present in glyoxysome are isocitrate lyase and malate synthetase. In the first few days after germination, glyoxysomes appear in the endosperm in association with lipid bodies.

Functions   One of the prime metabolic activities of glyoxysomes is to convert stored lipids into hexose sugars, the process known as gluconeogenesis. The metabolic role of glyoxysomes is diagrammatically represented in Figure 2.45. The property of glyoxisomes, to convert fat into sugar, does not occur in mammalian (including human) cells. In the seedlings, stored lipids in fat bodies are initially hydrolysed to glycerol and fatty acids. The fatty acids formed are then transported to glyoxysomes where they are converted to acetyl CoA by the process called β-oxidation. In animal and other plant tissues, acetyl CoA enters mitochondria where it is oxidized to CO_2 through Krebs' cycle. But in cells of seedling tissues, there is an alternate pathway in the glyoxysomes called glyoxylate cycle, in which the acetyl CoA is converted into C4 acids, namely, succinate and malate. In mitochondrial Krebs cycle, succinate is converted into malate and in the cytoplasm, malate is utilized to produce hexose sugars—glucose, the ultimate source of energy used by growing seedlings.

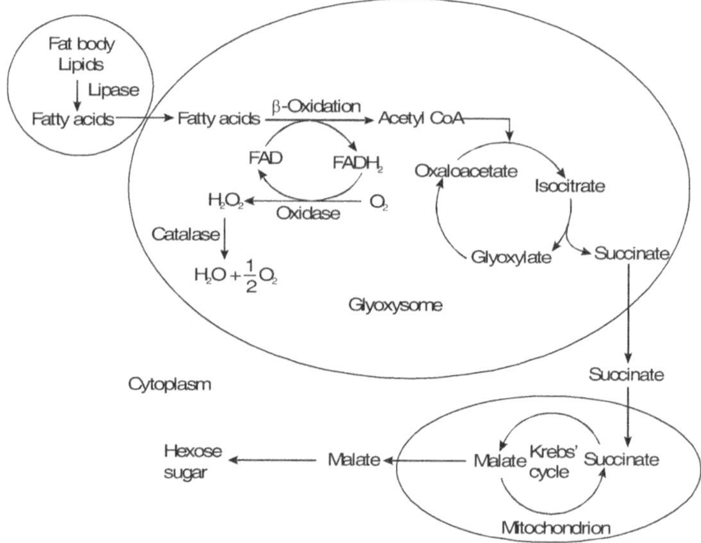

Figure 2.45 Functional interaction of glyoxysome with mitochondrion in conversion of fats to carbohydrates in the germinating seed

⊙ C.de Duve located the presence of hydrolytic enzymes in pericanalicular dense bodies and renamed them as lysosomes.

⊙ The primary lysosomes are pinched off from Golgi complex. The endocytotic vesicle enclosing foreign particle fuses with the primary lysosome to form secondary lysosome. The undigested and unabsorbed material present in residual body is expelled from the cell by exocytosis.

⊙ Lysosomes are present in cells that take up macromolecules or large substances from their environment, whereas they are absent in prokaryotic cells.

⊙ Lysosomes are bounded by a single lipoproteinous membrane enclosing hydrolytic enzymes like proteases, nucleases, acid phosphatase, glycosidase, acid lipase, heparin sulphamidase and phospholipase.

⊙ In most of the higher animals, phagocytosis is performed as a part of the defence mechanism, rather than as a mode of feeding, while in some cells lysosomes perform reverse phagocytosis to release its hydrolytic enzymes in extracellular space.

⊙ Lysosomal enzymes have the ability to break down or digest the cellular components, hence they are known as suicidal bags. Primary lysosomes are also concerned with autophagy, i.e., intracellular digestion of cell organelles like mitochondria and endoplasmic reticulum and thus serve for the destruction of organelles and their replacement.

⊙ When the activity of one or more of hydrolytic enzymes of lysosome is absent or very low, it can cause inborn errors of human metabolism. Lysosomal enzyme deficiencies leading to lysosomal storage disorders are Gaucher's disease, Tay-Sachs disease, Farber's lipogranulomatosis, Fabry's disease, etc.

⊙ Disturbances in lysosomal function have profound effect on human health. Some other disorders resulting from defects in lysosomal functions are silicosis, asbestosis, rheumatoid arthritis, and gout.

⊙ Lysosomal enzyme deoxyribo nuclease can cause breaks in DNA that transform the normal cell into cancer cells.

⊙ Microbodies are membrane-bound vesicles present in the cytoplasm of variety of cells. On the basis of their enzymatic composition, they are categorized as liver peroxisomes, leaf peroxisomes and seed glyoxysomes.

⊙ A typical hepatocyte contains approximately 1000 peroxisomes in the form of oval or spherical vesicles bounded by a single membrane. Liver peroxisomes contain enzymes like oxidase, catalase, glycolate oxidase, fatty acid CoA oxidase, amino acid oxidase, etc.

SUMMARY

- Peroxisome enzymes carry out the two-step reduction of molecular oxygen to water. In the first step, oxidase removes electrons from a variety of substrates and hydrogen peroxide (H_2O_2) is produced. In the second step, the enzyme catalase degrades hydrogen peroxide into water and oxygen.

- The peroxisomes are involved in β-oxidation of long chain fatty acids, and the synthesis of liver cholesterol and bile acids and also mediate the initial steps in synthesizing plasmologens.

- Peroxisomal oxidase converts uric acid into allantoin. The absence of this enzyme in primates results in uric acid accumulation in body fluid that forms crystals leading to gout as a genetic disease of primates. Proliferation of peroxisomes can be associated with liver cancer.

- Leaf peroxisomes are present in palisade cells of C3 leaves and bundle sheath cells of C4 leaves, where they help chloroplasts to perform photorespiration in association with mitochondria. One of the significant enzymes of leaf peroxisomes is glycolate oxidase.

- Glyoxysomes are found in plant seedlings of certain beans and nuts. In addition to oxidase and catalase, they contain isocitrate lyase and malate synthetase. Their prime metabolic activity is to convert stored lipids into hexose sugars.

REVIEW QUESTIONS

1. What are lysosomes? How are they differentiated on their morphological basis?

2. Explain the chemical composition of lysosomes with special reference to their enzyme content.

3. Describe the role of lysosomes in the processes of phagocytosis and autophagy.

4. Define inborn errors of metabolism. Enumerate various lysosomal storage disorders with their deficient enzyme and symptoms.

5. What are microbodies? Describe their types and morphological features.

6. Enlist the enzyme present in liver peroxisomes. What is impact of proliferation of peroxisomes on liver tissue?

7. Describe the position and function of leaf peroxisome. Trace out its role in photorespiration.

8. What are glyoxysomes? Explain their involvement in the conversion of stored fats into carbohydrates.

9. Write short notes on:

    i.   Primary lysosomes

    ii. Phagosomes

    iii. Residual body

    iv. Autophagy for mitochondrial replacement.

    v.   Tay-Sachs disease

    vi. Silicosis

    vii. Suicidal bags

    viii. Gout

    ix. Gluconeogenesis

    x.   Acatalasemia

CYTOSKELETON

In the body of animals, there is a skeletal system made up of hardened elements, which supports the soft tissues and assists in the bodily movements. The cell also has a "skeletal system"—a cytoskeleton that performs similar functions. The cells of plants and animals have the ability to move and change their shapes. Changes in the cell shape are brought about largely by changes in the orientation of cytoskeletal elements within the cell. During embryonic development, this happens as cells migrate before differentiation. There are certain specialized cells, which are in constant motion throughout their life, as in certain epithelia and blood cells in animals and pollen tubes and root hairs in plants. Even within most cells, cytoplasmic streaming (cyclosis) commonly occurs. All these movements are mediated by the internal cytoskeleton.

It has become clear from numerous electron microscopic photographs and from subsequent biochemical analysis that the cytoskeleton consists of at least three distinct structures classified on the basis of their diameter—**microtubules,** 24 nm; **intermediate filaments**, 10 nm; and **microfilaments**, 5 to 7nm; that together form an elaborate interactive network. Microtubules are hollow, cylindrical structures whose wall is composed of polymers of **tubulin,** acidic protein subunits, and present nearly in every eukaryotic cell including the core of cilia and flagella and the mitotic spindle of dividing cells. Microfilaments are solid, thinner structures, composed of protein **actin**. Intermediate filaments are tough, rope-like-fibres composed of a variety of proteins including **keratin, vimentin, desmin**, etc., having similar structure.

MICROTUBULES, CELL MOVEMENT, AND INTRACELLULAR TRANSPORT

Microtubules are made up of polymers of tubulin protein subunits, which are of two types, viz. α and β-tubulin molecules, the dimer being referred to as tubulin. Tubulin polymerizes to form a hollow tube bounded by 13 identical **protofilaments** arranged in longitudinal rows of subunits. The polymerization takes place in a head-to-tail fashion so that a microtubule has a definite polarity with the ends designated as plus (+) and minus (–) ends. This structural polarity is a significant factor in the assembly of these organelles and their ability to take part in directed mechanical activities. Microtubules rapidly collapse unless their ends are protected. The **microtubule organizing centre** (MTOC), a structure near the nucleus and containing (in animal cells) a pair of small bodies called centrioles (made of fused microtubules), protects the minus ends of microtubules. Microtubules radiating out from the MTOC pervade the cell as shown in the Figure 2.46.

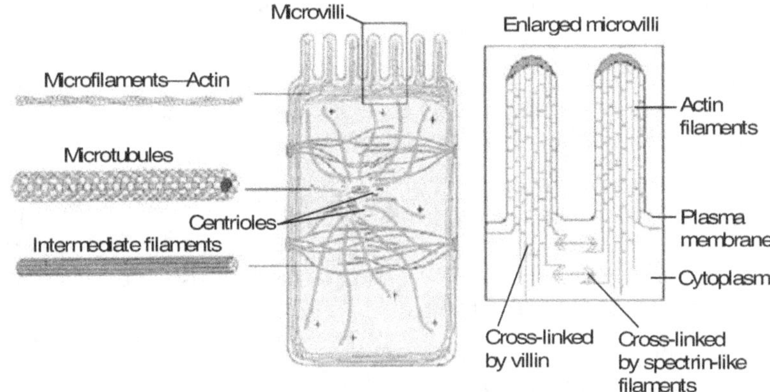

Figure 2.46   Schematic diagram of an intestinal epithelial cell showing each of the three major components of cytoskeleton. Note that actin filaments extending down from microvilli into the cell, the microtubules with (+) and (–) ends radiating out from MTOC, the centrioles are a pair of tube-like structures made of fused microtubules and intermediate filaments form an elaborate cage-like network around the nucleus and also ramify through the cytoplasm. The enlarged view of microvilli shows the arrangement of actin filaments cross-linked by its associated filaments.

Biochemical analysis of microtubules has been achieved by the treatment of cell with the drug **colchicine**, the alkaloid obtained from meadow saffron, which destroys the mitotic spindle and prevents the separation of chromosomes. Electron microscopy shows that colchicine causes the microtubules in the mitotic spindle and other cells to depolymerize. Under carefully controlled conditions of depolymerization, the protofilaments first dissociate from each other, and then subunits dissociate. In addition to the presence of tubulin protein as the principal constituent of microtubules, several **microtubule-associated proteins** (MAPs) have been found bound to tubulin that are isolated and identified in different tissues. MAPs 1 and 2 have been characterized from brain tissue as high molecular weight MAPs. Other MAPs are MAP 3, MAP 4, vesikin, kinesin, tau, dynein, and dynamin. Most of these MAPs occur in brain tissue except MAP 4 that has a widespread distribution.

Functions of Microtubules

1. Microtubules provide an internal skeleton or scaffold that provides structural support and helps to maintain the position of cytoplasmic organelles. The role of microtubules in maintaining cell structure can be inferred from electron micrographs from a wide variety of tissues. For instance, in the

neuronal axon, bundles of microtubules lie parallel to the longitudinal axon, and in growing plant cells, cortical microtubules lie parallel to the axis of cell elongation.

2. Microtubules form principal elements of cilia and flagella as motile cell projections that are discussed later in this chapter.

3. They are the part of the machinery that moves materials and organelles from one part of a cell to another. For example, the transport of vesicles from one membrane compartment to another is dependant on the presence of microtubules because disruption of these cytoskeletal elements often brings the movements to a halt.

4. Microtubules serve as the primary components of the machinery responsible for mitosis and meiosis. The movements of chromosomes and spindle formation during cell division are mediated by microtubules.

CILIA AND FLAGELLA

These are motile cell projections formed of microtubules and MAPs. These are hair-like, motile organelles projecting from the surface of a variety of eukaryotic cells. Cilia are short (5–10 μm), numerous (hundreds or thousands per cell), and occur in protozoa like *Paramecium* and echinoderm larvae, where they propel the organisms through their aqueous environment. Ciliated epithelia line the respiratory passage, where their action moves liquid and mucus along and helps in elimination of foreign bodies from the lungs. The fallopian tube and oviduct of mammals are lined by the ciliated epithelium that helps in transfer of ovum and process of fertilization. The beating of cilia is usually a coordinated activity.

Flagella are typically longer than 150 μm and are one or two per cell. They are present in flagellated protozoa such as *Euglena* and in the mammalian sperm tail as propulsive organelles. Bacteria also possess structures referred to as flagella, but prokaryotic flagella are simple filaments that bear no evolutionary relationship to their eukaryotic counterparts. Unlike a cilium, the beat of a flagellum is undulatory with more than one wave present along the length of the flagellum at one time. A beating flagellum generates a force that pushes or pulls a cell in a direction that is parallel to the long axis of the flagellum.

Structure of Cilia and Flagella

An electron micrograph of a cross section of cilium or flagellum shows that the entire projection of cilium or flagellum is covered by a membrane, which is continuous with the plasma membrane of the cell. The **axoneme** is the core of the cilium or flagellum that contains an array of microtubules running longitudinally the entire length of the organelle. The axoneme in nearly every case shows a "9 + 2" configuration, with nine peripheral microtubule doublets around the central pair of single microtubules (Figure 2.47).

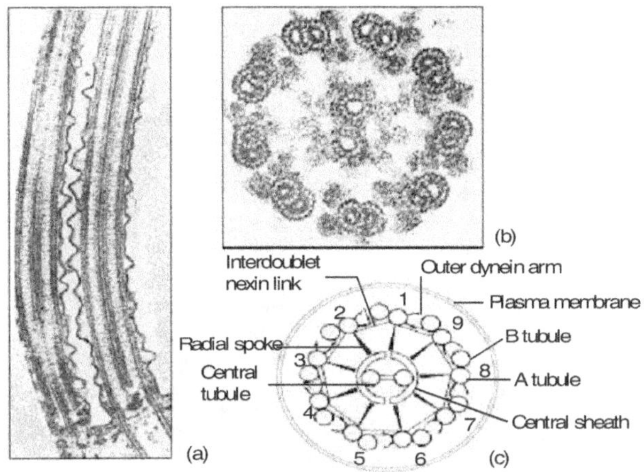

Figure 2.47 (a) Longitudinal section of cilia, (b) Transverse section of a cilium and (c) its corresponding diagrammatic representation showing nine peripheral doublets and two central microtubules.

In a cross section, each peripheral doublet shows one complete microtubule (A tubule) and one incomplete microtubule (B tubule). The doublets are interlinked by interdoublet bridge made of an elastic protein called **nexin**. The A tubule has a pair of projecting **dynein** "arms"—inner and outer arms in a clockwise direction. The central tubules are enclosed by projections that form the central sheath that is connected to the A tubule of peripheral doublets by a set of radial spokes.

A cilium or flagellum comes from the basal body which are specialized MTOCs that is often identical to the centrioles and have their own proteins in addition to tubulin. Like a cilium or flagellum, a basal body contains nine peripheral fibres, but each consists of three microtubules (triplets) instead of two. The A tubule is complete, while B and C tubules are incomplete. In addition, a basal body consists of a more complete radial spoke network. The central microtubules are absent in the basal body or centrioles.

Mechanism of Flagellar and Ciliary Movements

The generation of force and movement by flagellum and cilia have been studied extensively. The movements of flagellum captured by time-lapse cinematography have clearly shown that the beat occurs as a series of circular arcs, with the arc moving towards the tip. The beating of flagellum is undulatory and produces a force that propels or pulls the possessor (Figure 2.48a). It is difficult to visualize the movements of individual cilia because of their small size and large numbers. Ciliary movements are coordinated and also involve bending. Cilia beat in metachronal wave in which the cilia in a given row are in the same stage of the beat cycle, but those in adjacent rows are in different stage (Figure 2.48b). There is a rigid effective power stroke and a bending recovery stroke as shown in the Figure 2.48c. Factors like Ca^{2+} and cAMP (cyclic adenosine monophosphate) influence the rate and formation of a ciliary and flagellar beat.

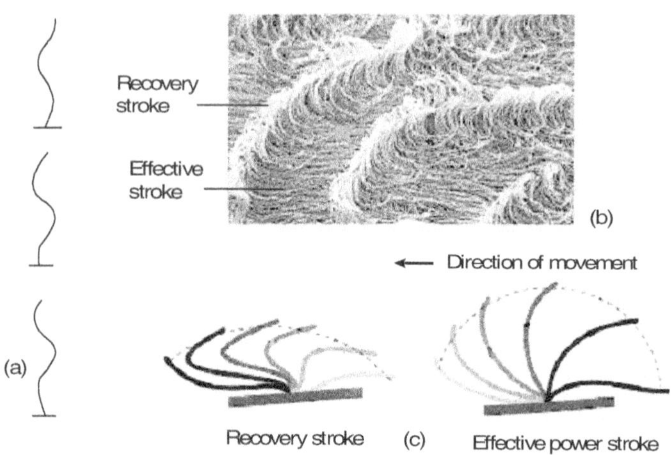

Figure 2.48 (a) Undulating movement of a flagellum, (b) Electron micrograph of beating cilia, and (c) The effective power stroke and recovery stroke during the beat cycle of a cilium.

Cilia and flagella perform locomotory functions in ciliates and flagellated protozoa, ctenophores, some platyhelminthes and nemertins, rotifers, some snails, and ciliated larvae of annelida, mollusca, and echinodermata. In addition to locomotion, cilia also help in feeding of ciliary filter feeders. Most of the sessile and sluggish animals are ciliary filter feeders, e.g. *Amphioxus*, *Herdamania*, rotifers and fresh water mussels where ciliary current assist in feeding. In coelenterates, like sea anemone and corals, cilia present on tentacles produce a cleansing effect. The ciliary lining in nephrons in kidney, respiratory tract, and genital duct help in the passage of materials through them.

MICROFILAMENTS

Cells have many shorter (1–$2\,\mu m$) and thinner (5–$7\,\mu m$) microfilaments in addition to containing microtubules. With the help of conventional electron microscopy and immunofluorescence (the technique that uses a fluorescent antibody to a muscle protein to localize the protein in a non-muscle cell), it became clear that the microfilaments of higher eukaryotes are composed of the protein **actin**. The human skin fibroblast stained with antibody to actin often shows parallel arrays of microfilaments (Figure 2.49). In tissue culture cells, microfilaments are called stress fibres, which play significant role in cell structure and motility.

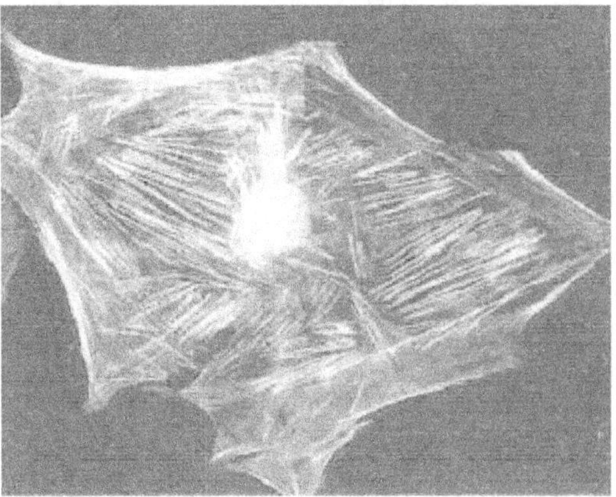

Figure 2.49   The parallel arrays of the actin filaments present in human skin fibroblast stained with the antibody

In some non-muscle cells, actin is the most abundant cellular protein. Although these cells have myosin, the ratio of actin to myosin is much higher than in muscle. Actin is a polymer of individual globular subunits (G-actin). In the presence of ATP, G-actin polymerizes in a head-to-tail manner to form a flexible filament composed of two strands of actin molecules wound around each other in a double helix as shown in Figure 2.46. The regulation of polymerization takes place via specialized actin-binding proteins, some of which are non-muscle myosin, gelsolin, α-actinin, filamin, fimbrin, profilin, spectrin, villin, and vinculin.

Functions of Microfilament

There are remarkable cell membrane movements, which involve actin filaments. For example, during cell differentiation in vertebrate embryo, the cells of the neural crest migrate across the entire length of the embryo and differentiate to form diverse tissues like chromatophores in the skin, and the cartilage of the jaws, etc. Similar movements of whole cells are executed by leucocytes that patrol the tissues of the body searching for debris and microorganisms to be cleared off from the body as a part of the defence mechanism. Microfilaments also mediate the cytoplasmic streaming or cyclosis in most of the cells. Such intracellular motions help in the migration of cell organelles, including mitochondria, lysosomes, vacuoles, transport vesicles, etc. at their targeted sites where these organelles have to perform specific functions.

Another good example of actin involvement in cell motility is the brush border of intestinal epithelium that lines the lumen of the gastrointestinal tract. The plasma membrane of each cell is folded at its apex into microvilli to increase the digestive and absorptive area of the gut. The intestinal microvilli contain parallel arrays of 20–30 microfilaments that run along the longitudinal axis and are cross-linked by the associated binding proteins such as fimbrin and villin (Figure 2.41). The "plus" ends of actin filaments are directed towards the tip of the villi and they are connected to the adjacent plasma membrane through protein cross bridges. There is an electron-dense region at the tip of microvilli, which may serve as the growing point for the polymerization of microfilaments. At the base of each microvillus, there is a second set of microfilaments that run perpendicular to the actin projections and are anchored by the spectrin-like filaments. The movements of microvilli are probably mediated by the sliding of actin filaments that are essential for circulating the fluid containing digested food around them and to increase its subsequent absorption.

Actin filaments are also involved in provision of propulsive force that is essential to release the acrosomal hydrolytic enzymes from the sperm of an animal into the jelly coat of the ovum. As a result, the coat is dissolved and the plasma membrane of sperm and the egg can fuse to form a zygote.

The cultured cells also migrate in a manner resembling that of amoebae, forming filamentous pseudopodia called filopodia, which are having microfilaments. Using immunofluorescence technique, it became clear that these structures largely contain actin filaments that directly anchor to the cell surface and orient themselves in the direction of cell movement.

INTERMEDIATE FILAMENTS (IFs)

These are smooth-surfaced, solid and unbranched filaments having a diameter of 7–11 nm and are intermediate between microtubules and microfilaments. Unlike microfilaments and microtubules, intermediate filaments are chemically heterogeneous in structure and composition. They also show wide variation in antigenicity and solubility. In humans, there are at least 60 different genes that code for the polypeptide subunits of intermediate filaments, which can be divided into six major classes on the basis of their tissue distribution (Table 2.6), as well as biochemical, genetic, and immunological properties.

Table 2.6 Major classes of the mammalian intermediate filaments

Filament class	IF protein	Average mol.wt.	Tissue distribution
I	Keratin (acidic)	40,000–65,000	Epithelia
II	Keratin (Basic, neutral)	50,000–67,000	Epithelia
III	Desmin	53,000–54,000	Muscle
	Vimentin	53,000–57,000	Mesenchyme
	Peripherin	54,000–57,000	Neurons
	Glial fibrillary	45,000–50,000	Glial cells
IV	Lemin proteins	60,000–70,000	Nuclear envelopes
V	Neurofilaments	62,000–110,000	Neurons
VI	Nestin	80,000–240,000	Neuronal stem cells

The various intermediate filaments are different but they show remarkable homology in their overall structure. Each polypeptide of IFs contains a central, rod-shaped, α-helical domain of similar length and homologous amino acid sequence. This polypeptide chain is flanked on each side by globular domains of variable size and sequence. Two such polypeptides wrap around each other forming a rope-like dimer (45 nm in length), which has polarity, with one end called C-termini of polypeptides and the opposite end the N-termini. Once the

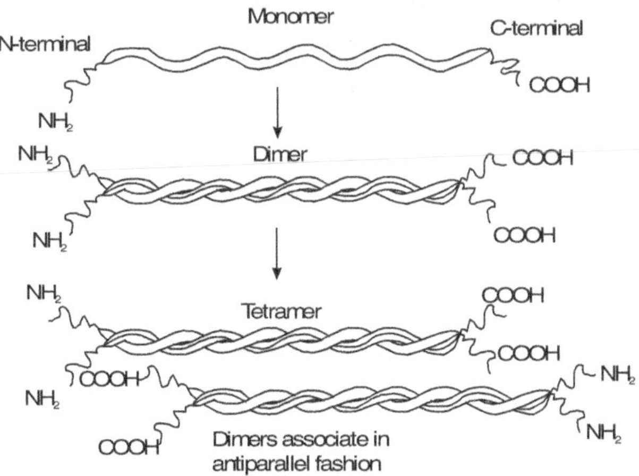

Figure 2.50 Sequential assembly of intermediate filament

dimer has formed, it interacts with a second dimer in an anti-parallel fashion to form a tetramer (Figure 2.50). Two of these units then associate to form a protofilament, an octamer of the basic chain. Four such protofilaments make up the filament of 10-nm of diameter seen in cells. In all, about 25,000 individual proteins associate to form a 40-μm filament.

Functions of Intermediate Filaments

Like that of microtubules and microfilaments, intermediate filaments perform a significant role as motility molecules within the cytoplasm of different classes of cells.

1. The keratin filaments of epithelial cells form an elaborate network around the nucleus and also ramify through the cytoplasm where their role is to hold the organelles in place. The cytokeratin filaments are typically organized into bundles (tonofilaments) that can be attached to cell junctions called desmosomes.

2. Desmin is found primarily in cardiac, skeletal, and smooth muscle. In skeletal muscles, desmin filaments form an organized network at the Z-line where they help to integrate muscle function.

3. Neurofilaments play an integral part in axonal transport of different macromolecules including synaptic proteins towards presynaptic membrane.

4. Vimentin filaments are present in mesenchymal-derived cells such as adipocytes, the endothelium of blood vessels, and fibroblasts, where they prevent the aggregation of cell structures.

5. Intermediate filaments provide the clues to clinicians with cell markers to determine the identity of the tumour. In immunofluorescence, use of specific monoclonal antibodies can differentiate the intermediate filaments.

SUMMARY

- Cytoskeleton is a "skeletal system" in the cells of plants and animals that helps them to move and change their shapes.

- Cytoskeleton is composed of microtubules, intermediate filaments, and microfilaments, which together form an elaborate interactive network.

- Microtubules are polymers of tubulin protein subunits, which polymerize to from a hollow tube bounded by 13 identical protofilaments.

- MAPs are microtubule-associated proteins like MAPs 1 to 4, vesikin, kinesin, tau, dynein, and dynamin that are isolated and identified in different tissues.

- Microtubules provide structural support and help to maintain the position of cytoplasmic organelles. They are a part of the machinery that moves materials and organelles from one part of a cell to another. They form principal elements of cilia and flagella as motile cell projections.

- The movements of chromosomes and spindle formation during cell division are mediated by microtubules.

- Cilia and flagella are motile cell projections made up of microtubules and MAPs. These are hair-like, motile organelles projecting from the surface of a variety of eukaryotic cells.

- The axoneme is the core of the cilium or flagellum that shows a "9 + 2" configuration, with nine peripheral microtubule doublets around the central pair of single microtubules. A basal body contains nine peripheral fibres, but each consists of three microtubules (triplets) instead of two.

- The beating of flagellum is undulatory and produces a force that propels or pulls the possessor. Cilia beat in metachronal wave in which there is a rigid effective power stroke and a bending recovery stroke.

- Microfilaments are shorter and thinner than microtubules and they are composed of parallel arrays of protein actin. In tissue culture cells, microfilaments are called stress fibres, which play a significant role in cell structure and motility.

- Actin is a polymer of globular subunits (G-actin), which polymerize to form a flexible filament composed of two strands of actin molecules wound around each other in a double helix. The actin-binding proteins are nonmuscle myosin, gelsolin, α-actinin, filamin, fimbrin, profilin, spectrin, villin, and vinculin.

- The movement of cells during differentiation in vertebrate embryos, the movements of whole cells as shown by leucocytes, cytoplasmic streaming or cyclosis in most of the cells, the cell motility in the brush border of intestinal epithelium, all are mediated by actin filaments.

SUMMARY

- Intermediate filaments (IF) are smooth-surfaced, solid and unbranched filaments and are intermediate between microtubules and microfilaments. In humans, there are at least 60 different genes that code for the polypeptide subunits of intermediate filaments that are divided into six major classes I to VI on the basis of their tissue distribution.

- The keratin filaments of epithelial cells hold the organelles in place. The cytokeratins are involved in cell junctions. Desmin in skeletal muscles helps to integrate muscle function.

- Neurofilaments are involved in axonal transport. Vimentin prevents the aggregation of cell structures. Intermediate filaments also help in diagnosis of cancer.

REVIEW QUESTIONS

1. What is cytoskeleton? List some of the generalized functions of the cytoskeleton.

2. Compare the structures of a fully assembled microtubule, actin filament and intermediate filaments.

3. What is an MTOC? Describe its structure and function.

4. Explain the structure of cilium. Compare it with the structure of flagellum.

5. Describe various classes of intermediate filaments with their tissue distribution.

6. Contrast the movement of *Paramecium* with that of *Euglena*.

7. Write short notes on:

    i. G-actin

    ii. MAPs

    iii. Dynein "arms" in A tubule

    iv. Neurofilaments

    v. Colchicine

    vi. Functions of cilia

    vii. Protofilament

    viii. Axonal transport

MICROSCOPY AND MICROMETRY

HISTORICAL EVENTS IN MICROSCOPY

The fields of cell biology, microbiology, medical parasitology, etc., have been inseparable from microscopy. The design of the present-day microscope used in biological laboratories has been developed over the course of several centuries. In the 14th century, Romans, Greeks, and Indians had known lenses of glass and quartz which they used as magnifiers. In the latter part of the 15th century, Leonardo da Vinci and Galileo Galilei of Italy stressed the importance of using lenses to view small objects. By the mid-1600s, a handful of pioneering scientists had used home-made microscopes to uncover a mysterious world that would never have been revealed to naked eye.

It was Anton van Leeuwenhoek (1632–1723), a Dutch (a cloth merchant) spending his spare time in grinding lenses and construction of microscopes of remarkable quality, who was the first to examine a drop of water and also the first to publish the descriptions of protozoa, bacteria, spermatozoa and red blood cells observed under these simple microscopes that could magnify the objects 200 to 300 times. These microscopes consisted of a biconcave lens enclosed in two metal plates (Figure 3.1). Robert Hooke (1665) confirmed these observations using a compound microscope in which a double-lens system is used, where the second lens magnifies the image from the first lens (Figure 3.2). The size of the observed image depends on the magnification of both lenses. The microscope developed by Hooke could be referred to as the forerunner of the modern-day microscope. The microscopic details of the cell, study of embryonic development, and the study of the microbial world became more apparent as the microscope was improved in design. Acromatic lenses began to be used by about 1830 and in the 1870s immersion lenses were available for higher magnification.

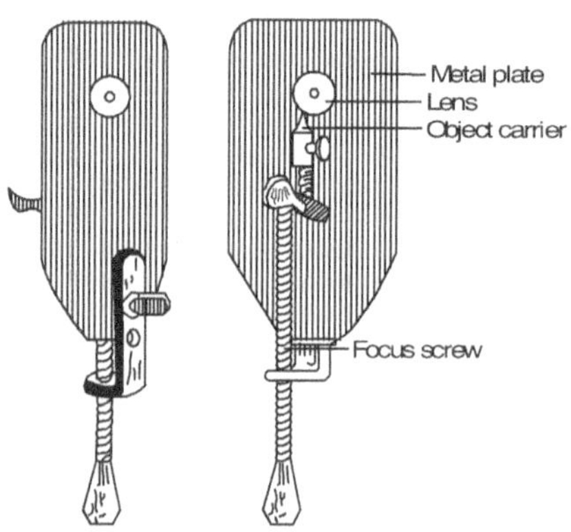

Figure 3.1   Leeuwenhoek's microscope

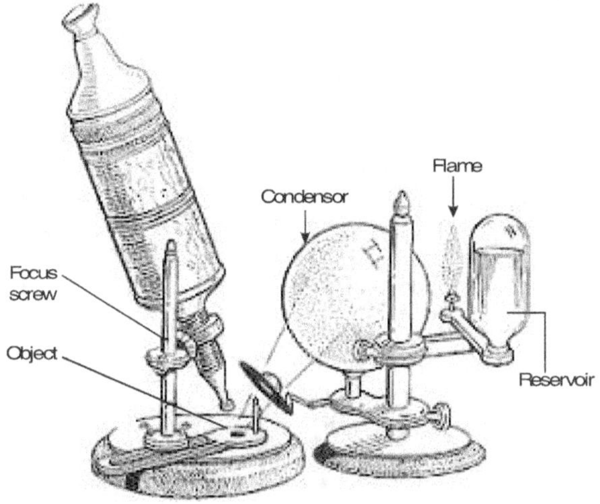

Figure 3.2 Hooke's compound microscope

SIMPLE MICROSCOPE

In this type of microscope a single convex lens that provides an enlarged image of an object to the eye is used. Since the image is larger, the eye is able to resolve finer details that cannot be distinguished without the lens. Simple microscope consists of the stage, the arm, the substage mirror, an adjustment knob, the foot and the optical system. A single convex lens constitutes its optical system that can be used to magnify the image up to 20×. Simple microscopes are also called dissection microscopes since most of them are used to magnify and study finer details of the dissected materials from biological specimens.

COMPOUND MICROSCOPES

These are modern microscopes that consist of a complex system of arrangement of lenses to have higher magnification and better resolution. There are basically two types of modern microscopes, light microscopes and electron microscopes. In light microscope, visible light is used to illuminate the object whereas in electron microscope, the source of illumination is a beam of electrons.

A typical compound microscope has two types of parts. The mechanical parts provide the structural framework to the microscope and support the optical parts that magnify the object. The mechanical parts include foot or base, arm, body tube, course- and fine-adjustment knobs, nose piece, stage, etc. and the optical system consists of two major lenses, the ocular or eyepiece and objectives, and condenser to gather the light and concentrate it onto the object (Figure 3.3a). The microscope with single ocular lens is called monocular microscope and that having two eyepieces is called binocular microscope.

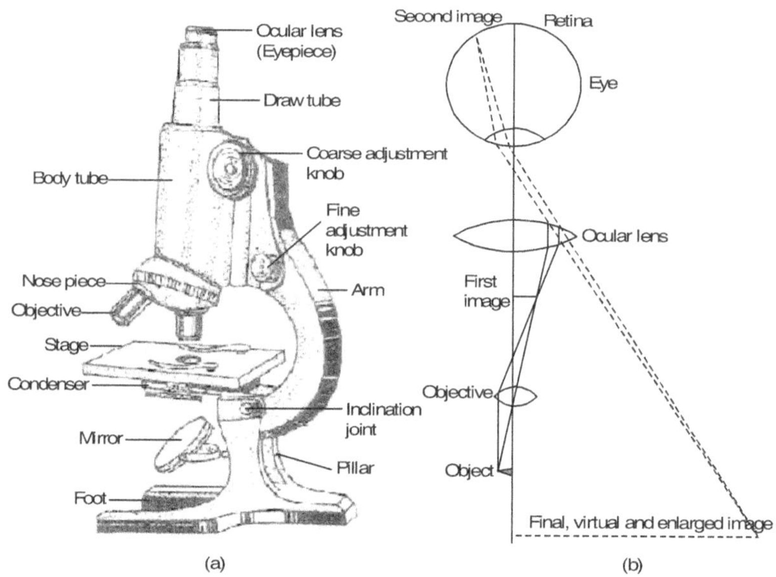

Figure 3.3 (a) The principal parts of the compound microscope, (b) The optical system of compound microscope showing the path of light rays. The objective focuses as a real, inverted image of the object. The image is focused on to the retina of the eye by the ocular lens and is seen as a virtual inverted image. Thus, a compound microscope functions on the principle that an image formed by one lens can serve as the object of the second lens.

The functions of different mechanical parts are mentioned in brief as follows:

❖ Base—provides support and stability to microscope

❖ Arm—used to carry the microscope and to set it in a comfortable position

❖ Stage—a flat platform that holds the slide in the position. Microscope may be equipped with mechanical stage that moves by means of adjustment knobs

❖ Body tube—used to transmit the magnified image

❖ Nose piece—holds the objectives and

❖ Adjustment knobs—the course (larger) and fine (smaller) knobs used to focus the image of the object

In the optical system of the compound microscope, two lenses of short focal lengths are used and the path of light rays passing through them to form a final, virtual and enlarged image of the object is diagrammatically shown in Figure 3.3b. The objective is a converging lens that forms a real, enlarged and inverted image of the object. This first image acts as the object for the ocular lens, which acts as a simple magnifying glass and helps to form a final, virtual and a still further enlarged image of the object. Our eye sees this image through the eyepiece.

In addition to the above-mentioned parts and their respective functions, the substage condenser, an iris diaphragm and a mirror assist in illumination of the object. The lenses present inside the condenser help in concentrating the light on the slide. The iris diaphragm controls the angle and amount of light used. The mirror reflects light from the source of illumination through the condenser. Natural day light (sunlight) or artificial light (lamp) can be used to illuminate the object. Modern microscopes have a built-in light source.

Magnification

It is the process of enlarging the image of an object which is being viewed and that cannot otherwise be seen by the naked eye. Magnification by a microscope is the product of the individual magnifying ability of oculars and objectives. For example, if the ocular is 10×, and objective is 45×, the specimen is magnified 450 times. If an oil immersion objective (100×) is used with 15× ocular, to visualize the object, the specimen is magnified 1500×. Effective magnification depends on the resolving power of the microscope.

Resolving Power of a Microscope

The resolving power of our eye is its ability to see closely placed objects as two separate objects. Theoretically, our eye can resolve two objects separated by a distance of at least 20 μm. But practically, due to factors like illumination, contrast, object size, the resolving power of our eye is about 100 μm. Therefore, human eye cannot see anything that is smaller than 0.1 mm. Similar to our eye, the resolving power of a microscope is the ability to distinguish two adjacent objects as separate and distinct images rather than as a single blurred image. The microscope's resolving power is dependant on the wavelength of the light used to illuminate the object and the numerical aperture of the objective.

In optical light (bright-field) microscopes, the wavelength of the light used falls in the visible range (400–750 nm). Within this range, if the light with shorter wavelength is used to illuminate the objects, it will give better resolution. For example, the blue light has shorter wavelength than red, as a result when object is illuminated with blue light, greater resolutions are obtained than that with the red light. In addition, when ultraviolet light or high voltage electron beams having still shorter wavelengths are used to illuminate the object, we get greater resolution. The resolving power of the microscope can be mathematically calculated by the following formula:

$$\text{Resolving power (RP)} = \frac{\lambda}{2\text{NA}}$$

where,

λ = wavelength of the light used to illuminate the object,

1/2 = a constant that depends on the contrast, and

NA = numerical aperture.

The numerical aperture (NA) is the second factor that affects the resolving power of microscope, which can be defined as the diameter of the objective lens in relation to its focal length. In other words, it is an index of its light-gathering capacity and is measured in terms of the angle subtended by the optical axis and the outermost ray entering into the objective. The NA can be expressed by the formula:

Numerical aperture (NA) = $\eta \sin \theta$

where,

η = refractive index of the medium through the light passes before entering into lens,

θ (Theta) = angular aperture that can be defined as the angle between the most divergent rays passing through the lens and the optical axis of the lens.

The divergent rays cannot enter the objective if their angle of divergence from the straight rays is greater than half the angular aperture.

The refractive index of air is 1.00. The maximum θ is 90° and sin 90 is 1.00. Therefore the maximum numerical aperture for any lens system working in air medium cannot be greater than 1, since $\eta \sin \theta = 1 \times 1 = 1$.

If a yellow light of the wavelength 580 nm were used to illuminate the object under a microscope having an objective with numerical aperture 1.0, the resolving power of the microscope would be calculated as follows:

$$\text{Resolving power (RP)} = \frac{\lambda}{2\text{NA}}$$

where,

λ = 580 nm and

NA = 1.0.

Then, RP of the microscope $= \dfrac{580}{2} = 290\,nm$

If blue light of wavelength 450 nm is used for illumination of the object, the RP of the microscope will be 225 nm indicating that the microscope can resolve objects smaller than 225 nm.

A practical way to increase the numerical aperture of the objective of a microscope above 1.0 is to increase the refractive index of the medium. Replacing the air space between the specimen and the objective lens with immersion oil can achieve this. The refractive index of the immersion oil (cedar wood oil or paraffin oil) is larger than that of air, and equal to that of glass. The higher refractive index of immersion oil helps all the reflected and refracted light rays that previously failed to enter into objective lens, to do so now. As a result, the numerical aperture of the objective is greatly increased. When the immersion oil is employed in the medium, η is 1.56, and if θ is 58° then,

$$NA = \eta \sin\theta$$

$$= 1.56 \times \sin 58°$$

$$= 1.56 \times 0.85 = 1.33$$

The numerical aperture for the lower power objective (10×) is 0.25, for the higher power objective (45×), 0.65 and for the oil-immersion objective (100×) 1.25.

The resolving power of the microscope depends on the numerical aperture of not only the objective but also that of the substage condenser. Taking into consideration both the numerical apertures, the resolving power of the complete microscope becomes:

$$NA \text{ (of complete microscope)} = \frac{\lambda \text{ (wavelength of the illuminating light)}}{\text{(NA of objective)} + \text{(NA of condenser)}}$$

Theoretically, when blue light having the wavelength 450 nm is used to illuminate the object and the lower power objective with 0.25 NA, higher power objective with 0.65 NA and oil-immersion objective with 1.25 NA, the resolving power of the compound microscope can be calculated as follows:

$$RP \text{ for lower-power objective} = \frac{450}{2} \times 0.25 = 900\,nm = 0.9\,\mu m$$

$$RP \text{ for higher-power objective} = \frac{450}{2} \times 0.65 = 346\,nm = 0.35\,\mu m$$

$$RP \text{ for oil immersion objective} = \frac{450}{2} \times 1.25 = 180\,nm = 0.18\,\mu m$$

Thus, one can say that when the objects smaller than and /or placed close to each other cannot be seen clearly with low-power, high-power or oil-immersion objectives, if they are 0.9 μm, 0.35 μm, and 0.18 μm, respectively.

There are numerous forms of light microscopy; some of them are bright-field microscopy, dark-field microscopy, phase-contrast microscopy, fluorescence microscopy, and polarizing microscope. Similarly, electron microscopes are of two types, transmission electron microscope (TEM) and scanning electron microscope. In addition to video microscopy, the comparative data of the above-mentioned microscopes is given in Table 3.1.

MICROMETRY

It is the measurement of microscopic objects, for example, the diameter of RBC, ova, nuclei, etc., the length of paramecium, intestinal villi, thickness of the epidermal layer, and even the size of microorganisms including bacteria. Since all these objects are small, their dimensions are usually expressed in units smaller than millimetres. The micrometre (μm) which is one-thousandth (10^{-3}) of a millimetre and one millionth (10^{-6}) of a metre is utilized to express the measurements of these microscopic objects that can be done under the microscope equipped with ocular and stage micrometers. One cannot measure dimensions of the objects smaller than 0.2 mm because it is less than the resolving power of a compound microscope. The ordinary compound microscope can be used to measure the dimensions of objects in the range of 0.2 mm to 25 mm.

The ocular micrometer (OM) is simply a glass disc with a diameter of 1 cm. It is engraved with an arbitrary scale of 100 divisions or less (Figure 3.4). Since it is fitted into the eyepiece of the compound microscope, it is more appropriately called ocular micrometer, which is to be used for all measurements. The distance between the graduations of an ocular micrometer does not have any standard value and varies depending on the objective used. Therefore it is necessary to calibrate (standardize) the ocular micrometer with the known standard scale of stage micrometer.

The stage micrometer (SM) is a special glass slide having in its centre a scale of defined length (Figure 3.4). Usually, the scale is 1 mm (1000 μm) divided into 100 divisions, so that one stage micrometer division is equal to 10 μm or 0.01 mm. The distance between micrometer's divisions become correspondingly enlarged when high-power and oil-immersion objectives are used.

Table 3.1   Comparison of various types of microscopes

Type	Magnification	Common uses/advantages	Requirements/disadvantages
Bright-field microscope	1500×	Allows visualization of stained or unstained specimens including microbes. Cytological and histological features of specimen can be studied.	Unable to resolve bacteria and viruses. Staining procedure may introduce artefacts.
Dark-field microscope	1500×	Allows viewing of unstained living microbes including very small bacteria. Specimens appear bright against dark background.	Staining reactions cannot be used for examining stained specimens. Modified substage condenser is essential.
Phase-contrast microscope	1500×	Used to observe all the features in stained living cells and tissues. Transparent protoplasmic components can be viewed without staining and killing the cells.	Special kind of objective with phase analyser and condenser with annular stop are essential. Staining reaction cannot be evaluated.
Fluoroscence microscope	1500×	Allows to identify infectious microbes in tissues. Widely used in immunology (immunofluorescent antibody technique), medical microbiology, and environmental microbiology.	Mercury vapour lamp, a dark-field condenser, 3 sets of filters and fluorescent dye are essential. Only specimens with natural fluorescence or that can be stained with a fluorescent dye can be observed. Such microscopy can be harmful and injurious if not used properly.
Polarizing microscope	--	Used to observe biological structures like fibres, tubules, and crystals that have high degree of molecular orientation. Common in geology laboratory to study minerals in thin sections of rocks.	A polarizer (prism) in a condenser and analyser in ocular are essential. Only objects with birefringence (isotropic and anisotropic material) can be observed.

(Contd.)

Table 3.1   (Continued)

Type	Magnification	Common uses/advantages	Requirements/disadvantages
Transmission electron microscope (TEM)	$500,000\times-$ $1000,000\times$	Allows visualization of ultrastructure of cells and microbes. The electrons are focused with magnetic lenses and transmitted through the specimen providing greater anatomical details, which cannot be observed by light microscopy.	Very expensive. Living biological materials cannot be observed. A variety of special techniques are used to prepare the material for observation that can introduce artefacts like breakdown of chemical bonds and alteration in conformation of the specimen.
Scanning electron microscopy (SEM)	$10,000\times-$ $200,000\times$	Used to study surface features of biological material at ultrastructural level. The image is created from electrons scattered off the surface of the coated specimen. SEM facilitates to produce three-dimensional image of the object.	As expensive as TEM. The resolution power of a SEM is lesser than TEM. The sample preparation involves fixing, drying and coating with a thin layer of metal that can alter some of the properties of the specimen.
Video microscopy	Images can be digitally magnified	Microscopic images of objects recorded by video camera are fed to computer where image-processing software can manipulate images with brightness, contrast, colour, etc.	Expensive. Trained persons having knowledge of video recording, computers and its application software can perform the operations.

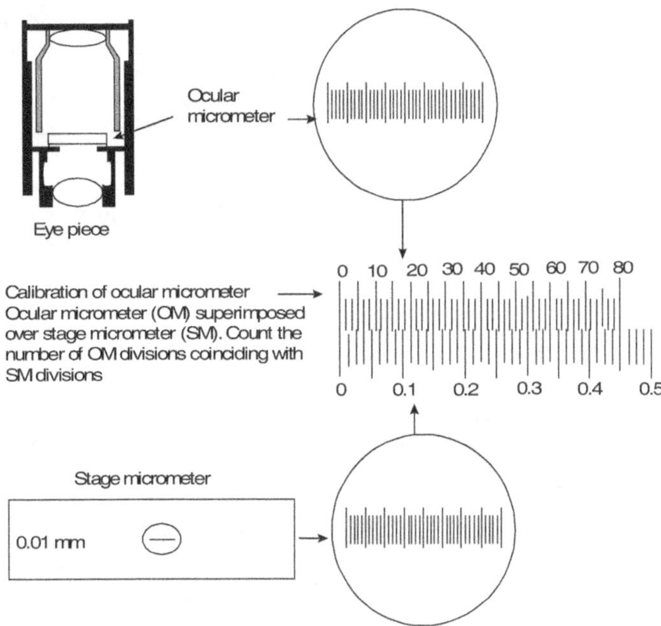

Figure 3.4   Calibration of an ocular micrometer with a stage micrometer

Calibration of Ocular Micrometer

It refers to comparison or superimposition of ocular scale with that of stage micrometer. This is necessary to determine how many graduations of ocular micrometer coincide with divisions on stage micrometer and thus the calibration factor for one division is calculated. This can be done by rotating the ocular lens until the parallel lines of ocular micrometer be parallel with those of the stage micrometer. The same procedure can be repeated for the high-power and oil-immersion objectives for calibration of ocular micrometer since ocular divisions are arbitrary while the divisions on stage are exactly 10 μm apart from each other. When the SM is magnified under different powers of the objectives, it is important to note that under low-power objective, the entire scale is visible and in higher magnification, only a part of it is seen. Under oil immersion, only a few lines of SM scale can be observed. However, whatever be the objective magnification, the size of the image of OM remains the same, magnified by the eye lens only.

Record of Observations

Once the calibration has been done, the stage micrometer is removed and the object is placed on the stage and its dimensions can be measured by counting the divisions of ocular micrometer the object occupies. The number of divisions of OM multiplied by the calibration factor will provide the size of the object. The calibration factor can be calculated as follows:

a) Low-power objective (10×10 =100 magnification)

Suppose,

20 divisions of ocular micrometer (OM) = 30 divisions of stage micrometer (SM)

Therefore,

One division of ocular micrometer $= \dfrac{30}{20} \times 10$ (Since 1 SM division$=10\mu m$)

$$= 15 \ \mu m$$

b) High-power objective (10 × 45 = 450× magnification)

Suppose,

14 divisions of ocular micrometer = 5 divisions of stage micrometer

Therefore,

One division of ocular micrometer $= \dfrac{5}{14} \times 10$

$$= 3.6 \ \mu m$$

c) Oil-immersion objective (10 × 100 = 1000× magnification)

Suppose,

12 divisions of ocular micrometer = 2 divisions of stage micrometer

Therefore,

One division of ocular micrometer $= \dfrac{2}{12} \times 10$

$$= 1.7 \mu m$$

When the stage micrometer is replaced by the freshly prepared or permanent slide of bacterial cell or blood smear to count the number of the ocular micrometer divisions occupied by the single cell, the calibration factor is taken into consideration. For example, if a bacterial cell occupies 4 divisions of ocular micrometer when examined using oil-immersion objective, then dimension of the bacterial cell would be:

$$4 \times 1.7 \mu m = 6.8 \mu m$$

This is because the calibration factor for 1 OM = 1.7 μm when oil-immersion objective is used. Similarly, a lymphocyte occupies 5 divisions of OM hence its size would be 8.5 μm.

SUMMARY

- Leeuwenhoek for the first time constructed the microscope having a biconcave lens enclosed in two metal plates that was used to examine protozoa, bacteria, spermatozoa and red blood cells.

- Robert Hooke designed a compound microscope in which a double-lens system is used, where the second lens magnifies the image from the first lens. The microscope developed by him is the forerunner of the modern-day microscope.

- The simple microscope consists of a single convex lens that provides an enlarged image of an object to the eye.

- Compound microscopes are modern microscopes having a complex system of arrangement of lenses with higher magnification and better resolution.

- A typical compound microscope has two types of parts. The mechanical parts provide the structural framework to the microscope and support the optical parts that magnify the object.

- In the optical system of the compound microscope, two lenses of short focal lengths are used and the light rays passing through them form a final, virtual and enlarged image of the object.

- Resolving power of a microscope is the ability to distinguish two adjacent objects as separate and distinct images rather than as a single blurred image. It is dependant on the wavelength of the light used to illuminate the object and the numerical aperture of the objective.

- The numerical aperture is defined as the diameter of the objective lens in relation to its focal length. It is an index of the light-gathering capacity of the objective.

- There are various forms of microscopy including bright-field microscopy, dark-field microscopy, phase-contrast microscopy, fluorescence microscopy, polarizing microscope, transmission electron microscope (TEM), scanning electron microscope, and video microscopy. Each type of microscopy has its own drawbacks.

- The dimensions of microscopic objects can be measured by micrometry in which the calibration factor is to be calculated in the presence of the stage micrometer and ocular micrometer. The size of the object is then expressed in micrometres taking into consideration the number of ocular divisions occupied and the calibration factor at different objectives.

REVIEW QUESTIONS

1. What is the contribution of Leeuwenhoek towards the development of microscopy?

2. Explain the compound microscope designed by Robert Hooke.

3. What are the different mechanical parts of the modern compound microscope?

4. Describe the principle of the bright-field microscope and its optical system.

5. Explain the resolving power of the microscope and the factors affecting it.

6. Calculate the resolving power of microscope when the yellow light is used for illumination of the object at low power, high power and oil-immersion objectives.

7. Give an account of various microscopes with their advantages and disadvantages.

8. What is micrometry? Explain the role of ocular and stage micrometer in measurement of microscopic objects.

9. Define calibration factor. Explain the method for its calculation using various objectives.

10. Write short notes on:

    i. Refractive index

    ii. Magnification

    iii. Condenser of the microscope

    iv. Numerical aperture

    v. Oils used in oil-immersion objective

    vi. Polarizer

    vii. Phase-contrast microscope

    viii. Stage micrometer

    ix. Calibration of ocular micrometer

VIRUS WORLD

Structural organization of viruses

Classification of viruses

Genomic organisation of viruses

DNA viruses

Retroviruses

Teminism

Reverse transcriptase and gene cloning

Bacteriophages as cloning vectors

Viruses and human diseases

Viruses and cancer

Retroviruses and AIDS

INTRODUCTION

Viruses cannot be classified as cells, because they have no nucleus or cytoplasm, and cannot proliferate outside a living cell. But they are the simplest particles to possess the fundamental properties of living systems—that is, their structure and function are determined by their genetic material, and they can produce copies of themselves by infecting suitable host cells and using the host's raw material and metabolic machinery for their own reproduction. Thus viruses are obligatory parasites that are capable of development only within the cells of the host organism.

Towards the end of the 19th century, it was accepted that many of the microorganisms are associated with diseases. In 1892, the Russian botanist Iwanowsky, in the study of mosaic disease in tobacco plants, discovered that juice extracted from affected plants contained the infectious agent. He assumed that the agent was a bacterium and tried to remove it by filtration but was unsuccessful even with the finest bacterial filter. Iwanowsky concluded that the agent must be smaller than any other type of bacteria then known. Other plant and animal diseases were later found to be caused by infectious agents that could pass through very fine filters and were not visible even under the highest-resolution of light microscopes. The agents came to be known as **filterable viruses**, or simply **viruses** (a Latin word which means poison), and were still considered to be very small bacteria. They could not, however, be grown on the usual media used for culturing bacteria and unlike bacteria, they retained their infective power after precipitation from alcoholic solution. It was not until 1935 that Stanley managed to crystallize the virus that causes tobacco mosaic disease, thus proving that it could not be cellular and must differ greatly from a bacterium. The virus crystals still retained their infective capacity and caused symptoms of mosaic disease when inoculated into healthy plants.

STRUCTURAL ORGANIZATION OF VIRUSES

The viruses are composed of a nucleic acid core surrounded by a protein coat. In present terminology, an intact virus unit is called the **virion**, and its protein coat is known as the **capsid**; the latter is composed of a number of subunits of a particular shape, known as **capsomeres**. The capsid serves to protect the nucleic acid (usually a single RNA or DNA molecule) core against attack by nuclease enzymes.

Most viruses are between 100 Å and 3,000 Å in size, the largest being the size of the small bacteria. Many of them have been examined by means of electron microscope. Figure 4.1 shows the shapes as well as sizes of the viruses which vary enormously. The structure of plant and animal viruses which is usually simpler than that of bacteriophages; the former are rod-shaped (such as tobacco mosaic) or polyhedral (such as adenovirus) and lack the tail of bacteriophage, which means that they cannot inject their nucleic acid core into the cells which they infect. Some viruses, such as the influenza and mumps viruses that are members of the group of myxoviruses, possess an outer membrane known as the **envelope**, which consists of lipid and protein and surrounds the capsid. Herpes also sometimes has an envelope. It has been reported that there is a plant virus, the potato spindle tuber virus, which consists only of RNA, without protein coat; although it would not be the first virus studied that could not make its own coat protein (the Rous virus, which is an avian tumour-producing virus, cannot do this), the potato spindle tuber virus would be the first observed to be transmissible and infectious without the protein coat.

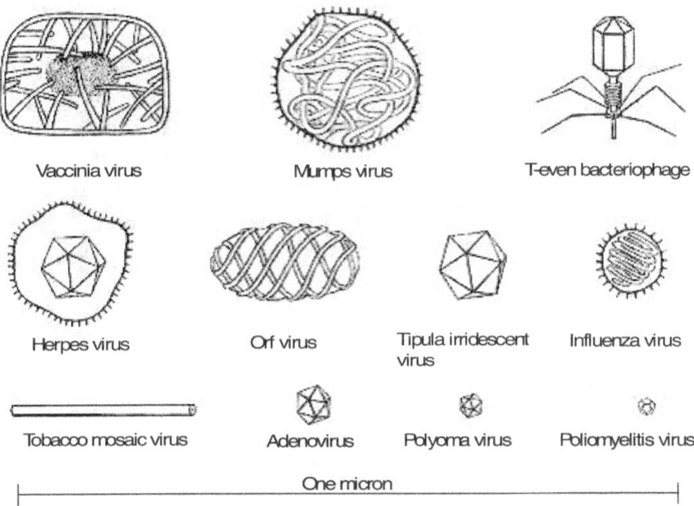

Figure 4.1 Relative sizes of viruses compared to one micron

As per the shapes, most viruses would fall in any one of the following categories: (1) spheroidal or cuboidal, (2) elongated or rod-like, and (3) mixed with shapes of first two types. Apart from these possible groups there are a few viruses with shapes that do not fall into any of the above categories, like for instance, the brick-shaped poxviruses. The most widely used and effective method for the determination of shape of the virus particles is electron microscopy. Except for certain "large" viruses (like smallpox and vaccinia), a major distinguishing property of viruses is their visibility under the best optical microscopes with magnifications up to 2000×. However, electron microscope can give resolution at magnifications of several hundred thousand times, making even the smallest viruses and their innermost structures readily visible.

Virions that appear approximately spherical in electron micrograph are called polyhedral because their protein building blocks are arranged to form a **polyhedral** shell, usually composed of 20 triangular faces (an **icosachedron**), that surrounds the viral nucleic acid and in some cases, protein. In some bacteriophages, the head is a polyhedral structure and the tail is helical, such combination is called binal structure. The binal structure is never found in animal or plant viruses. In such binal phages, the nucleic acid is located within the head and the tail acts as an organ of attachment to host cells. In most phages, tails are made of a single protein tube, but some also have an outer tube, called a sheath, surrounding an inner tube. The structure of the tail in T-even phages (for example, T2, T4 and T6) is remarkably complex. Their tail sheath contains 24 annular rings that form a tube, which is joined to the phage head by a thin collar. The distal end of tail is attached to a hexagonal base plate, which has a protein spike called a tail pin or prongs at each of its corners. Thus there are six tail fibres attached to the base plate, which is essential for adsorption to bacteria. The virus particles are in most instances composed of numerous identical protein subunits and one or rarely a few molecules of nucleic acid. The core of virion contains only one kind of nucleic acid; depending on the virus, it may be single or double-stranded DNA (ss or dsDNA) or single- or double-stranded RNA (ss or dsRNA), but in most cases it provides the genetic information required for the synthesis of viral components and their assembly into new virions by the infected host cell.

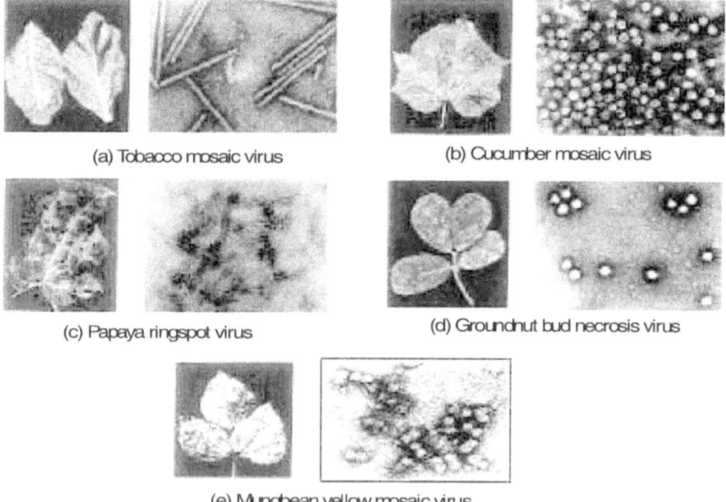

Figure 4.2   Symptoms and particle morphology of some plant viruses

Plant viruses also exist in a variety of forms; they contain either DNA or RNA as their genetic material, have either rod or polyhedral shapes, and are transmitted either by insect bites or by contact upon wounded regions. Most of the plant viruses contain single-stranded RNA as their genetic material. The morphological features of some of plant viruses and the symptoms of the diseases that are caused by them, observed under electron microscope are shown in Figure 4.2.

CLASSIFICATION OF VIRUSES

Viruses are classified as per the nature of nucleic acid, size and architecture of their capsid. In some group of RNA viruses, a virion contains several chromosomes, each coding one or two viral proteins. Virions that exhibit this highly unusual genetic organization are said to possess a **segmented genome**. In some of the single-stranded RNA viruses, the chromosomes also serve as viral RNA; in others, transcribed RNA complementary to the RNA chromosome, serves as mRNA. Virion RNA that can act as mRNA is called plus-strand RNA, while RNA that is complementary to virion mRNA is called minus-strand RNA. Viruses are separately classified as bacteriophages (that infect bacteria), plant viruses and animals viruses. The classification of viruses belonging to some of the families with the type of their nucleic acid and the host which they infect are mentioned in Table 4.1.

Table 4.1 Some of the families of viruses with their species and hosts

Type	Family	Genus	Type species	Host
Positive-stranded ssRNA virus (ss means single-stranded)	Narnaviridae	Narnavirus	*Saccharomyces cerevisiae* 20S narnavirus	Yeast
	Leviviridae	Levivirus	Enterobacteriophage MS2	Bacteria
	Picornaviride	Enterovirus	Poliovirus 1	Vertebrates
		Rhinovirus	Human rhinovirus 1A	Vertebrates
		Hepatovirus	Hepatitis A virus	Vertebrates
		Cardiovirus	Encephalomyocarditis virus	Vertebrates
		Aphthovirus	Foot-and-mouth disease virus O	Vertebrates
	Comoviridae	Comovirus	Cowpea mosaic virus	Plants
		Nepovirus	Tobacco ringspot virus	Plants
	Potyviridae	Potyvirus	Potato virus Y	Plants
		Bymovirus	Barley yellow mosaic virus	Plants
		Tritimovirus	Wheat streak mosaic virus	Plants

(Contd.)

Table 4.1   (Continued)

Type	Family	Genus	Type species	Host
	Caliciviridae	Vesiculovirus	Swine vesicular exanthema virus	Vertebrates
		Lagovirus	Rabbit haemorrhagic disease virus	Vertebrates
ssDNA virus	Inoviridae	Inovirus	Coliphage fd	Bacteria
	Geminiviridae	Mastrevirus	Maize streak virus	Plants
		Curtovirus	Beet curly top virus	Plants
		Begomovirus	Pea golden mosaic virus	Plants
	Circoviridae	Circovirus	Chicken anaemia virus	Vertebrates
	Parvoviridae	Parvovirus	Mice minute virus	Vertebrates
		Iteravirus	Bombyx mori densovirus	Invertebrates
		Brevidensovirus	Aedes aegypti densovirus	Invertebrates
dsDNA virus (ds means double-stranded)	Myoviridae	T4-like viruses	Coliphage T4	Bacteria
		P1-like viruses	Enterobacteriophage P1	Bacteria
		P2-like viruses	Enterobacteriophage P2	Bacteria
	Siphoviridae	λ-like viruses	Coliphage λ	Bacteria
		L5-like viruses	Mycobacterium phage L5	Bacteria
	Poxviridae	Orthropoxvirus	Vaccinia virus	Vertebrates
		Capripox virus	Sheeppox virus	Vertebrates
	Herpesviridae	Simplexvirus	Human herpesvirus 1	Vertebrates
		Muromegalovirus	Mouse cytomegalovirus 1	Vertebrates
	Adenoviridae	Mastadenovirus	Human adenovirus 2	Vertebrates
		Rhizidiovirus	Rhizidomyces virus	Fungi
	Papovaviridae	Polyomavirus	Murine polyomavirus	Vertebrates
		Papillomavirus	Cottontail rabbit papillomavirus	Vertebrates
DNA and RNA reverse transcribing viruses	Retroviridae	Lentiviruses	Human immunodeficiency virus 1	Vertebrates

(Contd.)

Table 4.1 (Continued)

Type	Family	Genus	Type species	Host
		Alpharetrovirus	Avian leukosis virus	Vertebrates
		Betaretrovirus	Mason-Pfizer monkey virus	Vertebrates
		Gammaretrovirus	Mouse mammary tumour virus	Vertebrates
		Deltaretrovirus	Bovine leukemia virus	Vertebrates
		Epsilonretrovirus	Walleye dermal sarcoma virus	Vertebrates
	Hepadnaviridae	Orthohepadnavirus	Hepatitis B virus	Vertebrates
	Caulimoviridae	Caulimovirus	Cauliflower mosaic vein virus	Plants
		SbCMV-like viruses	Soybean chlorotic mottle virus	Plants
		RTBV-like viruses	Rice tungro bacilliform virus	Plants
		Badnavirus	Commelina yellow mottle virus	Plants
	Metaviriade	Errantivirus	*Drosophila melanogaster* gypsy virus	Invertebrates

GENOMIC ORGANIZATION OF VIRUSES

Almost all plant and some bacterial and animal viruses have RNA genome. These genomes tend to be particularly small. For example, the genome of mammalian retroviruses such as HIV are about 9,000 nucleotides long, and that of the bacteriophages $Q\beta$ has 4,220 nucleotides. Both types of viruses have single-stranded RNA genome. Similarly, the genomes of DNA viruses vary greatly in size (Table 4.2). For example, a bacteriophage T2 has a linear DNA molecule with a molecular weight of approximately 1.2×10^8 daltons (180,000 base pairs) and measures 68 μm in length and as shown in electron micrograph, when the DNA is released from the ruptured head of the virus, its length is approximately 600 times longer than the phage head (Figure 4.3). Many viral DNAs are circular for at least part of their life cycle. During viral replication within a host cell, specific types of viral DNA called **replicative forms** may appear; for example, many linear DNAs become circular and all single-stranded DNAs become double-stranded.

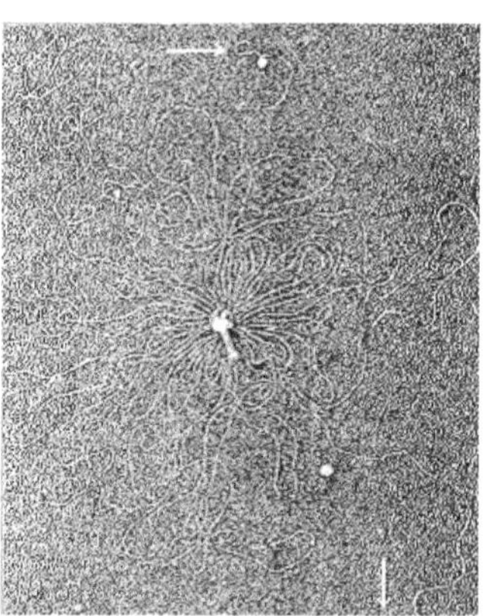

Figure 4.3   Electron micrograph of the DNA being released from the ruptured head of T2 bacteriophage. Note the two ends of the linear molecule.

A typical medium-sized DNA virus is bacteriophage λ (lambda), which **infects** *E. coli*. In its replicative form inside cells, λ DNA is a circular double helix. This double helix DNA contains 48,502 bp and has a counter length of 17.5 μm. Bacteriophage ΦX174 is a much smaller DNA virus; the DNA in the viral particle is a single-stranded circle, and the double-stranded replicative form contains 5,386 bp. Although viral genomes are small, the counter lengths of their DNA are much greater than the long dimensions of viral particles that contain them. The DNA of bacteriophage T4, for example, is about 290 times longer than the viral particle itself.

Table 4.2   Sizes of DNA and viral particles for some bacteriophages

Virus	Size of viral DNA (bp)	Length of viral DNA (nm)	Long dimension of viral particle (nm)
ΦX 174	5,386	1,939	25
T7	39,936	1 4,377	78
λ (lambda)	48,502	17,460	190
T4	168,889	60,800	210
T2	180,000	68,000	120

In most cases, linear DNA chromosomes are circularized within the host cell by one of two distinct mechanisms depending on whether the linear molecule has cohesive ends or terminal redundancy (Figure 4.4). Cohesive ends are short, single-stranded regions that are complementary to each other; within host cells, these anneal to form a circular molecule with a single nick in each DNA strand. Nicks are then sealed by the action of DNA ligase. Linear DNA chromosomes with terminal redundancy circularize by recombination within their homologous terminal regions. Once the circular DNA is formed, it undergoes the process of replication to form daughter DNA molecules. The process starts at a specific site and proceeds in both directions around the molecule similar to that by which the bacterial chromosome is replicated, and the method of replication of circular DNA is referred to as **rolling circle replication** (Refer Chapter 5).

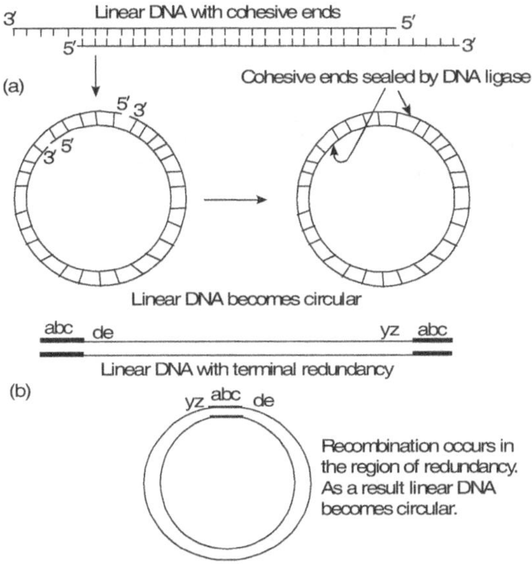

Figure 4.4 Alternate strategies for conversation of linear DNA into circular molecules

ACTION OF VIRUSES

The specificity of viruses appears to vary; some may attack only one specific cell type, while others may attack a number of different cells. They all possess one property in common, they cannot be grown on non-living material. Viruses grown in the laboratory (for production of vaccines, for example, and for research purposes) are usually cultured in fertilized hen's egg or in cells in tissue culture. The mode of action of viruses containing double-stranded DNA and the retroviruses having RNA as a genetic material is discussed.

DNA Viruses

In the late 1930s, geneticists began to investigate the various types of phages that attack *E.coli*. The best-known examples of these are the similar strains T2,

T4 and T6. As described earlier, structurally they have a polyhedral head enclosing a genetic material made up of DNA and a tail which is a hollow core, surrounded by a contractile sheath through which the DNA is injected into a bacterium. A bacteriophage infects a suitable *E.coli* host cell by adsorption, the tail fibres are used to place the virus in the correct position, and the prongs on the base plate anchor it there. A small hole is then produced in the bacterial call wall with the help of lysosomal enzymes that are present in the tail of the phage. The viral DNA is then injected through this hole, and generally it rapidly begins to reproduce inside the bacterial cell. About 100 to 1,000 new copies of viral DNA are produced which then acquire a newly synthesized protein coat and a tail, and finally form complete phage particles. The entire life cycle, which takes only a few minutes, finishes when the bacterium lyses, that is its cell wall breaks down, releasing the contents and freeing the many phage progeny. This process is known as a **virulent** infection and the phage that causes it is referred to as lytic virus (Figure 4.5).

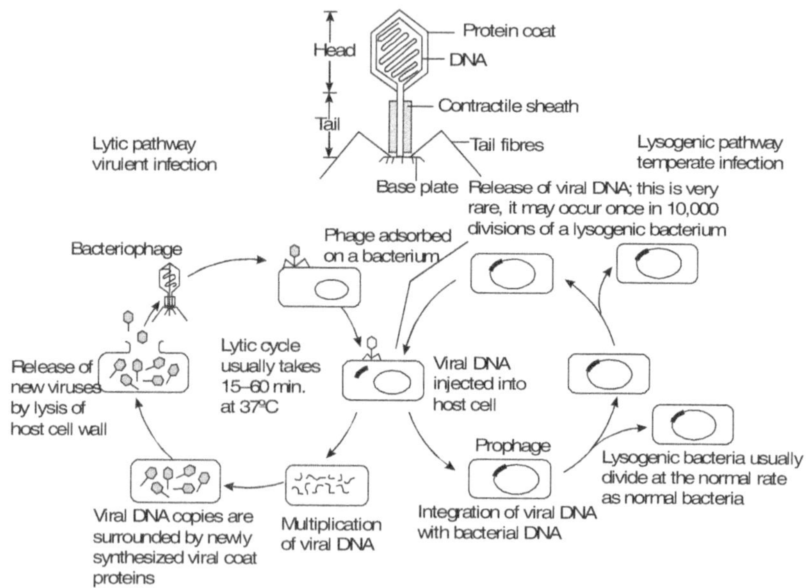

Figure 4.5   Life cycle of bacteriophage showing lytic and lysogenic pathways

There is another process, which may occur in certain cases, known as a **temperate infection**. Some bacterial viruses, called **lysogenic viruses**, do not always cause lysis in the usual way. Instead, the viral DNA becomes attached to the host cell genome, that is, the viral DNA can behave as an episome, either existing autonomously or as part of the host genome. When it is attached to the bacterial genome, it is known as the **prophage**, and appears to behave as part of the bacterial genetic material, duplicating once every cell generation. The prophage may be carried on the bacterial genome indefinitely, without destroying the host cell. Very rarely it becomes released from the host bacterium and multiplies as in a virulent infection, with resultant lysis.

Retroviruses

The mode of action of retroviruses is altogether different as compared to that of T-even bacteriophages. Retroviruses (retro is the Latin prefix for "backward") are the viruses in which single stranded RNA molecules serve as the genetic material. They infect animal-cells and carry within the viral particle an RNA-dependent DNA polymerase called **reverse transcriptase**. On infection, the single-stranded viral genome (approximately 10,000 nucleotides) and the enzyme enter the host cell. The reverse transcriptase first catalyses the synthesis of a DNA strand complementary to the viral RNA, then degrades the RNA strand of RNA–DNA hybrid and replaces it with DNA forming dsDNA (Figure 4.6).

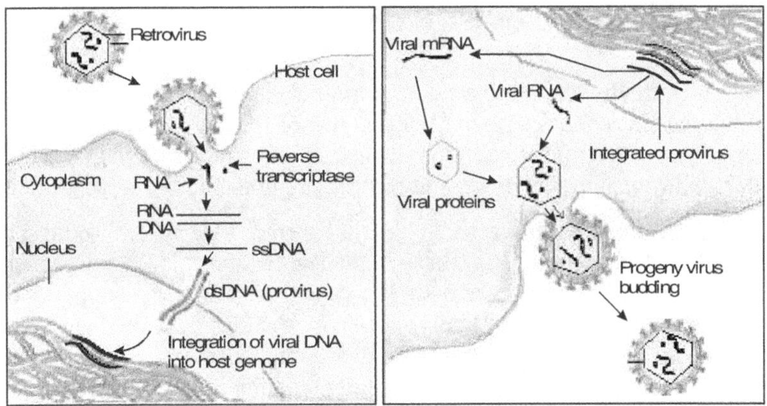

Figure 4.6 Retroviral infection. The viral enzyme reverse transcriptase converts its own RNA into double-stranded DNA, which later integrates into host genome. Provirus transcribes to form viral genomic RNA and mRNA in the normal cell division process. The viral mRNA transcribes to form viral proteins that assemble to form a virion enclosing its own genetic material. The progeny virions bud off from cell surface.

The resulting duplex DNA often becomes incorporated into the genome of the eukaryotic host cell. These integrated (and dormant) viral genes can be activated and transcribed, and gene products—viral proteins and the viral RNA genome itself—packaged as new viruses.

Teminism

In 1962, the existence of reverse transcriptase in RNA was predicted by Howard Temin and the enzymes were ultimately detected by Temin and, independently, by David Baltimore in 1970. Their discovery aroused much attention as it was a dogma-shaking proof that genetic information can flow "backward" from RNA to DNA by the process called reverse transcription catalysed by reverse transcriptase having two subunits, α (mol.wt. : 65,000) and β (mol.wt. : 90,000), specifically known as RNA-dependent DNA polymerase and the phenomenon is referred to as teminism. Thus single-stranded RNA is converted into double-stranded DNA in presence of the enzyme reverse transcriptase.

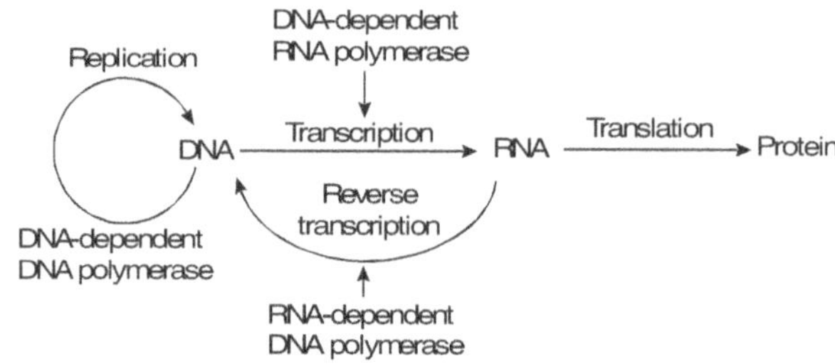

In retroviruses, there are long terminal repeat (LTR) sequences of a few hundred nucleotides present at each end of the linear RNA genome. These sequences are transcribed into duplex DNA facilitating integration of viral chromosome into host DNA, and contain promoters for viral gene expression.

There are three different activities that are catalysed by reverse transcriptase, viz. RNA-dependent DNA synthesis, RNA degradation, and DNA-dependent DNA synthesis. Like many DNA and RNA polymerases, reverse transcriptase contains Zn^{2+}. Each transcriptase is most active with the RNA of its own virus, but each can be used experimentally to make DNA complementary to a variety of RNAs. The DNA and RNA synthesis and RNA degradation activities use active sites on the protein. For DNA synthesis to begin, the reverse transcriptase requires a primer, which is a cellular tRNA obtained during as earlier infection and carried within viral particle. This tRNA is base-paired at its 3′end with a complementary sequence in the viral RNA. The new DNA strand is synthesized in the 3′ to 5′ direction, as in all RNA and DNA polymerase reactions. Reverse transcriptases, like RNA polymerases, do not have 3′ to 5′ proofreading exonucleases. They generally have error rates of about 1 per 20,000 nucleotides added. An error rate this high is extremely unusual in DNA replication and appears to be a feature of most enzymes that replicate the genomes of RNA viruses. As a consequence, there is higher mutation rate and faster rate of viral evolution, this is the reason why new strains of disease-causing retroviruses appear frequently.

REVERSE TRANSCRIPTASE AND GENE CLONING

The use of reverse transcriptase has become an important tool in DNA cloning technique. Synthesis of cDNA library is an essential activity in genetic engineering. Reverse transcriptase is used to synthesize a copy or complementary DNA (cDNA) from the mature mRNA template that is isolated from a eukaryotic cell. The reactions involved in the overall process are summarized in Figure 4.7.

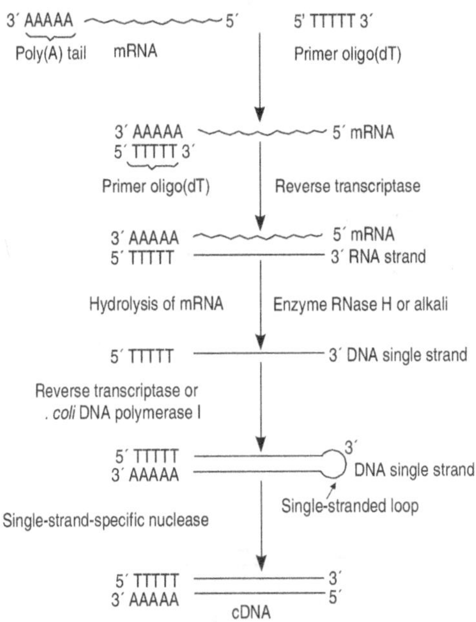

Figure 4.7 Synthesis of complementary DNA onto mRNA template with the help of reverse transcriptase

BACTERIOPHAGES AS CLONING VECTORS

Genetic engineering involves isolation of desired gene and its subsequent insertion into another piece of DNA of the cloning vector, which will allow it to be taken up by bacteria or any other suitable host and replicate within them as the cell grows and divides. The principle that governs the delivery of recombinant DNA in clonable vector form to a host cell, and its subsequent amplification in the host, is well illustrated by the use of bacteriophages as cloning vectors.

Bacteriophage λ has a very efficient mechanism for delivering its 48,502 bp of DNA into a bacterium, and it can be used as a vector to clone somewhat larger DNA segments. Two important features contribute to its utility: (1) About one-third of the λ genome is non-essential and can be replaced with the foreign

DNA, and (2) DNA is packaged into infectious phage particles only if it is between 40,000 and 53,000 bp long, a constraint that can be used to ensure packaging of recombinant DNA only. Attempt have been made to develop bacteriophage λ vectors that can be readily cleaved into three pieces, two of which together are only about 30,000 bp long. The third piece, "filler" DNA, is discarded when the vector is to be used for cloning and additional DNA is inserted between two essential segments to generate ligated DNA molecules long enough to produce viable phage particles. In effect, the packaging mechanism selects for recombinant viral DNAs.

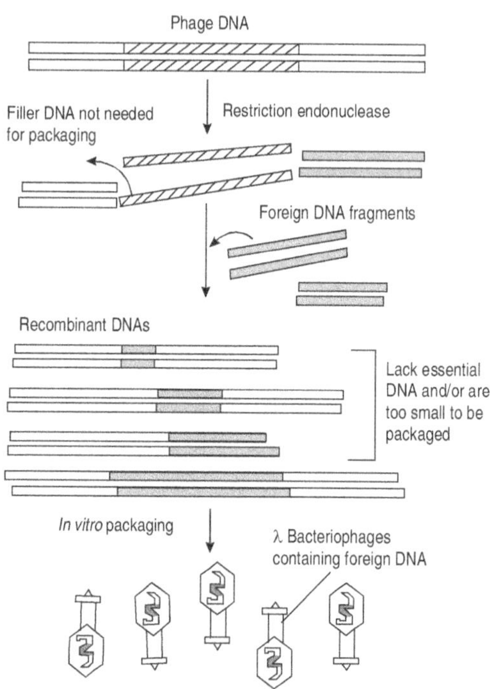

Figure 4.8   Recombinant DNA technology used to modify genome of bacteriophages λ that are to be served as cloning vectors

The λ phage vectors permit the cloning of DNA fragments of up to 23,000 bp. Once the bacteriophage is ligated to foreign DNA fragments of suitable size, the resulting recombinant DNAs can be packaged into phage particles by adding them to crude bacterial cell extracts that contain all the proteins needed to assemble a complete phage. This process is referred to as *in vitro* packaging and the steps involved in it are summarized in Figure 4.8. All the viable phage particles will contain a foreign DNA fragment. The subsequent transmission of recombinant DNA into *E. coli* cells is highly efficient.

VIRUSES AND HUMAN DISEASES

All of the major groups of the animal viruses cause human diseases. Some of important viral diseases are mentioned in Table 4.3.

Table 4.3 Some of human diseases caused by viruses

Disease	Pathogenic virus (Group)	Genetic material	Mode of transmission and symptoms
Smallpox	Variola, (Poxvirus)	Double-stranded DNA	Highly contagious and spread by respiratory route. Following an incubation period of 12 days, intense fatigue and high fever, spotty red rashes on skin. Spots become blisters that rupture, and scars are formed.
Chickenpox	Varicella (Herpesvirus)	Double-stranded DNA	Acquired by respiratory route or by direct contact with infected person. Following an incubation period of 10 to 21days, small blisters surrounded by reddened area of skin appear, then burst and covered by scab.
Mononucleosis	Epstein–Barr virus (Herpesvirus)	Double-stranded DNA	Acquired by direct contact or inhaling droplets from infected person. Following incubation period of 1 to 2 months, gradual onset of fatigue and fever, soar throat and enlargement of lymph nodes in neck.
Serum hepatitis	Hepatitis B virus (Orthohepadnavirus)	Double-stranded DNA	Acquired by contact with infected person, also transmitted from infected mother to foetus via placenta. Following incubation period of 1 to 3 months, fatigue and fever gradually begin, followed by jaundice that is caused by accumulation of bilirubin. Extensive liver damage may lead to death.

(Contd.)

Table 4.3 (Continued)

Disease	Pathogenic virus (Group)	Genetic material	Mode of transmission and symptoms
Infectious hepatitis	Hepatitis A virus (Picornavirus-Hepatovirus)	Plus-strand RNA	Acquired by oral–faecal route, mild intestinal symptoms, occasionally virus spreads to liver. Fatality rate is less than 1 per cent.
Polio	Poliomyelitis virus (Picornavirus-Enterovirus)	Plus-strand RNA	Acquired by oral–faecal route, replicating first in intestinal cells and spreading to lymph nodes. Virus spreads to spinal cord and brain where it infects and kills neurons, causing permanent paralysis.
Influenza	Influenza virus (Orthomyxovirus)	Minus-strand RNA	Acquired from infected person by respiratory route. Following incubation period of 1 to 3 days, headache, fever, and muscular pains begin with cough and sore throat.
Rabies	Rhabdovirus	Minus-strand RNA	Acquired by dog or cat bite. Incubation period 2 to 16 weeks. Symptoms begin with loss of appetite, fatigue, sensation of tingling or burning at the site of bite, intense and painful spasms of throat muscle, hydrophobia, and 100 per cent fatal due to destruction of breathing centres in brain.
Yellow fever	Togavirus	Plus-strand RNA	Disease transmitted by mosquito (*Aedes*) bite. Monkeys are primary reservoirs. After in-cubation period of 3 to 6 days, headache and chills occur. Fever lasts several days. The characteristic yellowing of skin (jaundice) occurs in only severe cases.

Retroviruses and AIDS

Retroviruses are animal viruses with DNA or RNA as their genetic material. RNA viruses have the enzyme reverse transcriptase. In addition, they have a highly unusual replicating cycle; upon entering the cytoplasm of a host cell, this enzyme synthesizes a strand of DNA (ssDNA) that is complementary to viral (plus-strand) RNA. The ssDNA is then converted into dsDNA (double-stranded) with the help of the same enzyme. This leads to formation of the chromosome of the virus called **provirus**, which migrates to the nucleus and becomes integrated into a host chromosome. Then the RNA polymerase of the host transcribes the integrated viral genes, producing viral RNA for incorporation into virions (Figure 4.6).

The human immunodeficiency virus (HIV) that causes acquired immunodeficiency syndrome (AIDS) is a retrovirus [Figure 1.6c]. In early 1983, Luc Momtegnier and his colleagues at the Pasteur Institute, Paris, isolated a virus from lymph tissue of man who had lymphadenopathy syndrome that normally precedes full-blown AIDS. Later on several workers gave various names to the same virus like LAV (lymphadenopathy associated virus), IDAV (immuno deficiency associated virus), ARV (AIDS associated virus), LAV/HTLV III-CDC-151 (Human T-cell Leukemia virus—the name given by Center for Disease Control, USA), and finally in May 1986, International Committee on Taxonomy of Viruses gave a single name to the virus, i.e., "HIV".

It is generally accepted that HIV is included under Lentiviruses since it shows similarities to lentiviruses in the following characteristics: budding, antidrift, antigen shading, sophisticated self regulation of replication involving self inhibitory action, genetic structure and also presence of membrane-bound glycoproteins gp120 and gp41. Two strains of HIV, I and II are now known, of which HIV-II is less common and perhaps less dangerous. The genome of HIV-I contains the genes *gag, pol, vif, vpr, tat, rev, vpu, env* and *nef*. The product of these genes and their functions are summarized in Table 4.4.

Table 4.4 The HIV genes with their functions

Gene	Product of gene (protein with mol. wt. in kDa)	Function
Gag	p24	Capsid structural protein
	p17	Matrix protein
	p9	RNA-binding protein (?)
	p6	RNA-binding protein helping in virus budding

(Contd.)

Table 4.4 (Continued)

Gene	Product of gene (protein with mol. wt. in kDa)	Function
pol	Reverse transcriptase (RT, p55)	Reverse transcription
	RNase-H 63 (p63)	Ribonuclease (inside core)
	Protease 15 (p15)	Post translation processing of viral protein
	Integrase (p11)	Viral cDNA integration
vif	p23	Increases viral infectivity and cell–cell transmission
vpr	p18	Helps in viral replication and perhaps transactivation
tat	p14	Transactivation
rev	p19	Regulation of viral mRNA
vpu	p15	Helps in viral budding
env	gp120 and gp41	Envelope surface and envelope *trans*-membrane proteins
nef	p27	Probably involved in virus suppression

Viruses and Cancer

Cancer is a diverse class of diseases that results from changes in cellular behaviour caused due to changes in the genetic material of the cell. Because of these changes, cancer cells divide in an uncontrolled manner and form an **abnormal mass of tissue. Such masses are called neoplasms or tumours** that are classified by their pattern of growth into two groups: those that do not invade surrounding tissue are referred to as **benign**. They grow by displacing adjacent cells but rarely kill the organism unless they occur in the brain. On the other hand, **malignant tumours**, called cancers, invade and destroy surrounding tissue as they grow. One of the prominent characteristics of malignant tumours is their tendency to release the cells that break away from the parent mass, enter the lymphatic or vascular circulation, and spread to distant sites in the body where they establish secondary tumours, the phenomenon called **metastasis**.

Cancers are generally classified into four groups, depending upon the type of cells originally involved. Two types of cancer lead into overproduction of white blood cells. **Leukemias** are diseases of the bone marrow that cause excessive production of leucocytes. **Lymphomas** are diseases of lymph nodes and spleen that cause excessive production of lymphocytes. **Sarcomas** are tumours of tissue that arise from embryological mesoderm such as muscle, bone, and cartilage. Most common form of cancers is **carcinomas**, tumours arising from epithelial tissue such as glands (including hepatomas—the malignant tumours in liver cells), breast, skin (including melanomas, the malignant tumours arising in melanocytes, the pigment cells of skin), and linings of the urinogenital (for example, cancers of kidney, ovarian, uterus, cervix, prostate, etc.) and respiratory systems (including throat and lung cancer).

One of the important behavioural changes in human cancer cells, whether residing in the body or on a culture dish, is its loss of growth control. The capacity for the growth and division is not drastically different between a cancer cell and most normal cells. When normal cells are grown in tissue culture under conditions that promote cell proliferation, they will grow and divide at a rate similar to that of their malignant counterparts. However, when the normal cells proliferate to the point where they cover the bottom of the culture dish, their growth rate decreases markedly and they tend to remain as a single layer (monolayer) of cells as shown in Figure 4.9a. The drop in growth rate occurs

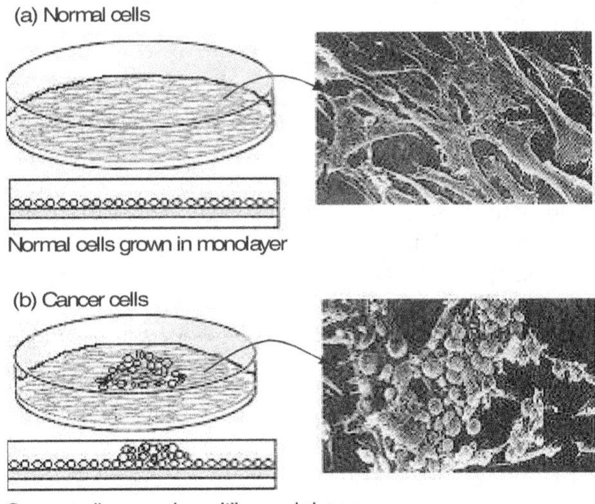

Figure 4.9 Differences in growth pattern of (a) normal cells, and (b) cancer cells

as the cells respond to inhibitory influences from their environment. These growth-inhibiting influences may arise as a result of depletion of growth factors in the culture medium or contact with surrounding cells on the dish (this is referred to as contact inhibition of growth). On the other hand, when malignant cells are cultured under the same conditions, they continue to grow, piling one on top of another to form clumps (Figure 4.9b). It clearly indicates that malignant cells are not responsive to the type of influences that cause their normal cells to stop growth and division.

In 1775, Percival Pott, a British surgeon, made the first known correlation between an environmental agent and the development of cancer when he concluded that high incidence of cancer of nasal cavity and skin of the scrotum that occurred in chimney sweeps was due to their chronic exposure to soot. In addition to several chemical mutagens, a number of other types of agents have been shown to be carcinogenic, including several types of ionizing radiation and a variety of DNA- and RNA-containing viruses. The first evidence of a causal relationship between viruses and cancer was obtained in 1908 when V. Ellerman and O. Bang demonstrated that a type of leukemia that affects chicken could be transmitted to healthy birds by injecting them with a cell-free filtrate of the blood of a leukemic bird. In 1911, F. Peyton Rous demonstrated that a chicken sarcoma could be similarly transmitted and he established that the active agent in the filtrates was a first retrovirus, called Rous sarcoma virus (RSV) containing RNA as genetic material.

The interest in the phenomenon that demonstrated the relationship between viruses and cancer was sharply increased in 1936 when J. Bittner experimentally proved that a mammary tumour virus (MTV) transmitted from a female mouse to her offspring can cause mammary cancer. Although studies of carcinogenic-induced tumours have yielded much information on the chemical mechanisms involved in alteration of DNA, until recently they revealed nothing about the specific gene responsible for transforming normal cells into cancer cells. Far more insight into these genes came first from the study of viruses that cause cancer in experimental animals and transform cells in culture. This is because the viruses usually carry the cancer-inducing genes, or **oncogenes**, as part of their genome.

A number of viruses are capable of infecting vertebrate cell, transforming them into cancer cell. The tumour viruses are able to transform cells because they carry with them genes whose products interfere with the cell's normal growth-regulating activities. These viruses are broadly divided into two large groups: **DNA tumour viruses** for example papovaviruses [Simian virus (SV)40, polyoma, and papilloma viruses], adenoviruses, and herpesviruses and the other group of **RNA tumour viruses** include retroviruses. In general, the oncogenes of DNA viruses are virus-coded gene products that are also required for the normal replication of the virus. In contrast, most of the retrovirus oncogenes that have been studied are not required for replication. Rather, they are cellular genes that were transduced by chance onto the viral genome. Normal cellular genes with potential to become oncogenes are called **proto-oncogenes** or **cellular oncogenes** (c-oncs).

Although the use of tumour viruses in laboratory has proven invaluable in identifying the genes involved in carcinogenesis, such viruses are responsible for only a few minor types of human cancers that are mentioned in Table 4.5.

Table 4.5   Viruses that cause human cancer

Virus type	Malignant tumor
Epstein–Barr virus	Burkitt's lymphoma, nasopharyngeal carcinoma and B-cell lymphomas
Hepatitis B virus	Hepatoma
Human papillomaviruses	Skin cancer, tumours in cervix, vulva, penis, perianal and anal regions
Human T-cell leukemia virus	Adult T-cell leukemia

Oncogenes   These are genes whose activity has the potential to transform a normal cell into a cancer cell. In most cases studied so far, transformation results from the presence of a single viral gene called oncogene. This gene encodes proteins that promote the loss of growth control and the conversion of a cell to a malignant state. The existence of oncogenes was discovered through a series of investigations on RNA tumour viruses. The turning point in these studies came in 1976, when it was discovered by F.P. Rous that an oncogene called *src* was carried by an RNA tumour virus called avian sarcoma virus or Rous sarcoma virus (RSV).

The *src* encodes a protein of MW 60 kDa (denoted pp 60^{src}), which is largely associated with the cytoskeleton, a network of protein microfilaments underlying the cytoplasmic membrane, and that phosphorylates tyrosine residues in certain proteins. One of these, vinculin, is a membrane protein associated with zones, termed adhesion plaques, where the membrane establishes contact with a surface. It is hypothesized that phosphorylated vinculin cannot function in establishing these contacts. The class of retroviruses called **acute transforming viruses**, includes the viruses having oncogenes with the ability to induce tumours *in vivo* after a short latent period, sometimes as little as a few days and they also transform cells in culture, a property that is almost essential to the molecular analysis of a transforming gene. About 50 different transforming retroviruses have been isolated from different animal species, and about 20 to 25 different oncogenes have been discovered in the genomes of these viruses. Some of the acute transforming viruses with their oncogenes are given in Table 4.5.

Table 4.5   Some of the acute transforming viruses with their oncogenes and the encoded proteins with their functions

Retrovirus and host	Oncogene	Protein (MW in kilodalton)	Function
Rous sarcoma virus (chicken)	*src*	pp60 ^{src}	Tyrosine kinase
Simian sarcoma virus (monkey)	*sis*	P28$^{env\text{-}sis}$	Platelet-derived growth factor
Avian myoblastosis virus (chicken)	*myb*	P45^{myb}	Nuclear protein
Abelson murine leukemia virus (mouse)	*abl*	P90–P160$^{gag\text{-}abl}$	Tyrosine kinase
UR2-avian sarcoma virus (chicken)	*ros*	P68$^{gag\text{-}ros}$	Tyrosine kinase
Harvey murine sarcoma virus (mouse)	*H-ras*	pp21^{ras}	GTP-binding protein
Avian erythroblastosis virus (chicken)	*erbA*	P75$^{gag\text{-}erb\text{-}A}$	Membrane receptor for growth hormone
Gardner–Arnstein feline sarcoma virus (cat)	*fes*	P110$^{gag\text{-}fes}$	Tyrosine kinase

Proto-oncogenes   The non-acute retroviruses lack oncogenes, they encode only *gag* (the gene for capsid protein), *pol* (gene for synthesis of reverse transcriptase), and env (gene that codes the protein spikes of the viral envelope) as shown in Figure 4.10a. These viruses, which are frequently found in chickens, mice, and cats, usually cause different forms of leukemia that typically take much longer to appear than tumours induced by transforming retroviruses. Because they never transform cells in culture, for a long time it was a mystery as to how these viruses induce tumours. The answer came from studying B-cell leukemias induced by avian leucosis virus (ALV) in chickens. In this case, the provirus integrates with host genome adjacent to cellular oncogens. As a consequence, protooncogenes become activated and induce tumour formation in the host.

The proto-oncogenes encode proteins that have various functions in the cell's normal activities. Several processes appear to be involved in the activation of proto-oncogenes (Figure 4.10b). For example,

1. The gene can be mutated in a way that alters the properties of the gene product so that it can no longer carry out its normal activity. This can be exemplified by a single base-pair substitution mutation (GC into TA transition) that converts a normal cellular gene into an oncogene, which in turn causes some human carcinomas;

2. A mutation in a nearby regulatory sequence can alter the expression of the gene, so that too little or too much of the gene product is produced; and

3. A chromosome rearrangement occurs that brings a DNA sequence from a distant site in the genome into close proximity of the gene, which can either alter the expression of the gene or the nature of the gene product. Most of the leukemias and lymphomas are characterized by chromosomal abnormalities, called translocations; where an arm of one chromosome has been broken and rejoined to the arm of another chromosome. Such abnormality was first discovered in **chronic myelocytic leukemia**, where an arm of chromosome 9 is translocated next to the gene for antibody light chain synthesis on the chromosome 22.

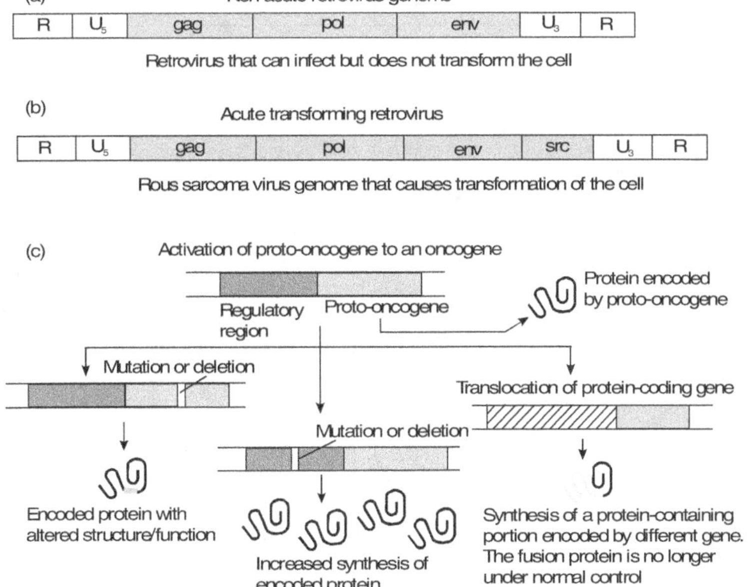

Figure 4.10 (a) Genome of non-acute retrovirus. (b) Genome of RSV (Rous sarcoma virus) containing *src*, the oncogene. (c) Various ways of activation of proto-oncogene to an oncogene.

HIV INFECTION

The human immunodeficiency virus consists of the RNA molecule and an enzyme reverse transcriptase surrounded by protein coat, which in turn is enclosed in a lipid coat derived from the plasma membrane of a T cell of the immune system. HIV selectively infects the T4 helper cells containing CD4 molecules on their surface. HIV binds CD4 cells via gp120. When the infected T cells (HIV provirus) participate in an immune response, the cellular and viral DNAs are transcribed. The viral RNA transcript is translated into viral proteins, and new viral particles are formed. These particles bud off from the surface of the T cell, rupturing and killing the cell and setting off a new round of T-cell infection. Gradually over the course of HIV infection, there is decrease in the number of helper T4 cells thereby decreasing the immune power (Figure 4.11). This ultimately results in an increased susceptibility to pathogens and increased risk of certain forms of cancer. AIDS patients often succumb to opportunistic infections such as pneumocystis pneumonia. HIV has a long incubation period and symptoms of the disease make their appearance in a phased manner at different rates in different persons. Between 4 weeks to 4 months after infection, 60–70% patients may develop transient symptoms similar to influenza or glandular fever.

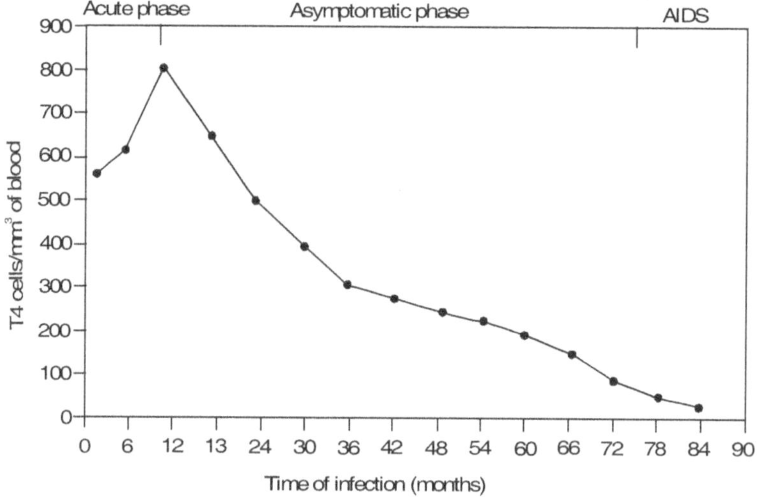

Figure 4.11 The progressive loss of T4 cells during the HIV infection. Usually, when levels of T4 declines below 200/mm³ of blood, the patient shows clinical symptoms of AIDS.

When the infection is 3 years old, 30–40% patients show fever, sweat, aches, fatigue, weight loss, sickness, diarrhoea and swelling in lymph nodes in neck and armpit. About 10–15% patients infected 3 years ago show symptoms of severe infection or full-blown AIDS. The symptoms are lung diseases, fever, skin tumour, severe fungal infection of mouth and esophagus, severe diarrhoea and unexplained weight loss. Once full-blown AIDS is developed, 80% patients die within 2 years.

Possible Routes of HIV Infection

The disease is transmitted primarily by direct contact between infected homosexuals or heterosexuals, from HIV positive mother to her child, via unprotected sex, use of unsterilized hypodermic needles and syringes, and also through blood or blood products obtained from infected persons. The infection does not spread through saliva or through insect bites (mosquitoes, ticks, etc.) social or casual contacts. The method for diagnosis of the disease is ELISA (Enzyme-linked immunosorbant assay) that detects anti-HIV antibodies followed by western blotting; it is the most practicable and inexpensive method.

The enzyme reverse transcriptase of HIV is even more error prone than other known reverse transcriptase, ten times or more, resulting in high mutation rate in this virus. One or more errors are generally made every time the viral genome is replicated, so any two RNA molecules are likely to differ. Many modern vaccines for viral infections consist of one or more coat proteins of the virus. These proteins are not infectious on their own but stimulate the immune system to recognize and resist subsequent viral invasion. Because of the high error rate of the HIV reverse transcriptase, the viral genome undergoes very rapid mutation, which makes the development of an effective vaccine for HIV a difficult and complicated task. Blocking the action of enzyme activity has prime importance in development of AIDS therapy. A class of drugs that are structure analogues of thymidine, cytidine, and inosinic acid, when incorporated into DNA, prevent the extension of the DNA by reverse transcriptase. The drugs are azidothymidine (AZT, zidovudine), dideoxycytidine (ddC) and dideoxyinosine (ddI). However, these drugs are highly toxic and their benefits are transient.

SUMMARY

- Viruses are obligatory parasites having the capacity to develop only within the cells of the host organism.

- In 1892, Iwanowsky discovered a filterable agent that caused mosaic disease in tobacco plants.

- Stanley (1935) crystallized tobacco mosaic viruses which retained their infective capacity and caused mosaic disease when inoculated into healthy plants.

- Viruses are composed of a nucleic acid made up of DNA or RNA core surrounded by a protein coat.

- Electron microscopic observations indicate enormous variation in the shapes and the sizes of the viruses. The plant and animal viruses are usually simpler than bacteriophages; the former are rod-shaped or polyhedral.

- The polyhedral shell is usually composed of 20 triangular faces (an icosachedron), which surrounds the viral nucleic acid.

- T-even phages have remarkably complex structures with polyhedral head enclosing nucleic acid and a tail is attached to a hexagonal base plate having tail pins or prongs and six tail fibres.

- Viruses are classified as per the nature of nucleic acid, size and architecture of their capsid. On the basis of the type of the host which they infect, viruses are separately categorized as bacteriophages, plant viruses and animals viruses.

- The genomic organization of RNA and DNA viruses vary greatly in size. For example, the genome of mammalian retroviruses such as HIV is about 9,000 nucleotides long, and that of the bacteriophage $Q\beta$ is about 4,220 nucleotides.

- The life cycle of T-even bacteriophages (DNA phages) can follow a lytic pathway, which takes only a few minutes, finishes with lysis of the bacterium and release of many phages. This process is known as a virulent infection. But in a rare lysogenic pathway, there is no lysis of bacterial cell and the process is known as a temperate infection.

- Retroviruses have single-stranded RNA molecules as the genetic material. They infect animal cells and carry an RNA-dependent DNA polymerase called reverse transcriptase, the enzyme that was predicted by Temin and Baltimore.

- On retroviral infection, the single-stranded RNA genome and the enzyme enter the host cell. The enzyme first catalyses the synthesis of a DNA strand complementary to the viral RNA, then degrades the RNA strand of RNA–DNA hybrid and replaces it with DNA- forming dsDNA, which is then incorporated into the genome of the eukaryotic host cell. On activation, the integrated viral genes are transcribed to form viral RNA genome and then viral proteins that are packaged as new viruses.

SUMMARY

- Reverse transcriptase is used to synthesize copy or complementary DNA (cDNA) library from the mature mRNA template isolated from a eukaryotic cell.

- Genetic engineering exploits bacteriophage Z as cloning vectors. It has a very efficient mechanism for delivering its 48,502 bp of DNA into a bacterium, and it can be used as a vector to clone somewhat larger DNA segments.

- The major groups of animal viruses cause human diseases like smallpox, chickenpox, serum hepatitis, infectious hepatitis, influenza, rabies, AIDS, cancer, etc.

- Cancer results from changes in cellular behaviour caused due to changes in the genetic material of the cell and hence these cells divide in an uncontrolled manner and form an abnormal mass of tissue. Such masses can be called benign tumours if they do not invade surrounding tissue or malignant tumours if they grow, invade and destroy surrounding tissue. Malignant tumours also release cells, which enter the circulation and spread to distant sites in the body and form secondary tumours, the phenomenon called metastases.

- Cancers are classified into four groups, depending upon the type of cells originally involved such as leukemias, lymphomas, sarcomas and carcinomas.

- In a culture medium, the normal cells proliferate and tend to remain as a single layer (monolayer) whereas malignant cells are cultured under the same conditions, they continue to grow, piling one on top of another to form clumps.

- DNA tumour viruses, for example papovaviruses, Simian virus (SV40), polyoma, and papilloma viruses, adenoviruses, and herpesviruses and RNA tumour viruses include retroviruses that transform a normal cell into cancer cell.

- Acute transforming retroviruses carry oncogenes as key elements in the transformation of normal cells into a cancer cell just like that of src in Rous sarcoma virus.

- The non-acute retroviruses lack oncogenes but when the provirus integrates adjacent to the cellular oncogenes or proto-oncogenes, it induces tumour formation in the host.

- The proto-oncogenes encode proteins that have various functions in the cell's normal activities. The cellular oncogenes can be converted into oncogenes by several ways that include a single base-pair substitution mutation, a mutation in regulatory sequence that can alter the expression of the gene or a chromosomal translocation.

- HIV infection causes decrease in the number of helper T4 cells thereby decreasing the immune power (AIDS) that ultimately results in an increased susceptibility to pathogens and increased risk of certain forms of cancer.

REVIEW QUESTIONS

1. What are viruses? Explain the contribution of Iwanowsky and Stanley in the discovery of viruses.

2. Describe the structural organization of plant and animal viruses and bacteriophage.

3. Explain the criteria for classification of viruses. Describe various classes of viruses with examples for each class.

4. Describe the genomic organization of few significant DNA viruses.

5. Explain the life cycle of a bacteriophage showing the lytic and lysogenic pathways.

6. Define retroviruses. Explain the role of reverse transcriptase in retroviral infection.

7. How does reverse transcriptase help in making a cDNA library?

8. Explain the role of bacteriophages as cloning vectors.

9. Enlist the human diseases caused by viruses, with their symptoms and mode of transmission.

10. What is cancer? How does a normal cell differ from a malignant cell when grown in separate culture media?

11. Explain the mode of transmission of HIV infection with the symptoms of its disease.

12. Describe the genomic organization of the Rous sarcoma virus.

13. What are oncogenes? How do they differ from proto-oncogenes?

14. Enlist the acute transforming viruses with their oncogenes, encoded proteins and functions.

15. Write short notes on:

i. Filterable viruses	ii. Capsomeres
iii. Lentiviruses	iv. Teminism
v. Viruses that cause human cancer	vi. Contact inhibition of growth
vii. SV40	viii. T4 helper cells
ix. Avian leucosis virus	x. Benign tumours
xi. Metastasis	xii. Activation of proto-oncogenes
xiii. *src* oncogene	

BACTERIAL GENETICS

INTRODUCTION

Bacteria are the smallest self-contained living entities governed by the genetic information of DNA. Although the cells of most bacteria are smaller than other types of cells, there is considerable size variation between the smallest and largest bacteria. Since their genetic material is not organized in nuclei, they are called prokaryotes. There exists an enormous number of different types of bacteria, and they vary not only in size and shape but also in the nutritional conditions best suited for their growth and survival. Some bacteria, for example, are aerobes and grow only in the presence of oxygen; others are anaerobes and multiply only in the absence of oxygen; and still others, facultative anaerobes, can change their exact mixture of enzymes to allow growth in both environments. While most bacteria derive their energy from breaking down externally derived food molecules, others have evolved photosynthetic pigments to let them use sunlight to make ATP.

Generally, it is easy to grow bacteria in the laboratory once their nutritional requirements have been provided in the growth medium. For example, *Escherichia coli* will grow in an aqueous solution containing glucose and several inorganic substances like NH_4Cl, $MgSO_4$, KH_2PO_4, Na_2HPO_4 and traces of other ions. If *E. coli* grows only on glucose, about 60 minutes are required at 37°C to double the cell mass. But if glucose is supplemented by various amino acids and nitrogen bases, then only 20 minutes are necessary for the doubling of the cell mass. *E. coli* is the most intensively investigated and correspondingly best-known bacterium at both biochemical and genetic levels.

BACTERIAL GENOME

The average *E. coli* cell is rod-shaped and about 2 μm in length and 1 μm in diameter. Its genetic information is stored in a single large circular molecule of DNA, called bacterial chromosome of about 1 millimetre in circumference with a molecular weight of about 3.0×10^9 daltons, representing some 4.6×10^6 base pairs, the average molecular weight of a single base pair being 660 daltons. Most bacteria contain chromosomes that are about the same size as the one in *E. coli* but in some cases significant variations occur (Table 5.1). Usually the size of the bacterial chromosome correlates with the cell's physiological or morphological complexity. For example, *Mycoplasma pneumoniae*, which lacks a cell wall and is metabolically simple, has a chromosome that is 5 to 6 times smaller than the *E. coli* chromosome.

On the other hand, the chromosome of *Pseudomonas aeruginosa* (the largest prokaryote sequenced as yet) has about one and half times larger DNA than *E. coli* chromosome.

Table 5.1 Details of the chromosomes, of some bacteria

Bacteria	Mol. wt. of bacterial chromosome (dalton)	Number of base pairs	Number of genes
Mycoplasma pneumoniae	5.3×10^8	8.1×10^5	680
Rickettsia prowazekii	7.2×10^8	1.1×10^6	878
Helicobacter pylori	1.1×10^9	1.6×10^6	1589
Haemophilus influenzae	1.2×10^9	1.8×10^6	1738
Archaeoglobus fulgidus	1.4×10^9	2.2×10^6	2437
Synechocystis	2.3×10^9	3.5×10^6	4003
Vibrio cholerae	2.6×10^9	4.0×10^6	3890
Mycobacterium tuberculosis	2.9×10^9	4.4×10^6	4275
Escherichia coli	3.0×10^9	4.6×10^6	4406
Myxococcus xanthus	3.8×10^9	5.7×10^6	5103
Pseudomonas aeruginosa	4.1×10^9	6.2×10^6	5570

GENE EXCHANGE MECHANISM IN BACTERIA

Bacteria divide by simple fission, with an equal distribution of their genetic material to the two progeny cells. They are usually haploid but become partly diploid when the nucleoid of a bacterium contains two or more identical copies of the chromosome (plus plasmid, when present). The chromosomes of bacteria do not go through the mitotic and meiotic condensation cycles associated with cell division and gametogenesis in eukaryotes. In eukaryotes, genetic exchange involves union of two haploid gametes to form a zygote. Such an event never occurs during the exchange of genes among the prokaryotes. Recombination, however, is undoubtedly important in the evolution of bacteria just as it is in the evolution of eukaryotes.

In all those cases that have been studied sufficiently to reveal the molecular basis of gene exchange, only a small portion of the genome from one bacterium (the **donor**) is transferred to another bacterium (the **recipient**), thus forming an incomplete zygote. The genetic exchange in prokaryotes is an occasional process that mediates transfer of genetic material from donor bacterium to recipient one by three quite different mechanisms called transformation, transduction and conjugation. These three modes of genetic transfer can be distinguished by three simple criteria.

1. sensitivity to the presence of the enzyme deoxyribonuclease (DNase)

2. need of an external agent, and

3. dependence on cell contact (Table 5.2).

Table 5.2   Criteria to distinguish different modes for transfer of genes from one bacterium to another

Comments	Transformation	Transduction	Conjugation
Mechanism	Donor cells release naked DNA into solution, which is then incorporated into recipient's genome.	The gene from donor bacterium is transferred to recipient cell by a bacteriophage.	The process during which DNA from donor (male) cell is transferred to recipient (female) cell via conjugation tube.
Cell contact required?	Direct contact between donor and recipient cells is not necessary.	There is no need of cell-to-cell contact between donor and recipient.	It is a sort of sexual union that requires cell-to-cell contact to transfer the gene.
External agent required?	No need of any external entity or mediator.	A bacteriophage is essential to transfer the gene from donor to recipient.	There is no mediator required to transfer the gene from donor to recipient cell.
Sensitive to DNase?	Yes	No	No

TRANSFORMATION

The phenomenon of genetic transformation is the change of one genetic type to the other, which was first reported by British bacteriologist F.Griffith (1928) while working with the bacterium *Diplococcus pneumoniae*. This was the first study on bacterial transformation that has had special impact on bacterial genetics in particular, and on biology in general.

Pneumococci, like all other living organisms, exhibit genetic variability that can be recognized by the existence of two different phenotypes such as 1) the presence or absence of a surrounding polysaccharide (complex sugar polymer) capsule and 2) the type of capsule, that is, the specific molecular composition of the polysaccharides present in the capsules. When grown on appropriate media (such as blood agar) in petri dishes, pneumococci with a capsule form large, smooth colonies and are designated type 'S'. These encapsulated pneumococci are pathogenic to most mammals for example they cause pneumonia in humans. These virulent (disease-causing) type S pneumococci mutate to a non-virulent or non-pathogenic form that has no capsule (at a frequency of about one in 10^7). Such non-encapsulated, non-virulent pneumococci form small, rough-surfaced colonies when grown on blood agar medium and are designated type R. The pathogenicity lies in the capsule and synthesis of its components is under the control of genes. The capsule protects the bacterial cell against phagocytosis by leucocytes.

Griffith's Experiments

Griffith had done a series of experiments by injecting the type S and R strains of pneumococci to experimental mice and discovered the phenomenon of transformation (Figure 5.1). In the first stage, he injected experimental mice with type S (virulent) strains of pneumococci and found that mouse died because of the disease. When he injected type R (non-virulent), the mice survived. It was also found that if type S bacteria are killed by heating to 65°C, they become non-virulent. Griffith mixed heat-killed type S with living type R and injected into mouse. Contrary to expectation, the animals did get pneumonia. On autopsy, these mice were found to contain live virulent cells of type S. How did this happen? Evidently some heat-stable component present in dead bacteria of type S could confer the characteristics of this strain on living type R cells.

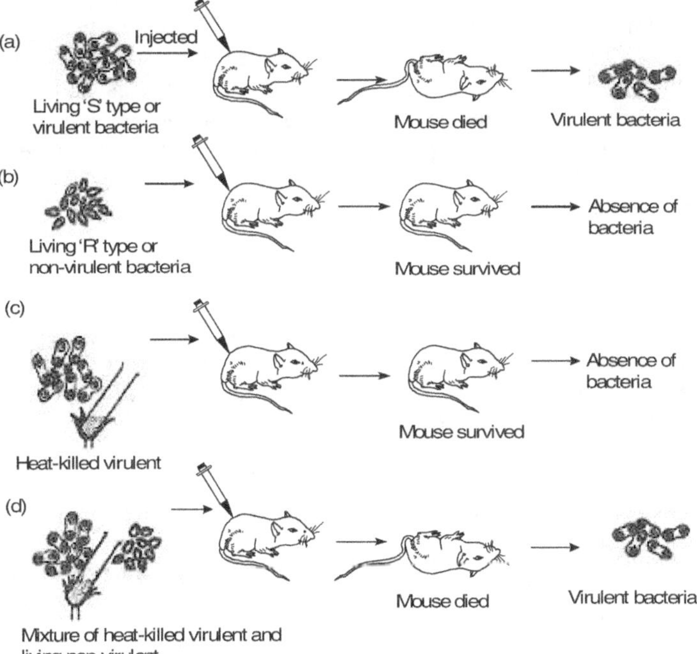

Figure 5.1 Diagrammatic representation of Griffith's experiment to demonstrate transformation in *Diplococcus pneumoniae*

The component released was hypothesized by Griffith to be a **transforming principle** (the cellular component mediating transformation), which was released by heat-killed cells of type S and taken by type R cells, thus transforming their hereditary properties into those of type S. Although Griffith did not identify the exact chemical nature of the transforming principle, yet his experiments set the stage for the work of Avery *et al.* (1944) who then discovered that Griffith's transforming principle was DNA.

Proof that the "Transforming Principle" is DNA

In 1944, Avery, MacLeod, and McCarty, on the basis of a set of extensive and laborious experiments, proved that the transforming principle is DNA. They showed that if highly purified DNA from type S pneumococci was present with type R pneumococci, some of the type R cells were transformed to type S. But how could one be sure that DNA was pure? Proving the complete purity of any macromolecule is extremely difficult. The DNA preparation may contain a few molecules of protein and these contaminating proteins were responsible for transformation. The most definite proof that DNA was the transforming principle involved the use of enzymes that degrade DNA, RNA or proteins. In separate experiments, highly purified DNA from type S cells, was treated with 1) deoxyribonuclease or DNase which degraded DNA, 2) ribonuclease or RNase that degrades RNA, or 3) proteases (which degrade proteins) and then tested for its ability to transform type R cells to type S. Only DNase treatment destroyed the transforming activity of DNA while RNase or protease had no effect on the ability of purified DNA preparation to transform type R cells to type S. This clearly proved that the transforming principle was DNA.

Mechanism of Transformation

A reasonable complete picture of the overall process of transformation has been established. The uptake of DNA by the recipient bacteria is an active, energy-requiring process. Transformation does not occur "naturally" in all species of bacteria, but only in those species possessing the enzymatic machinery involved in the active uptake and recombination process. The phenomenon of transformation has been studied in *D. pneumoniae*, *Bacillus subtilis*, and *Haemophilus influenzae*. Even in all these species, all cells in a given population are not capable of active uptake of DNA. Only "competent" cells that possess a so-called competence factor (probably a cell-surface protein or enzyme involved in binding or in uptake of DNA) are capable of serving as recipients in transformation.

The process of transformation can be divided into several stages that are diagrammatically represented in Figure 5.2. The stages are 1) reversible binding of double-stranded DNA molecule to receptor sites on the recipient cell surface, 2) irreversible uptake of donor DNA by recipient cell, 3) conversion of double-stranded donor DNA molecules to single-stranded molecule by nucleolytic degradation of one strand, 4) integration of all or a part of the single strand of donor DNA into the chromosome of the recipient, and 5) segregation and phenotypic expression of the integrated gene or genes in transformed cells.

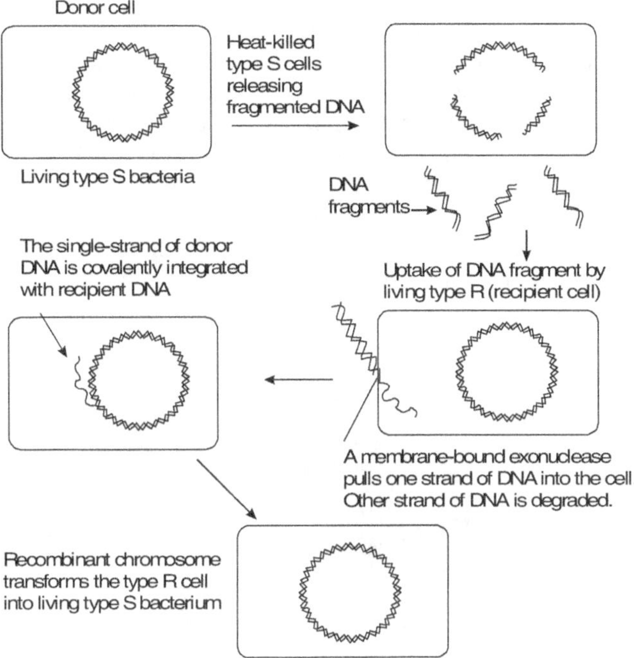

Figure 5.2 Diagrammatic representation of two important steps, uptake and integration, in transformation of *Diplococcus pneumoniae*

TRANSDUCTION

The transfer of genetic material from donor to recipient cell mediated by bacteriophages is referred to as transduction. The phenomenon was first discovered by N. Zinder and J. Lederberg (1952) in *Salmonella typhimurium* and subsequently by other workers in *Escherichia, Shigella, Staphylococcus*, etc. Transduction occurs when a bacteriophage particle carries a segment of the chromosome from one bacterium (the donor) to another bacterium (the recipient), facilitating subsequent recombination. Bacteriophages that occasionally produce transducing particles and therefore are capable of effecting transductional genetic exchange are commonly encountered in the microbial world.

The classical investigations of Zinder and Lederberg were on the mouse typhoid bacteria *Salmonella typhimurium*. Several mutant auxotroph strains were used in the experiments. The set-up for the experiment consists of a 'U' tube in which one auxotroph strain that required methionine ($met^- thr^+$) was placed in one ('A') arm of the 'U' tube and other auxotroph strain that required threonine ($met^+ thr^-$) was placed in the other ('B') arm of the 'U' tube.

Both strains were separated by a sintered glass filter, which was placed in the centre of the 'U' tube as shown in Figure 5.3. One strain of the bacterium was infected with bacteriophage. A little later when bacteria in both arms were analysed, it was noticed that the strains in both arms had become recombinants. That is when two strains were mixed and plated on a medium deficient for methionine and threonine, some wild types (prototrophs) occurred and gave rise to colonies of prototrophs. Since the sintered glass filter does not allow the migration of bacteria from 'A' arm to 'B' arm of the 'U' tube, the only way for transfer of genetic material from one auxotroph (the donor) to the other auxotroph (the recipient) was mediated by the bacteriophage that can pass through the glass filter.

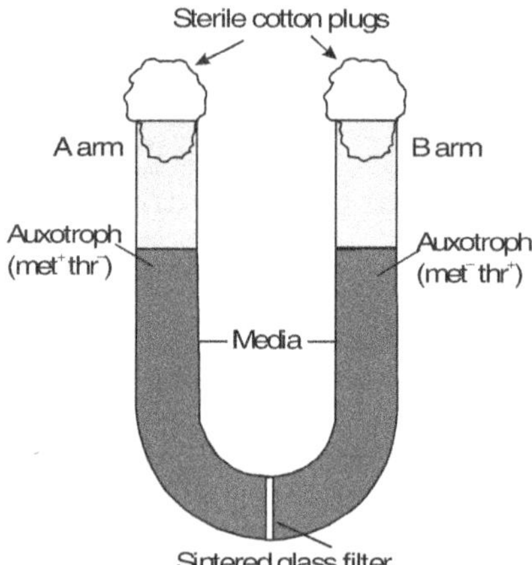

Figure 5.3   The 'U' tube experiment used to demonstrate the phenomenon of transduction. Two arms having different strains of bacteria are separated by sintered glass filter containing pores of a size that permit the passage of bacteriophages, but not bacteria.

Types of Transduction

Two very different types of transduction are known.

❖ In **generalized or non-specific transduction**, a random segment of bacterial DNA is picked up during maturation of the bacteriophage in the place of, or along with, the phage chromosome in a few "progeny" phage particles, called transducing particles. Generalized transducing phages can therefore transport any gene of the donor cell to the recipient cell.

❖ In **specialized or restricted transduction**, the phage transduces or transfers only a certain segment of the bacterial genome.

Generalized transduction   Bacteriophages are classified into two types on the basis of their interaction with the host (bacterial cell) as described in chapter 4. Virulent phages always multiply and lyse the host cell after infection. Temperate phages have a choice between two lifestyles after infection. They can either enter the **lytic cycle**, during which they reproduce and lyse their host just like virulent phages, or alternately, they can enter the **lysogenic pathway**, during which their chromosomes are integrated into the chromosomes of the host and replicate like any other segments of host chromosomes. Generalized transduction is mediated by some virulent bacteriophages whose chromosomes are not integrated at specific attachment sites on the host chromosome. Generalized transducing particles are produced during the lytic cycles of these phages but they are produced at a low frequency. Only one out of 10^5 to 10^7 of the progeny particles present in lysate contains bacterial DNA. The generalized transducing particles P1 of *E. coli*, P22 of *Salmonella* and PBS1 and SP10 of *Bacillus subtilis* are extensively involved in generalized transduction. These phages are used to construct the genetic fine structure mapping (mapping mutant sites within individual genes or short segments of the chromosome).

At the end of the lytic cycle when phage particles are packaged into protein coat, a small segment of the bacterial (donor) DNA is incorporated into the phage DNA. When such transducing phage infects a second bacterium (recipient cell), it injects a fragment of host DNA into a recipient cell. The donor DNA may either 1) be integrated into recipient chromosome (as in lysogenic pathway) in a manner similar to the integration of transforming DNA, except that the integrated segment is double-stranded, or 2) remain free in the cytoplasm. If it is not integrated, it will not replicate and will be transmitted to only one progeny cell during each cell division. The genes located on the transduced chromosome fragments may be expressed, even if they are not integrated. Cells carrying non-integrated transducing fragments are called **abortive transductants**. They are partially diploid and can be used to carry out complementation tests, the test which provides the operation definition of the gene. It helps to determine whether different mutations are in the same genes or different gene. The significance of the transduction lies in the fact that they provide a convenient way to construct bacterial strains with new combinations of mutations. In addition, they also help in construction of linkage maps of the bacterial genome.

Specialized transduction   In contrast to generalized transduction, specialized transduction is mediated only by a temperate bacteriophage whose chromosome is able to integrate at one or a few specific attachment sites on the host chromosome. The chromosomes of temperate phages of this type are thus capable of both autonomous replication (replication independent of the replication of host chromosome) and integrated replication (replication as a segment of the host chromosome) as they are examples of genetic elements called **episomes**.

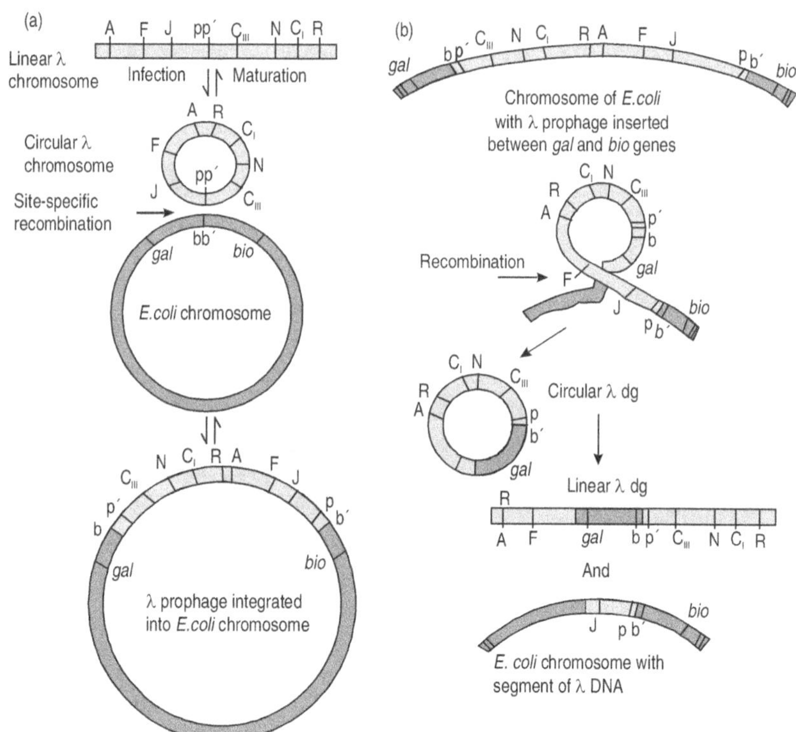

Figure 5.4 Specialized transducing phage λ integrated and latter excised from *E. coli* chromosome. (a) On infection to host, phage DNA is converted into circular form and enters into lysogenic pathway. It integrates with bacterial DNA at a specific site between the *gal* genes (required for utilization of galactose as an energy source) and *bio* genes (essential for synthesis of biotin). In the reverse of integration (excision), the prophage forms figure-8. In the normal process, intact circular λ DNA and *E. coli* chromosome are produced. (b) Formation of the λdg (lambda defective *gal*) as a specialized transducing particle. Occasionally the excision event occurs at a site other than the original attachment site. When this happens, a portion of the phage chromosome (gene 'J') is left in bacterial chromosome and a portion of host chromosome is excised with the phage DNA. This phage is defective because genes required for growth and maturation under lytic conditions are replaced by bacterial DNA. Thus, λdg can reproduce only in presence of a wild-type "helper" phage. The λdg can recombine with *gal⁻* endogenote (recipient chromosome) and transfer the *gal* genes to recipient bacteria to produce stable *gal⁺* transductants.

Integration of the chromosome of a specialized transducing phage, such as the lambda (λ) phage of *E. coli* involves a recombination event between the circular intracellular form of the phage chromosome and the circular bacterial chromosome at specific attachment sites on the two chromosomes. The site-specific recombination event results in the covalent linear insertion of the phage chromosome into the chromosome of the bacterium. In its integrated state, the phage chromosome is called a **prophage**. The bacterium harbouring a prophage is said to be lysogenic; the prophage–host relationship is called lysogeny.

Occasionally, the viral genome gets separated from its integrated site from the bacterial genome; the process is called **excision**, which is site-specific, like the integration process. Once the viral genome is excised from host chromosome, it starts replicating autonomously. Such a transition of viral genome, which causes the lysogenic state to switch over to lytic growth, is referred to as **induction**. The processes of site-specific integration of viral genome into host chromosome, its induction and formation of defective phages are summarized in Figure 5.4. Similar to lambda phage, the other specialized transducing phage Φ80 integrates near the *E. coli trp* genes (required for the synthesis of the amino acid tryptophan) and help in transduction of *trp* genes. The experimental utility of specialized transducing phages is considerable since they help in obtaining a high concentration of specific genes.

CONJUGATION

It is a fascinating way to transfer the DNA from donor bacterium to the recipient cell through a specialized protoplasmic connection or conjugation tube, that forms between them. In 1946, J. Lederberg and E.L.Tatum discovered a genetic exchange occurring between certain strains of *E. coli*. The phenomenon is eventually different from transformation with respect to the direct cell-to-cell contact that is essential for transfer of genetic material from donor to recipient bacterium. It is observed that in bacterial cultures, conjugation did not occur randomly between cells, rather it was the selective process. The genetic material is transferred selectively from a particular cell to certain other specific cells. This selectivity is due to the presence of a small circular molecule of DNA or "minichromosome" called F factor (for "fertility" factor, also called "sex factor" and "F plasmid"). The bacteria carrying F factor (donor cells or male bacterium) are called F$^+$ strain and the cells without F factor are referred to as F$^-$ strain (recipient cells or female bacterium). The conjugation takes place between F$^+$ and F$^-$ strains.

F Factor

Like the main *E. coli* chromosome, the F factor is a small circular, self replicating and double-helical DNA molecule present in one copy per cell. It replicates independently during cell division and is inherited by both the daughter cells. Therefore, it function is not governed by the bacterial chromosome. It only carries approximately 94,500 base pairs, which is 1/40 the amount of genetic information contained in the main chromosome. About one-third of the F^+ DNA consists of 19 transfer (*tra*) genes specifically involved in the transfer of male genetic material into the female cell. Among these genes, twelve genes are responsible for synthesis of the sex-specific F pili. When male cells are mixed with female cells, conjugal pairs form by the attachment of a male sex pilus to the surface of a female cell (Figure 5.5).

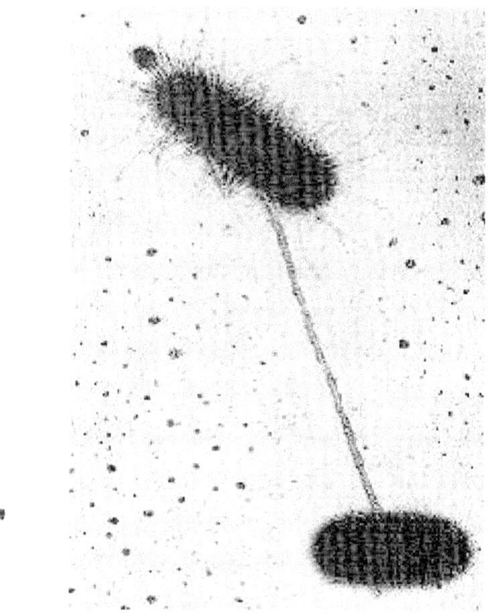

Figure 5.5 The attachment of a male *E. coli* through the F (sex) pilus to the surface of a female cell. This is followed by the replication of F factor and its subsequent transfer to the recipient cell, which then becomes F^+.

Genetic Map of F Plasmid

As mentioned earlier, the genomic size of the F plasmid of *E. coli* is made up of 94.5 kilobases. Figure 5.6 is the genetic map of the F plasmid that shows relative positions of genes coding for autonomous replication, sex pili formation and conjugal transfer functions. The F plasmid contains the transfer (*tra*) region and non-transfer related genes. The transfer genes are about 32 kb long and contain about 19 known transfer genes. Twelve genes are involved in F pilus formation are *tra* A, L, E, K, B, V, W/C, U, F, H, and G. Genes involved in the regulation are *fin*P and *tra*J. The non-transfer-related markers are the insertion sequences (IS3, γ, δ, and IS2), stable DNA degradation (*srn* B), inhibition of replication by T7 and II phages *(pif)* and a region for replication (*rep*), incompatibility (*inc*) and the origin of vegetative replication *(oriV)*.

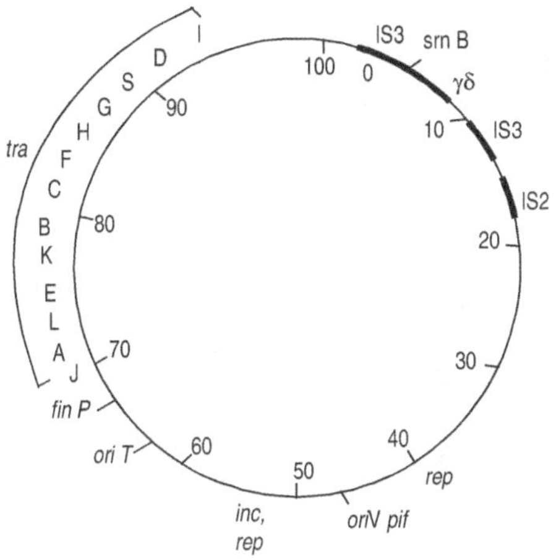

Figure 5.6   Genetic map of F plasmid

Mechanism of Conjugation

In *E. coli* strain K12, the cells that carry the sex factor F are known as F⁺ or male. During conjugation, direct cell-to-cell contact is established between an F⁺ cell and an F⁻ cell through a long appendage (sex pilus) that extends from the surface of F⁺ cell. Once the contact is established, the pilus is believed to be modified and serve as a protoplasmic channel called conjugation tube between two cells. At the *ori*T site of the plasmid, a nick is made by endonuclease producing 5′ and 3′ ends. The 5′-terminus of the single strand of plasmid DNA binds with pilot proteins and travels gradually through the conjugation tube as shown in the Figure 5.7. Ordinarily, the single strand of plasmid DNA passes from F⁺ to an F⁻ recipient while retains one strand for itself. Once the transfer of single strand of plasmid DNA is completed, the conjugation tube breaks off separating the partners (exconjugants). The transfer of F element from donor to recipient cell eventually leads every cell in the population to become an F⁺ cell. After entering into the recipient cell, the 5′ undergoes replication. Transfer of DNA is associated with synthesis of a replacement strand in the donor cell and of a complementary strand in the recipient cell. In both cells, the process of replication is mediated by DNA polymerase III and the replication is brought about by **rolling circle method**.

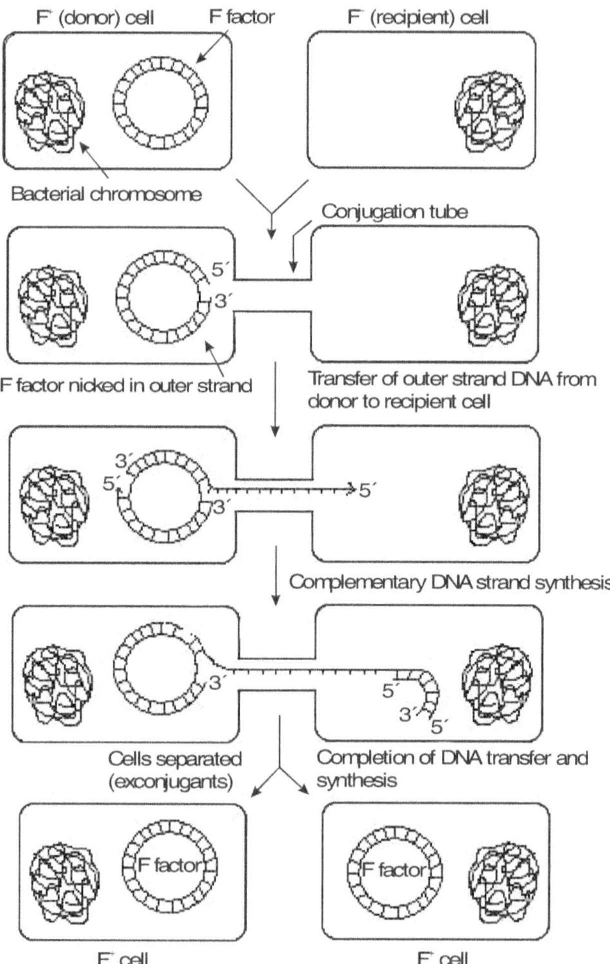

Figure 5.7   Steps involved in transfer of F plasmid from F⁺ cell to F⁻ cell during the conjugation in *E. coli*.

HIGH FREQUENCY RECOMBINATION (Hfr) STRAIN

When the bacterial chromosome of the F⁺ strain integrates with the F plasmid, it leads to the formation of high frequency recombination (Hfr) strain (Figure 5.8). The phenomenon occurs at about once in every 10,000 F⁺ cells. Even after integration of F plasmid into bacterial chromosome, it remains as a single, circular and helical DNA molecule. When an Hfr cell conjugates with an F⁻, the frequency of recombination becomes high and that of transfer of F factor becomes low. It takes approximately 2 minutes to transfer the F plasmid, but 100 minutes to transfer the entire bacterial chromosome from donor to recipient cell. This difference is mainly due to the relative size of F and the integrated chromosome.

The integrated F plasmid of the Hfr strain ordinarily replicates along with the bacterial chromosome. The mode of transfer of F element from Hfr to F⁻ cell is similar to that of the conjugation between F⁺ cell and F⁻ cell. The only point of difference is that usually the F⁻ cell receives an incomplete copy of F element and remains F⁻ cell. And if the transfer is complete, the recipient cell acquires a complete F plasmid and Hfr property; as a result, its progeny acts as a Hfr strain. The recipient cells that have received a part of the donor chromosome are called **partial zygotes or hemizygotes**, since they possess copies of one or more genes that are derived from two different parents. Such cells do not remain diploid but soon they participate in non-reciprocal genetic exchange. This property helps in the construction of chromosomal maps of bacteria.

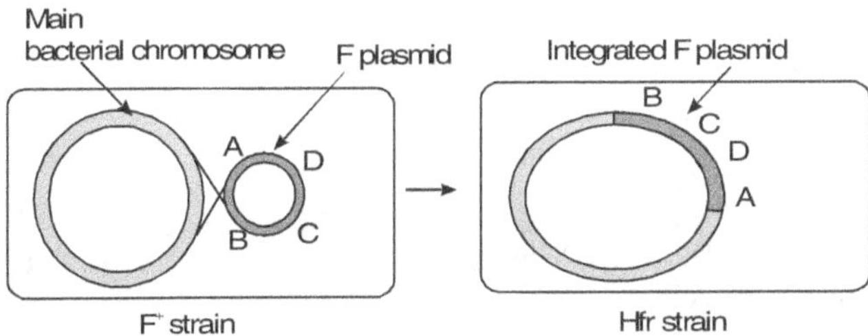

Figure 5.8 Integration of F plasmid in main bacterial chromosome to form Hfr strain

Interrupted Conjugation and Gene Mapping

In 1957, F. Jacob and E. Wollman used the technique of "interrupted conjugation" between an Hfr strain and an F⁻ strain and produced the first genetic map of a bacterial chromosome. In their experiments, male cells of Hfr strain were placed in the same culture with the female cell, and the two kinds of cells came together in conjugation. While the two cells of each conjugating pair were joined by a protoplasmic bridge, a linear group of genes migrated from donor cell across the bridge into the recipient cell. This was accomplished as follows: the ring shaped *E. coli* linkage group opened at the place where the F genome was integrated. The end opposite the F end entered the conjugation tube and moved across into the recipient cell. About two hours after the beginning of the transfer, the entire linkage group was across the tube with the F end entering the last. This was a one-way transfer of genetic material that always moves from donor to recipient cell.

When, for example, a cross was made between genetically different Hfr and F⁻ strains, the sequence of gene exchange during conjugation was demonstrated. Hfr organisms used in the experiment were capable of synthesizing the amino acids threonine (*thr⁺*), and leucine (*leu⁺*); they were sensitive (s) to the metabolic inhibitor sodium azide (azi-s), and sensitive to the bacteriophage T_1 (T_1-s); they could utilize lactose (*lac⁺*) and galactose (*gal⁺*); and they were sensitive to the antibiotic streptomycin (*str-s*). Genotypically, the Hfr bacteria were: *thr⁺ leu⁺ azi-s T_1-s lac⁺ gal⁺ str-s*, and the F⁻ strain were *thr⁻ leu⁻ azi-r* (r for resistant) T_1*-r lac⁻ gal⁻ str-r*. The cross was made by mixing them to remain together in liquid for 25 minutes. At the end of this period, the mixed cells were plated on the first selective medium, a minimal medium containing streptomycin.

On the selective medium, the streptomycin killed the Hfr parent cells, and the F⁻ parent cells were unable to grow because the essential amino acids threonine and leucine, which they could not synthesize, were not supplied in the medium. The only cells that could grow on the medium were those that could synthesize the necessary amino acids and were resistant to streptomycin. Since the genes *(thr⁺* and *leu⁺)* for synthesizing these amino acids came from the Hfr parent and streptomycin resistance *(str-r)* came from the F⁻ parent, the cells that survived were recombinants.

Jacob and Wollman periodically interrupted the process of conjugation between Hfr and F⁻ cells. The conjugation is stopped at any desired time by subjecting the mixture to sharing forces of a waring blender (violent agitation) that artificially disrupts the conjugation tube and thus conjugants were separated. This separation interrupted the transfer of genetic material from donor to recipient and presumably broke the linkage group. After separation, the cells were plated and tested for the genes from the donor that had been integrated into the recipient, which revealed the presence of recombinants. For example, at 0 minute from the time of mixing, 0 recombinants had occurred. The *thr⁺* gene became integrated in 8 minutes, *leu⁺* in 8½, *azi-s* in 9, T_1-s in 11, *lac⁺* in 18, and *gal⁺* in 25 minutes (Table 5.3). These studies showed that gene transfer from Hfr donor to F⁻ recipient cell is not a random process. Rather, the genes move across the conjugation tube in a regular order, as if they are tied in a linkage group in a linear sequence as shown in the Figure 5.9. Depending on the strain, it takes about 90 to 100 minutes to transfer the complete chromosome from Hfr to an F⁻ strain. The time interval between the transfer of any two genes, which is determined by interrupted conjugation experiments, is a good estimate of the physical distance between the genes on a chromosome. It is convenient to use the minute, representing the time interval between the transfer of genes in interrupted conjugation experiments, as the standard unit for measuring linkage in *E. coli*. A map distance of 1 minute corresponds to the length of the segment of the chromosome transferred in 1 minute during conjugation. On the basis of the interrupted mating experiment, the standard linkage map of *E. coli* is thus divided into minute intervals from 0 to 100 minutes with genes present at definite loci (Figure 5.10).

Table 5.3 Time required for the transfer of different genes from Hfr to F⁻ strain

Minutes	Genes transferred from Hfr strain
0	0
8	*thr⁺*
8½	*thr⁺ leu⁺*
9	*thr⁺ leu⁺ azi-s*
11	*thr⁺ leu⁺ azi-s T_1-s*
18	*thr⁺ leu⁺ azi-s T_1-s lac⁺*
25	*thr⁺ leu⁺ azi-s T_1-s lac⁺ gal⁺*

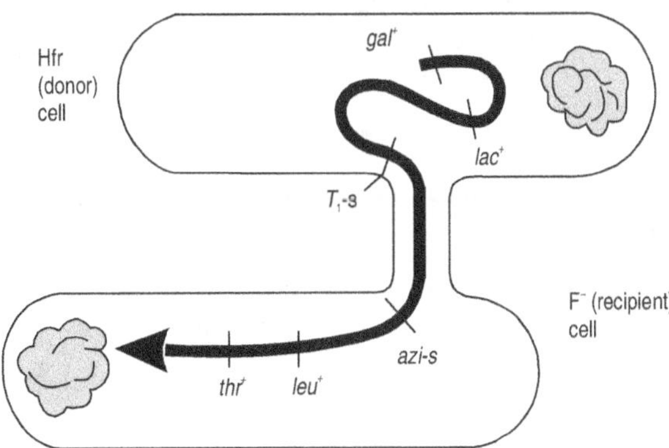

Figure 5.9 Diagrammatic representation of linkage group transfer from an Hfr cell to an F⁻ cell

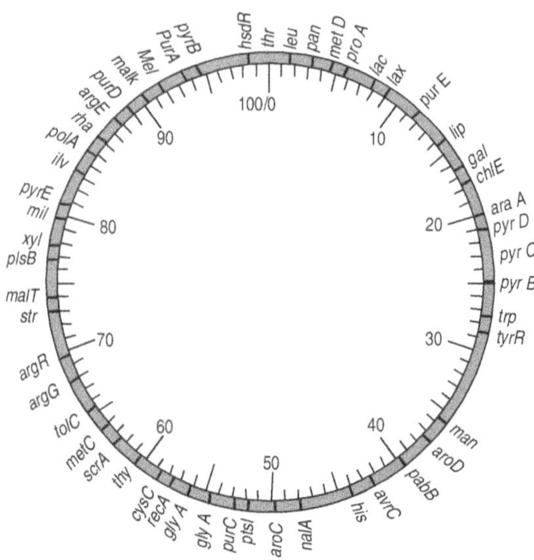

Figure 5.10 The genetic map of *E. coli* strain K12. The map is divided into one-minute intervals from 0 to 100 minutes based on the time required to transfer the gene during conjugation

CLONING EUKARYOTIC DNAs IN BACTERIAL PLASMIDS

Bacterial plasmids play indispensable role in the genetic engineering or recombinant DNA technology (rDNA technology) or gene cloning that essentially involves:

1. Isolation of the gene of interest to be cloned using restriction enzymes.

2. Insertion of desired gene into plasmid DNA, which acts as a cloning vector.

3. Transfer of the recombinant vectors into bacterial cells (hosts) either by transformation or by infection using viruses.

4. Selection of transformed (transgenic) cells, which contain the desired recombinant vectors.

5. Growth of the bacteria, which can be continued indefinitely, to give as much cloned DNA as needed.

6. Expression of the gene to obtain the desired product, which may be a polypeptide, enzyme, hormone, vaccine, etc. used for human welfare.

Recombinant DNA technology has important applications to gene mapping, detection of inherited diseases, cancer research, immunology, enzymology, novel biochemical synthesis and industrial manufacturing of commercial products. Vectors are the DNA molecules having the ability of self-replication and into which foreign DNAs or genes of interest are integrated to form recombinant DNAs. Then they are introduced into appropriate hosts for gene cloning or expression of the gene of interest.

Types of Vectors

There can be two types of vectors, namely cloning vector, and expression vector.

Cloning vector   These vectors are used for propagation of DNA inserts in a suitable host. There is no transcription or translation of cloned gene in such vectors. Cloning vectors are useful in creation of genomic library or Gene bank, in preparation of molecular probes and in genetic engineering experiments.

Expression vector   These vectors are designed to express the cloned gene, i.e., they allow the transcription and translation of cloned gene to produce a protein. This can be achieved by adopting certain strategies if a eukaryotic gene is to be expressed in a prokaryote. Expression vectors are used in producing transgenic plants or animals so that the cloned gene expresses to give the product (protein)which can be utilized for human welfare.

Plasmids

As mentioned earlier, plasmids are extra-chromosomal, small, circular, double-stranded DNA molecules capable of self-replication and present in bacterial cells. Plasmids can integrate themselves with the main bacterial chromosome to form **episomes**. The independent plasmids are not essential for bacterial cells except under specific environments. There are several types of bacterial plasmids, some of them are

Fertility (F) plasmid   It is a type of plasmid that is responsible for bacterial conjugation as   explained earlier.

Resistant (R) plasmid   It is also a type of plasmid that imparts to the bacteria (e.g. Shigella, causing dysentery) resistance to different antibiotics. Similar to F factors, R plasmids may be transferred from one bacterial strain to another. The small circular DNA of R factor carries genes responsible for resistance against antibiotics like penicillin, streptomycin, tetracycline, chloramphenicol, etc.

Colicin (Col) plasmid   It is the factor responsible for synthesis of colicins that are toxic antibiotics of complex chemical nature killing bacteria other than those producing them. *Col* plasmids can also pass through the conjugation tube from between donor and recipient bacteria.

The scope of genetic research, even with the simplest bacteria and viral systems, has dramatically improved as a result of the development of recombinant DNA technology. These procedures make the isolation and cloning of any gene essential for cellular growth and for which mutations exist that block its proper functioning. With genetically well-characterized microorganisms, the isolation of the desired gene is no longer a major challenge. The isolation of *E. coli* genes starts with the fragmentation of its chromosome by using an appropriate DNA restriction enzyme that recognizes a specific nucleotide sequence (Figure 5.11). These chromosomal fragments are then mixed with excess DNA of an *E. coli* plasmid that has been converted into a linear form by the same sequence-specific restriction enzyme. Addition of DNA ligase, the DNA joining enzyme, randomly links the two sets of fragments together, leading to a generation of many recombinant plasmid DNA circles into which have been inserted discrete fragments of the *E. coli* chromosome. Found within this recombinant DNA "library" are some thousand different recombinant plasmids, collectively containing all the DNA of the *E. coli* chromosome.

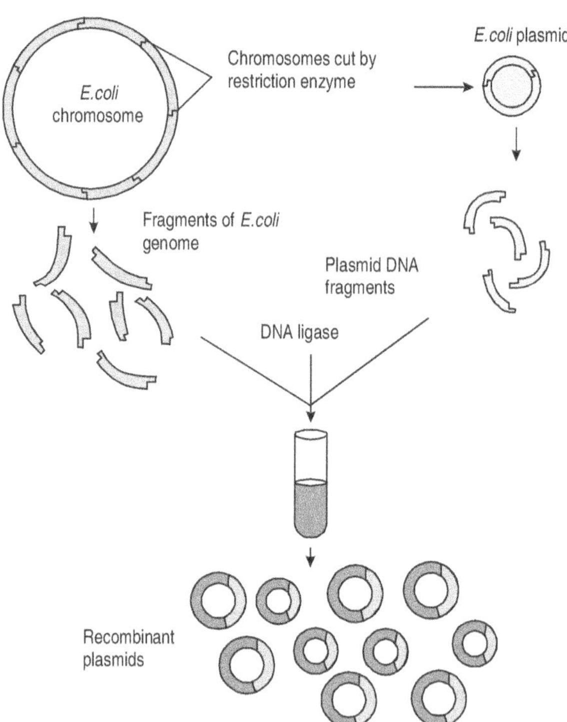

Figure 5.11   Formation of recombinant DNA "library"

Plasmid Vectors

Most of the plasmids which are used as cloning vectors in *E. coli*, are derived from natural plasmids. The natural plasmids are modified to form cloning vectors. Plasmid DNA used as vector can be cleaved at a site with restriction enzyme where foreign DNA can be inserted. Some of the reconstructed plasmid cloning vectors are as follows.

pSC 101   It is the earliest plasmid vector containing the replication module (*ori*) for replication in *E. coli*, *tet*[r] gene for resistance to tetracycline and single recognition site for restriction endonuclease *Eco* RI, *Hind* III, *Bam* HI, *Sal* I as shown in Figure 5.12.

pBR 322   Its name is derived as follows: p is for plasmid, B and R for Boliver and Rodriguez, the scientists who developed the plasmid. Its structural features include *Col* E1, origin of replication that allows production of several copies per cell and it has two genes for different antibiotic resistance, i.e., ampicillin and tetracycline resistance (*amp*[r] and *tet*[r]) as shown in Figure 5.13.

pUC   It is the most popular and widely used plasmid. Its name is derived from the place of its initial preparation, University of California. It has 2686 bp ampicillin resistance (*amp*[r]) gene and *lacZ* gene, Vectors of pUC family have several restriction sites in a small stretch of DNA, such region is called multiple cloning site (mcs) as shown in Figure 5.14.

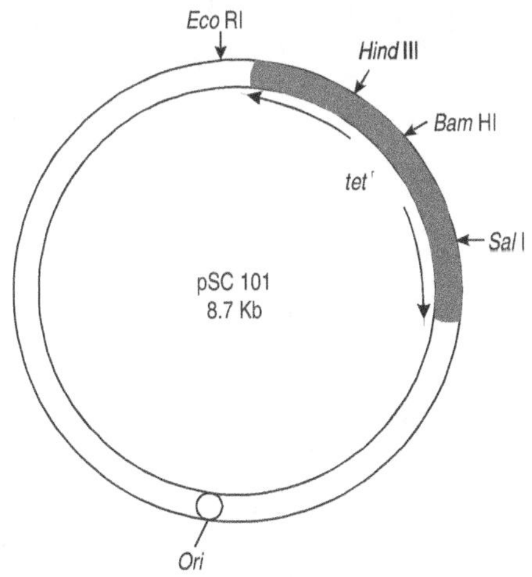

Figure 5.12   The plasmid vector, pSC 101

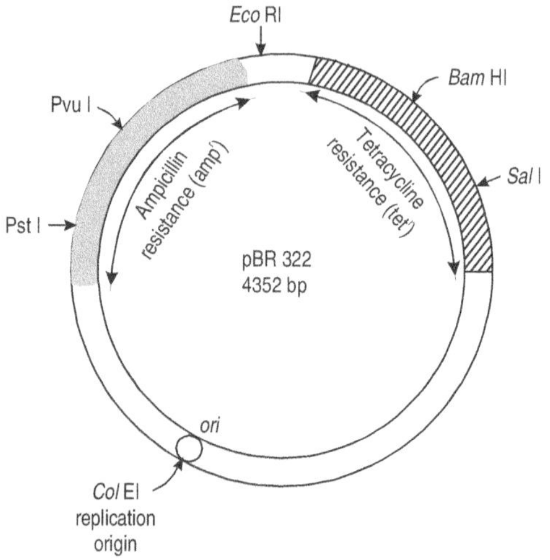

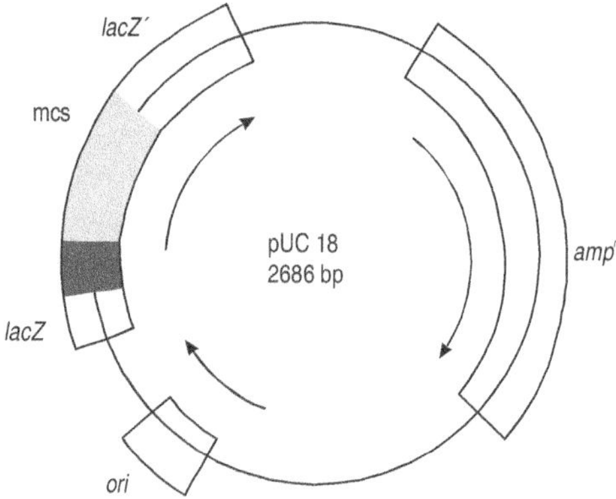

Figure 5.13   The plasmid vector, pBR 322

Figure 5.14   The plasmid vector, pUC 18

Cosmids as Cloning Vectors

Cosmids are cloning vectors having the properties of both plasmid and phage. They are essentially plasmids having a minimum of 250 bp of λ phage DNA. A typical cosmid has a replication origin, unique restriction site and selectable markers from plasmid as shown in Figure 5.15.

Structural features of cosmid vector made up of pBR 322 and λ segment with *cos* sites, a Pst I and Pvu I recognition for restriction enzymes within

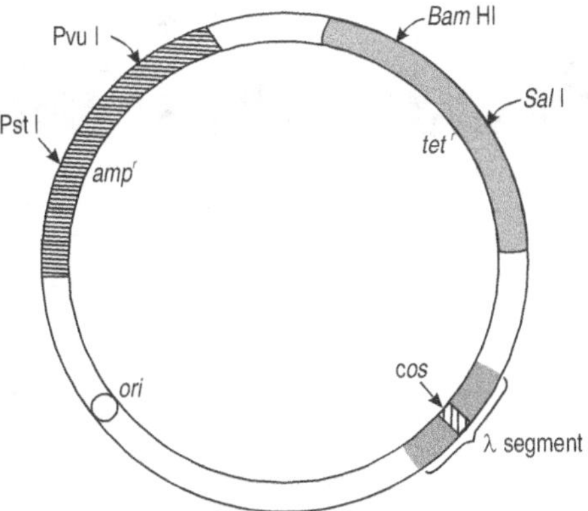

Figure 5.15 The structure of a cosmid vector

ampicillin resistance gene and *Bam*HI and *Sal*I sites within tetracycline resistance gene. Barbara Hohn and John Colling developed cosmids in 1978. The use of cosmid vectors for cloning of DNA insert is advantageous because cosmids can accommodate up to 45 bp long DNA inserts. Their efficiency is high enough to produce a complete genomic library of 10^6–10^7 clones from just $1\,\mu$g of DNA insert. Cosmids have been successfully used in producing gene banks of *Drosophila*, mouse and several other organisms.

SUMMARY

- Bacteria are the smallest self-contained living prokaryotes governed by the genetic information of DNA.

- The exclusively studied bacteria are E. coli in which the genetic information is stored in a single large circular molecule of DNA. It is called bacterial chromosome of about 1 millimetre in circumference with molecular weight of about 3.0×10^9 daltons, representing some 4.6×10^6 base pairs.

- Bacteria are usually haploid and divide by simple fission, with an equal distribution of their genetic material to two progeny cells. They become partly diploid when the nucleoid of a bacterium contains two or more identical copies of the chromosome.

- Genetic exchange in bacteria is an occasional process in which the genetic material is transferred from donor bacterium to recipient one by three different mechanisms called transformation, transduction and conjugation.

SUMMARY

- F. Griffith (1928) *studied bacterial transformation while working with the bacterium* Diplococcus pneumoniae.

- *In the pioneering experiment of Griffith, a transforming principle was released by heat-killed cells of type S (virulent)* D. pneumoniae *that was taken by type R cells (non-virulent) and thus transformed their hereditary properties into those of type S without any cell-to-cell contact or mediator.*

- *In 1944, Avery et al., by treating enzymes that degrades DNA, RNA or proteins proved that the transforming principle in Griffith's experiment was DNA.*

- *The process of transformation involves reversible binding of dsDNA molecule to receptor sites on the recipient cell surface, irreversible uptake of donor DNA by recipient cell, conversion of donor's dsDNA to single-stranded molecule, its integration into the chromosome of the recipient, and finally its phenotypic expression in transformed cells.*

- *Zinder and Lederberg discovered (1952) transduction in* Salmonella typhimurium *where a bacteriophage particle carries a segment of the chromosome from donor to the recipient bacterium and facilitates recombination.*

- *Transduction may be generalized or non-specific transduction, a random segment of bacterial DNA is transduced by bacteriophage from donor cell to the recipient cell or specialized or restricted transduction where phage transfers only a certain segment of the bacterial genome.*

- *The generalized transducing particles are* P1 *of* E. coli, P22 *of* Salmonella *and* PBS1 *and* SP10 *of* Bacillus subtilis.

- *Specialized transducing phages are lambda (λ) phage of* E. coli, *and the phage* F80 *that integrates near the* E. coli trp *genes.*

- *In 1946, Lederberg and Tatum discovered conjugation as a sort of mating between two different strains of* E. coli *where the genetic material from donor bacterium to recipient bacterium is transferred through the conjugation tube.*

- *The genomic size of the F plasmid or sex factor of* E. coli *is made up of 94.5 kilobases and it consists of genes coding for autonomous replication, sex pili formation and conjugal transfer functions.*

- *In* E. coli *strain K12, the cells with F factor are called F^+ or male and strains without F plasmid are known as F^- or female. During conjugation, direct cell-to-cell contact is established between a male and a female cell through a long sex pilus and a one-way transfer of outer strand of plasmid DNA is transferred to F^- cell within 2 minutes, thereby converting the recipient cell to F^+ cell.*

SUMMARY

- High frequency recombination (Hfr) strains are formed when the bacterial chromosome of the F+ strain integrates with the F plasmid.

- In 1957, Jacob and Wollman produced the first genetic map of a bacterial chromosome using the technique of "interrupted conjugation" between Hfr and F− strains.

- Similar to F plasmid, R (resistant) and Col (colicine) plasmids are transferred from donor to recipient cell.

- The genetic engineering or gene cloning utilizes plasmids, as cloning vector to propagate desired DNA inserts in an appropriate host. Plasmid DNA can be cleaved at a site with restriction enzyme where foreign DNA can be inserted and are then cloned in a suitable host.

- The pSC 101, pBR 322 and pUC are some of the reconstructed plasmid cloning vectors in E. coli and are derived from natural plasmids.

- Cosmids have the properties of both plasmid and phage and are also used as cloning vectors. They are essentially plasmids having a minimum of 250 bp of λ phage DNA.

REVIEW QUESTIONS

1. Comment on the bacterial genome and distinguish between three different mechanisms of bacterial genetics.

2. Define transformation. Explain the Griffith's experiment to demonstrate the role of the transforming principle.

3. What is the contribution of Avery *et al.* in providing proof that DNA is the genetic material?

4. Explain the mechanism of transformation in *Diplococcus pneumoniae*?

5. What is transduction? Give an account of the phenomenon of transduction.

6. Explain the types of transduction with their significance.

7. Comment on the 'U' tube experiment conducted by Zinder and Lederberg.

8. Describe the lytic and lysogenic pathways in the life cycle of viruses.

9.   How do viruses help in transduction?

10.   What is plasmid? Explain the types of plasmids with their importance.

11.   Define conjugation. Explain the mechanism of conjugation.

12.   How does the conjugation in bacteria differ from transformation and transduction?

13.   Explain the genomic organization of F plasmid with its role in conjugation.

14.   What is high frequency recombination (Hfr)? How does it help in construction of genomic map of bacteria?

15.   Define genetic engineering. Explain the role of plasmid as cloning vector.

16.   What are the different types of modified plasmids involved in recombinant DNA technology?

17.   In what respect does plasmid differ from cosmid? Give an account of the structural organization of a cosmid.

18.   Write short notes on:

    i.   Competence factor in recipient bacteria

    ii.   Episomes

    iii.   Prophage

    iv.   Transfer genes

    v.   Partial diploid bacteria

    vi.   P22 phage of *Salmonella*

    vii.   Protoplasmic connection or conjugation tube

    viii.   Lambda (λ) phage of *E. coli*

    ix.   Interrupted conjugation

    x.   Cloning vector

    xi.   Colicin

    xii.   Significance of gene cloning

    xiii.   *Ori* gene in plasmid

    xiv.   Induction and excision of viral genome

    xv.   Expression vector

CELLULAR REPRODUCTION AND DEATH

CELL CYCLE

The cell is a unit of biological structure and function. Every living cell is also the unit of biological continuity due to its reproducing ability. Any given cell can be said to have life, just as any given organism. The life of a cell begins with its formation by cell division of a mother cell and ends either with the formation of daughter cells or with its death. Eukaryotic cell divides for one of three reasons—reproduction, growth, and replacement.

In human adults, there are regions of constant cell division, mostly to replace cells that are lost. For example, 200 million erythrocytes are destroyed each day, and in the intestine, epithelial cells are constantly sloughed off to the extent of 1 kg cells/day. These cells are replaced by constantly dividing **stem cells** that are present in bone marrow (for erythrocytes) and in the crypts at the base of intestinal villi (for intestinal epithelial cells). Under normal circumstances, the demand of these and other activities result in the occurrence of 25 million cell divisions at any time in the human body.

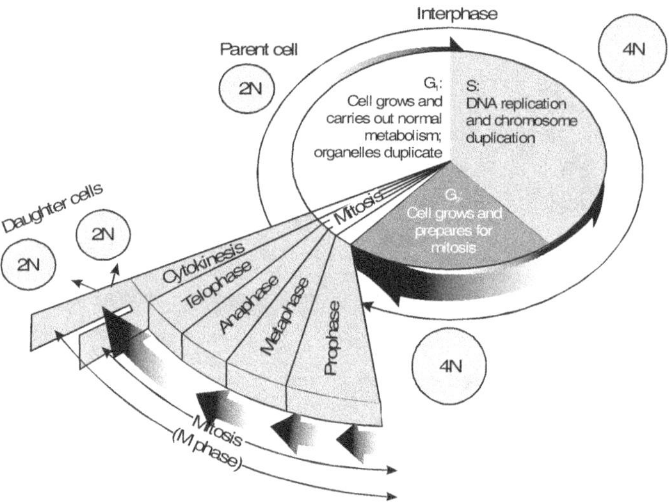

Figure 6.1   The eukaryotic cell cycle. M phase includes the successive stages of mitosis. Interphase is divided into G_1, S, and G_2 phases. During S phase, chromosomes are duplicated hence the parent cell becomes 4N.

Whatever the reason, the stages through which a cell passes from one cell division to the next constitute the cell cycle (Figure 6.1). All cell cycles have two basic phases that can be readily visible under the light microscope; these are M (Mitotic) phase and interphase. M phase includes the following:

1. the process of mitosis, during which the duplicated chromosomes are separated into two nuclei, and

2. cytokinesis, during which the entire cell is physically divided into two daughter cells. Vast majority of living cells spend their life in interphase during which synthetic activities of the cell increases so as to prepare the cell for the mitotic cycle. While M phase usually lasts only for 30 minutes to an hour, interphase may extend for hours, days, weeks, or longer, depending on the cell type.

M phase is a time when the cell is focused on the activities necessary for cell division—a time when the synthesis of macromolecules is generally shut down. In contrast, interphase is a period when cells often grow in volume and carry out active metabolic functions such as glucose oxidation, transcription, translation, and replication of DNA. Although M phase is the period when the contents of the cell are actually divided, numerous preparations for an upcoming mitosis are made during interphase, including the important events involved in duplication of its chromosomes. Interphase is divided into G_1 (for the first gap or the period prior to synthesis of DNA), S phase (period of DNA synthesis), and G_2 phase (for second gap or period between DNA synthesis and mitosis).

DURATION OF CELL CYCLE AND G_0 STAGE

The analysis of cell cycles of a large variety of cells has revealed great variability in the lengths of their cell cycles, particularly when examined at different stages of development. The duration of cell cycles can range from less than 30 minutes in very rapidly dividing cells of cleaving embryos (other than mammals, which cleave very slowly) to cycles lasting several months in slowly growing tissues, such as the mammalian liver. Rapidly growing adult cells typically divide every 12 to 36 hours. In a typical adult mammalian cell cultured from human tissue, G_1 lasts 8 hours, DNA synthesis occurs in 6 hours in the S phase, G_2 continues for about 4–5 hours, and mitosis is completed within 1 hour. With a few notable exceptions, cells that have stopped dividing, whether temporarily or permanently, whether in the body or in culture, are found in a stage prior to the initiation of DNA synthesis. Cells that are "locked" in this state are usually said to be in the G_0 **state** to distinguish them from the typical G_1 **phase** cell that soon enters S phase. Once the cell receives a signal to proceed from G_1 into S phase, the cell starts DNA synthesis and invariably completes that round of DNA synthesis and continues through mitosis.

The phenomenon of **contact inhibition** in the normal cells which stopped dividing once a monolayer is formed in culture medium has been already discussed in Chapter 4. Examination of fibroblasts that have stopped dividing upon formation of a confluent monolayer reveals that all the cells are blocked in early G_1. This resting point for quiescent cells is referred to as G_0. Similarly, when the growth of such cells is prematurely blocked by removal of essential growth factors from culture medium, the starved cells do not stop randomly in the cell cycle, but always proceed through mitosis, to yield cells with the 2N amount of DNA. In each case, growth is stopped very early in G_1, for DNA synthesis is commenced about 8 hours after reintroduction of the limiting growth factors. When so blocked in early G_1, quiescent cells remain healthy, even if the needed nutrients are withheld for many generation times. In contrast, if toxic chemicals like colchicine or hydroxyurea are added to the cells, they don't block the cells in G_1; colchicine stops cells in M phase, while hydroxyurea interrupts DNA synthesis in S phase. Cells so inhibited do not remain viable for long and usually die within 24 hours after exposure to these toxic chemicals.

In addition to hydroxyurea, there are a number of anticancer drugs like fluorouracil, cytarabine, doxorubicin hydrochloride, mercaptopurine, methotrexate sodium, thioguanine, etc., that specifically inhibit DNA synthesis thereby interfering with nucleotide pools. Because malignant cells are not necessarily rapidly dividing, the drugs also affect those normal tissues that are likely to be in the S phase.

CONTROL OF THE CELL CYCLE

The cell cycle is an orderly series of events, as is the development of an organism from a fertilized egg was discovered by conducting a series of experiments on the oocytes and early embryos of frogs and several invertebrates. The nature of the agent(s) that trigger DNA replication and entry of a cell into mitosis (or meiosis). The findings of these experiments have shown that the entry of a cell into M phase was triggered by the activation of a protein kinase called MPF (**maturation-promoting factor**). Isolation and purification of MPF reveals that it consists of two subunits, one with a molecular weight of 45,000 (p45) and one with a molecular weight of 32,000 (p32). The larger subunit is a catalytic subunit that adds phosphate groups to serine and threonine residues

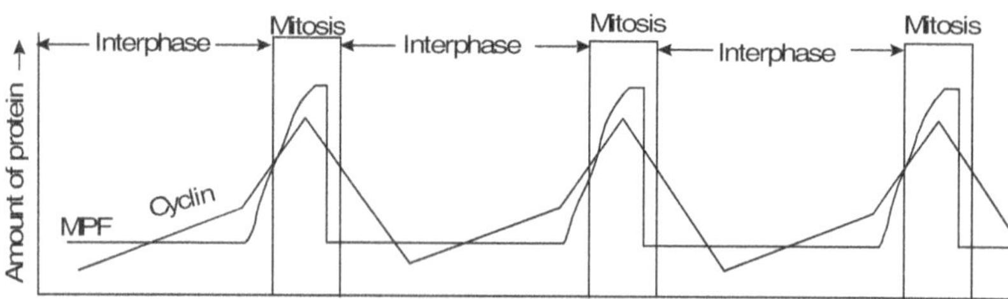

Figure 6.2   Cyclical fluctuations in the level of cyclin and maturation-promoting factor (MPF) during early frog development when mitotic divisions occur synchronously in all cells of the embryo.

on cellular proteins. The smaller regulatory subunit consisting of protein is referred to as **cyclin**. The term cyclin was coined to indicate that concentration of these regulatory proteins rises and falls in a predictable manner as the cell cycle progresses (Figure 6.2).

The activities of MPF and cyclin fluctuate with the cell cycle. As interphase proceeds from S and G_2, a cyclin is made and accumulates in the cell until there is enough of it to bind MPF, which up to then is inactive. Cyclin binding causes the kinase to become active, and mitotic induction follows. During mitosis, cyclin is degraded, and MPF is inactivated, and this allows the cell to exit mitosis and enter interphase. The degradation of cyclin essentially completes the mitotic induction cycle. After mitosis, the cells enter G_1, and a new cycle starts (Figure 6.3).

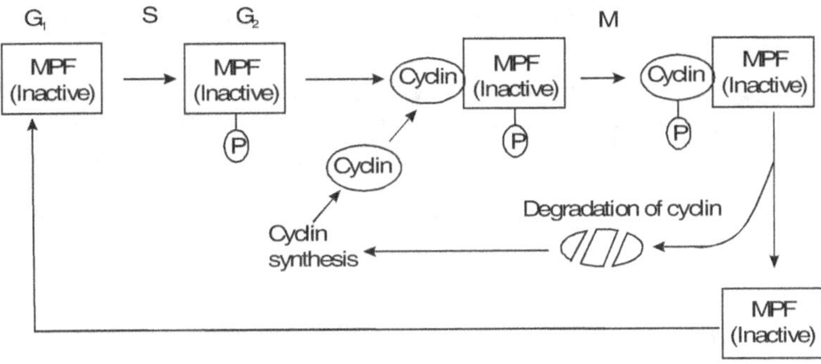

Figure 6.3   The cyclin-MPF cycle during a yeast cell cycle. Cyclin accumulates during G_1 and S to reach a peak at G_2. After M, it is broken down. Phosphorylated MPF is inactive, while dephosphorylated MPF is active and is suitably bound to cyclin.

Leland H. Hartwell of the University of Washington pioneered the genetic line of the cell cycle research in experiments with brewer's yeast, *Saccharomyces cerevisae*. The unicellular organism differs from many cells in that it does not divide by pinching in half. Instead, after starting to replicate its DNA in interphase, it sprouts a bud, thereby initiating mitosis. This bud grows continuously, as does the parent cell, and then separates, marking the end of the cycle in that yeast. He began by identifying mutants from the yeast that became "stuck" at specific points in the cell cycle. He reasoned that the deranged cells each reflected an alteration in a gene whose product was critical to passing the point of arrest. Those crucial genes are collectively called cell-division cycle (*cdc*) genes. The product of the *cdc2 gene* was found to be homologous to the catalytic subunit of MPF. In other words, it was a cyclin-dependent kinase (Cdk). Subsequent research on yeast, as well as many different mammalian cells, has supported the concept that the progression of a cell through its cycle is regulated primarily at two distinct stages of the cell cycle, one occurring just before the end of G_1 (the G_1–S transition) and the other near the end of G_2 (the G_2–M transition). Passage through each of these points commitment requires the transient activation of Cdks by specific cyclins (**G_1 cyclins** at first point and the **mitotic cyclins** at the second point).

MITOSIS (MITOS=THREAD)

The process of mitosis begins with chromosome condensation and ends with decondensation. In a precise manner, the entire nuclear matter of a mother cell has been divided into two equal and similar halves into two daughter cells. Therefore mitosis is an equatorial division that is divided into four sequential periods: prophase, metaphase, anaphase, and telophase. The overall process of mitosis and interphase is diagrammatically represented in Figure 6.4.

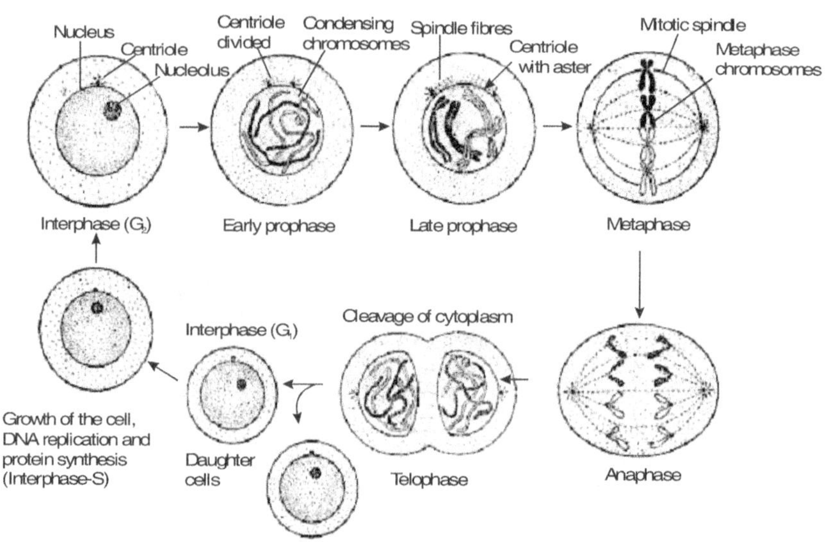

Figure 6.4 Diagrammatic representation of various stages of mitosis

In general, mitosis takes one to two hours, with prophase being the longest and anaphase the shortest.

Prophase (*Pro*=before, *Phase*=appearance)

It is characterized by three major events, coiling of chromosomes, nuclear disintegration, and breakdown of nuclear envelope. In early prophase the chromosomes make their appearance as thin, delicate, coiled thread-like structures, which then shorten to form chromosomes, the process called **condensation**. As chromosomes continue to condense, the distinction between heterochromatin and euchromatin becomes less distinct and finally disappears. Each shortened, thickened chromosome is divided longitudinally into two chromatids that are held together by centromere. Nucleolus decreases gradually and disappears towards late prophase. The nuclear membrane breaks down.

In the cytoplasm, near the nuclear membrane lies the centrosome containing one or two centrioles. The centrosome is the major microtubule-organizing centre (MTOC) (see chapter 2) during interphase. It consists of protein tubulin that precipitates out to form the spindle. If there is one centriole then it replicates into two. Short microtubules radiate out in a star-like structure (astral) from centrioles, which move apart and finally reach the opposite poles of the cell. In animal cells, the stage is **centric** and division is **astral** since the centriole is usually present and participates in cell division. In organisms that do not have centrioles the division goes on in the absence of astral rays. This is the **ecentric** stage and the division is the **anastral** division.

Cell organelles like mitochondria, lysosomes, peroxisomes and chloroplasts of a plant cell traverse mitosis relatively intact, while others, such as endoplasmic reticulum, and Golgi complex, undergo extensive fragmentation to form large number of cytoplasmic vesicles.

The dissolution of the nuclear envelope marks the beginning of next phase called **prometaphase**, during which the definite mitotic spindle is formed and the chromosomes are moved into position at the centre (equatorial plate) of the cell.

Metaphase

In this stage, the chromosomes align themselves to the spindle equator. The mitotic spindle contains a highly organized array of microtubules that can be functionally divided into three groups (Figure 6.5).

1. astral microtubules, radiating from the centrosome and probably help the cell to determine the plane of cytokinesis,

2. chromosomal (or kinetochore) microtubules, extending from the centromere of each sister chromatid to opposite poles. These microtubules are essential for the movement of chromosomes towards the poles during anaphase, and

3. polar microtubules that run from one centriole to the other. They form a structural basket and maintain the integrity of the spindle.

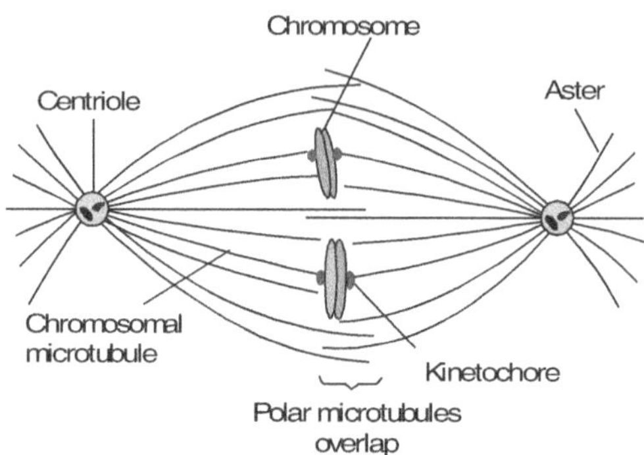

Figure 6.5   The components of eukaryotic mitotic spindle

Metaphase is a time of intense activity, even though there is very little evidence of movement. It is one of the checkpoints of the cell cycle where the cell decides whether to go ahead with or delay the progression of mitosis.

Anaphase

It starts when the sister chromatids of each chromosome suddenly separate from each other. The sister chromatids in each metaphasic chromosome are held together by strong forces. The loss of attraction in the presence of the enzyme topoisomerase II triggers the separation of sister chromatids (now called chromosomes) during anaphase. The mutation in the gene coding for this enzyme fails to separate chromosomes in yeast. The chromosomes move towards opposite poles due to shortening of microtubules attached to its kinetochore (anaphase A). At the same time, the polar microtubules expand further. Activities of these polar fibres are thought to be responsible for causing separation of the poles (anaphase B). When chromosomes travel towards the poles they become bent in the form of a "V" and arms of the V trail behind, pointing their arms towards the equator of the spindle.

The movement of chromosomes in anaphase is a key event in accurately separating the two copies of the nuclear genetic material. If both chromatids of a pair migrate to the same pole, it leads into an error called non-disjunction. As a consequence, one daughter cell carries an extra copy of the chromosome and cannot survive in animals but can in plants.

Telophase

It is the process exactly reverse of prophase since the chromosomes near their respective poles tend to collect in a mass. Telophase marks the beginning of the final stage of mitosis and initiates the preparatory stage for the cell to return to the interphase condition. The union of membranous vesicles reforms the double-membrane nuclear envelope; chromosomes become more and more uncoiled and dispersed until they disappear from view under the microscope; nucleoli and other membranous organelles like endoplasmic reticulum, Golgi complex, etc. reappear. Spindle dissolution proceeds during telophase and progressively becomes shorter, and finally centrioles in animal cells remain at the perinuclear position.

Cytokinesis

In all eukaryotes except land plants, a cleavage furrow appears at the equatorial plate that deepens and finally divides the cell contents into two separate, membrane-bound cells. The process begins at telophase or even late anaphase. As the furrow deepens, the cell assumes a dumbbell shape, and finally, the invaginating membranes fuse and two new cells are formed that have the same genetic constitution as the parent one. The mechanism of furrowing involves microfilaments (actin filaments), and their active growth is essential for the progression of the furrow. Plant cells, which are bounded by cell wall, divide their cytoplasmic contents into daughter cells by a very different type of mechanism. Plant cells build a cell membrane and cell wall called the **cell plate**, which grows outward to the existing lateral walls. Inside the dividing cell at the **phragmoplast** the first sign of cell plate formation appears, where the cluster of interdigitating microtubules arrange themselves perpendicular to the future plate. ER–Golgi system-derived vesicles further provide the material to construct the cell plate, which is the forerunner of the mature cell wall.

MEIOSIS

It is the reduction division in which the diploid number of chromosomes is reduced to haploid and occurs in organisms that reproduce sexually. The term meiosis was coined by J.E. More and J.B. Farmer. In humans, a typical mitotic cell (e.g. in the bone marrow) has a diploid number of chromosomes (23 pairs = 22 pairs of autosomes + a pair of sex chromosome), and its progeny also has 46 chromosomes. The product of meiosis is the formation of gametes in the process of oogenesis and spermatogenesis, each having 23 chromosomes (22 autosomes + one sex chromosome either X or Y). Figure 6.6 shows the summarized events that occur during meiosis.

As mentioned above, meiosis includes two sequential divisions that distribute the haploid set of chromosomes among four nuclei (Figure 6.7). The first meiotic division is reduction division since one member of each pair of homologous chromosomes of a diploid mother cell are pulled apart and distributed to form two haploid daughter cells. Subsequently, in the second meiotic division, the chromatids from each chromosome present in the two daughter haploid cells are separated and distributed to the four daughter cells.

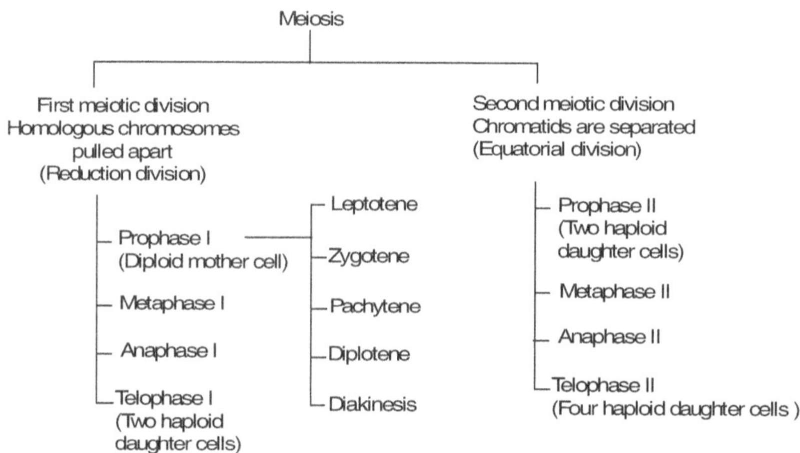

Figure 6.6   Two successive divisions in meiosis

First Meiotic Division

The first meiotic division is studied under four phases namely prophase I, metaphase I, anaphase I, and telophase I. The first prophase is typically the longest phase as compared to the prophase of mitosis. In the human female, for example, every oocyte in the ovary has entered meiotic prophase I at approximately the time of birth. Since a female produces no additional oocytes in her lifetime, many of these oocytes will remain in the same stage of prophase for several years. Since this phase is long and quite complex, it is conveniently divided into five sub-stages as follows:

Leptotene   The pre-leptotene stage resembles the early mitotic prophase. Chromosomes start to make their appearance in the form of thin, long, and highly coiled chromatin. Later on in leptotene, the already duplicated chromosomes appear as single threads with bead-like thickenings called **chromomeres**. The chromosomes at this stage show a definite orientation and a tendency to be polarized. This peculiar arrangement is called the **bouquet stage**.

Zygotene   During this stage, the paternal and maternal homologous chromosomes begin to attract and joined to each other to form pairs or **bivalents**. This event constitutes the process called **synapsis**. The pairing of homologous chromosomes starts from one point and proceeds onwards in a zipper-like fashion. Electron micrograph indicates that chromosome synapsis occurs by the formation of a complex structure called **synaptonemal complex** (SC). The SC is a ladder-like structure and a key element in the lining up of homologous chromosomes. Once formed, the complex is required for continued pairing and chiasma formation during pachytene.

Pachytene   The end of synapsis marks the end of zygotene and beginning of pachytene, which may last for up to several weeks. In this stage, each paired chromosome splits longitudinally to form two sister chromatids and thus each pair contains four chromatids (**tetrad**). At this stage, the nucleoli are clearly visible and are associated with a particular chromosomal region. The significant events that occur in this stage are exchange of paternal and maternal genes and chiasma formation. The exchange event takes place first and appears visually as chiasma formation. The synaptonemal complex facilitates the exchange between non-sister chromatids and genetic crossing over. The centre of SC contains a number of electron-dense bodies called **recombination nodules**, since they contain the machinery that help the genetic recombination.

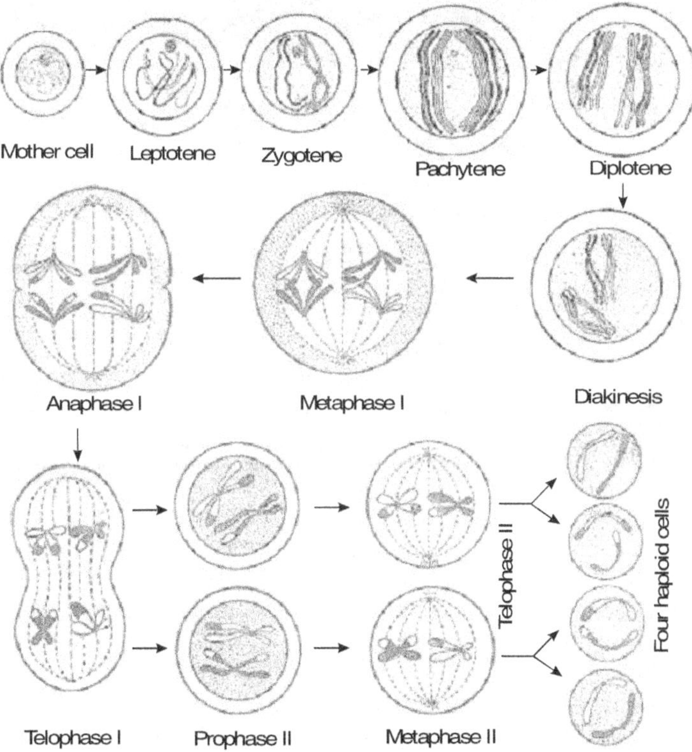

Figure 6.7   Stages of meiosis

Diplotene   This stage is marked by the beginning of separation of non-sister chromatids and the tight pairing of the homologue is relaxed. The centromere is not split, and longitudinal separation of the homologous chromosomes is incomplete. The chromosomes remain united at their points of interchange or chiasma (Figure 6.8). The frequency of chiasma formation is proportional to the length of the chromosomes. There may be one to several chiasmata in each pair of homologous chromosome.

The chiasmata are generally located at the sites on the chromosomes at which genetic exchange during crossing over had previously occurred. The presence of chiasma provides striking visual evidence for the extent to which genetic exchange has taken place. This event of reshuffling of genes provides chances for new combinations of characters, thereby producing variations leading to speciation, which plays an indispensable role in organic evolution.

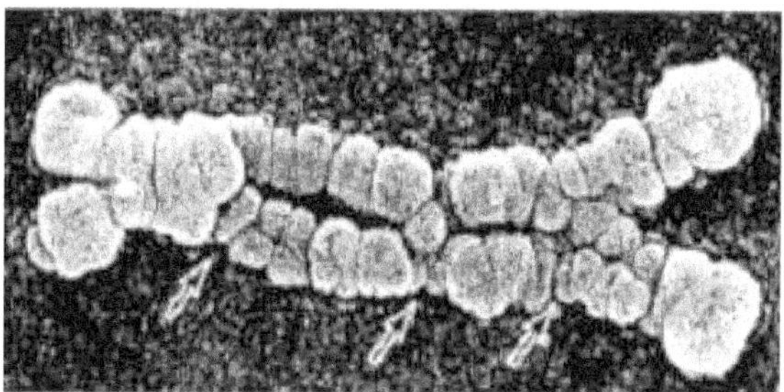

Figure 6.8   The evidence of chiasmata formation in a pair of diplotene bivalents indicated by arrows

Diakinesis   This stage prepares the chromosomes for attachment to the meiotic spindle fibres. As a sort of preparatory action, chiasma moves from centromere towards the end of the chromosomes, and it continues till the interstitial chiasma diminishes constituting the process called **terminalization**. Diakinesis ends with the disappearance of the nucleolus, the breakdown of nuclear membrane, and finally the movement of the tetrads to the metaphase plate.

As the prophase I of meiosis is completed, the cell enters metaphase I in which the tetrads move towards the centre of the cell and arrange themselves to the metaphase plate. The next step is the separation of the tetrad so that each homologous double-chromatid chromosome moves to one of the daughter cells (anaphase I). This is the reduction division step and results in two daughter cells that have one of each homologous chromosome and therefore are haploid with respect to chromosome number (telophase I). However, because each of these chromosomes has two sister chromatids, the cell quantitatively has the normal diploid amount of DNA but only half the number of chromosomes.

Second Meiotic Division

Once the first meiotic division is completed, the cell subsequently enters into the stage called interkinesis, which is generally short-lived. The second meiotic division is further divided into prophase II, metaphase II, anaphase II, and telophase II. In this division, the double-chromatin chromosomes line up themselves in the centre of the cell and chromatids separate and move to daughter cells as in a normal mitosis. The final product is a true haploid cell with respect to both chromosome number and DNA content. Thus, four cells containing haploid set of chromosomes are formed at the end of the second meiotic division. Few points of differences between mitosis and meiosis are mentioned in Table 6.1.

Meiotic error   As described earlier, meiosis is a complex process of cell division, where things occasionally go wrong. A pair of homologous chromosomes may fail to separate during meiosis I, or sister chromatids may fail to separate during meiosis II or during mitosis. This phenomenon is called non-disjunction. Conversely, homologous chromosomes may fail to remain together. These problems can result in the production of aneuploid cells. Aneuploidy is a condition in which one or more chromosomes are either lacking or present in excess causing hereditary disorders that are explained in chapter 7.

Table 6.1   Differences between mitosis and meiosis

Mitosis	Meiosis
It is the division by which diploid number of chromosomes in mother cell is equally distributed to two daughter cells.	It is the division by which diploid number of chromosomes in mother cell is reduced to half and distributed to four daughter cells.
It occurs in somatic as well as germ cells of organisms.	It is restricted to germ cells of organisms.
The entire process is completed in one sequence.	It consists of two successive divisions: the first reduction division and second equatorial division.
Prophase is of short duration.	Prophase is prolonged and occurs double.
Prophase is not divided into sub-stages.	Prophase I is divided into five sub-stages: leptotene, zygotene, pachytene, diplotene, and diakinesis.
There is no synapsis of homologous chromosomes.	Homologous chromosomes form synapsis.
There is no chiasma formation between non-sister chromatids.	Non-sister chromatids twist around each other to form chiasma.
No reshuffling of genes since there is no crossing over.	There is recombination of genes since there is crossing over.

(Contd.)

Table 6.1 (Continued)

Mitosis	Meiosis
Metaphase occurs once where chromatids arrange themselves at equatorial plate.	Metaphase occurs twice. In the first, chromosomes line up and in the second, chromatids arrange themselves on the equatorial plate.
During single anaphase chromatids move towards the opposite poles.	During first anaphase, homologous chromosomes and in second anaphase, chromatids move towards opposite poles.
Telophase is necessary after every division.	Telophase I may or may not be postponed to telophase II that is regular.
Spindle fibres disappear completely in telophase.	Spindle fibres do not disappear completely during telophase I.
Two daughter cells are formed from a single parent cell.	Four daughter cells are formed from a single parent cell.
Newly formed cells have identical genetic composition as compared to parent cell.	Newly formed cells have different genetic composition as compared to parent cell.
No role in speciation and evolution since there is no chance for recombination of genes to occur.	Plays a significant role in speciation and evolution since there is recombination of genes during prophase I.

GAMETOGENESIS

Sexually reproducing organisms produce gametes that have half the number of genetic material as compared to the normal diploid cell. As a result, when the nuclear material of the sperm and the egg (in the case of animals) unite during the fertilization process, the normal diploid amount of genetic material is restored. The newly formed zygote (the diploid mother cell) will divide by mitosis into two cells, the two will divide into four, and so on until an organism with thousands of cells is formed. In higher plants, the haploid pollen grains as male gametes and haploid ovules as female gametes unite to form a diploid sporophyte, which undergoes sporogensis (that includes meiosis) to produce haploid spores. The spores on germination produce haploid gametophyte that can be either an independent stage or, it is retained in the form of a tiny structure within the ovules. Regardless of the type, the gametophyte produces the gametes by mitosis.

As mentioned above, gametogenesis in sexually reproducing animals includes **spermatogenesis** (the process of formation of spermatozoa from the primordial germ cells in testes) and **oogenesis** (the process of formation of oocytes/egg from the primordial germ cells in the ovary). At the end of spermatogenesis, four haploid motile spermatozoa are formed, whereas in oogenesis, only one haploid non-motile and mature egg or ovum is formed. The overall process of gametogenesis in male and female is summarized in Figure 6.9.

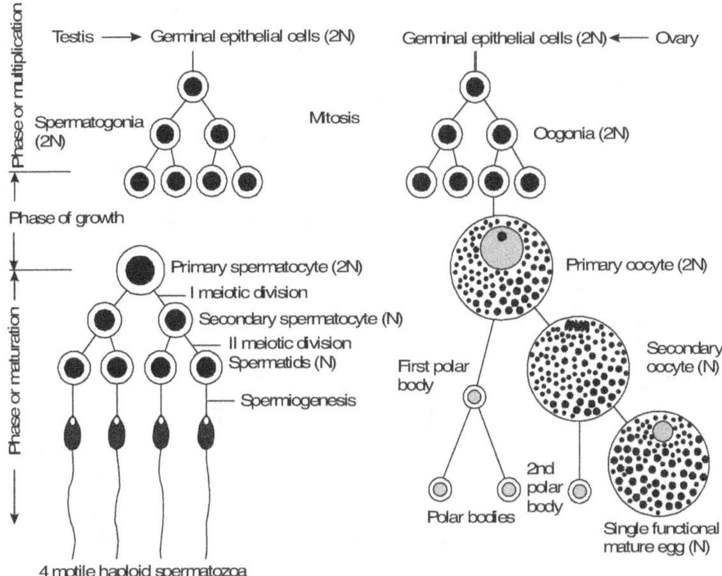

Figure 6.9   Comparative stages in spermatogenesis and oogenesis to form spermatozoa and egg

APOPTOSIS

It is a phenomenon of programmed **cell death** in a differentiated tissue and it is different from the death of cells due to tissue injury (an event called necrosis; points of differences are mentioned in Table 6.2). Apoptosis requires the activation of a specific set of genes that have evolved to destroy the cell in a way that has least effect on its surrounding environment. Once a higher organism attains a mature size, it stays at that size by carefully regulating the balance between cell proliferation and cell death. For example, erythrocytes in humans are constantly produced by pluripotent stem cells in the bone marrow to replace the "old" red blood cells that lack nuclei and have a finite life of about 120 days in the blood. Phagocytic cells in the liver and spleen destroy red blood corpuscles. Similarly, stem cells at the base of the mucosal lining of the intestine constantly divide to replace cells that are sloughed off into lumen of gut due to damage caused by the contents of the alimentary canal. Other examples of apoptosis are death of excess nerve cells, death of T lymphocytes capable of reacting with the body's own tissues, and the death of potential cancer cells.

One way in which tissues and organs take shape during development is through differential cell death. A classic set of studies demonstrating this principle has been made on chick limb development. It has been shown that the shaping of chick wing is in large part brought by localized zones of cell death. Specifically, the shaping of upper forearm, arm, and elbow and elimination of tissues between the digits are all caused by selective cell death. These specific regions are programmed to die at specific times during development. Even if the cells are removed from the embryo and placed in a different region of another embryo, or just maintained in culture, cells still die on schedule. In addition, to above-mentioned examples of apoptosis, the formation and differentiation of the proper gonad (testis or ovary) also involve the death of embryonic tissue. There is the massive degeneration (death) of tissue that was present in the indifferent stage, depending on which sex differentiates. These phenomena occur because the reproductive systems develop from primordial structures that contain presumptive tissues for both male and female reproductive organs. As illustrated by the examples above, apoptosis is a common occurrence during embryonic development, but it also occurs in adult tissues of organisms. Plant cells also have programmed cell death, for example, in leguminous plants, cells of the nitrogen-fixing nodules in the plant root die; they are no longer needed by the plant because the need for fixed nitrogen is reduced at a particular time.

Table 6.2 Difference between apoptosis and necrosis

	Necrosis	Apoptosis
Stimuli	Low O_2, toxic, ATP depletion, accidental damage	Specific genetically programmed physiological signal
ATP required	No	Yes
Plasma membrane	Bursts	Blebbing
DNA breakdown	Random fragmentation	Chromatin condensation, and nucleosome-sized fragments
Fate of dead cell	Ingested by phagocytic cells	Endocytozed by neighbouring cells
Tissue reaction	Inflammation	No inflammation

Apoptosis is a type of orderly, or programmed cell death in which the cell responds to certain signals by initiating a normal response that leads to the death of the cell. The molecular signals that stimulate apoptosis have been described in only a few cases. Generally, the cells that are stimulated to die are not actively cycling but are in G_0. In humans, glucocorticoids are thought to stimulate the death of lymphocytes. On the other hand, when young mammals are weaned from the mother's breast, her serum prolactin levels go down, and lack of a growth-promoting hormone promotes destruction of mammary gland tissue. In another experiment, when lymphocytes are cultured in the absence of the cell division promoter interleukin-3, they quickly become apoptotic. Epithelial cells of the prostate become apoptotic when deprived of the male sex hormone testosterone. This is the reason why prostate cancer that has been spread to other tissues can be treated with drugs that interfere with production of testosterone.

Death by apoptosis is characterized by overall compaction of the cell and its nucleus, which is brought about by several stages that include the breakdown of chromatin into pieces by special DNA-splitting endonuclease, shrinkage of cell from rest of the tissue, and fragmentation of cell into several membrane-bound bodies (containing broken pieces of nuclear material, organelles, and cytoplasm) that are endocytozed by neighbouring cells (Figure 6.10).

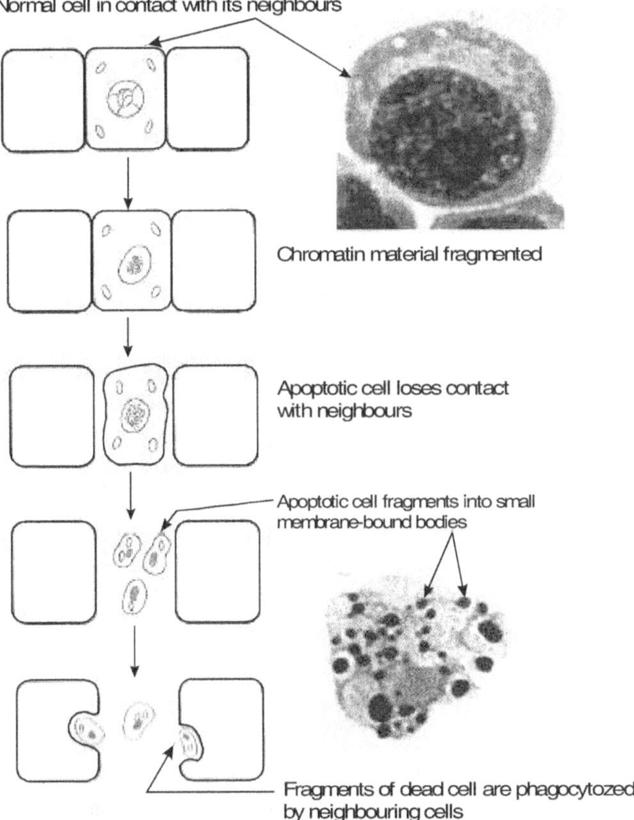

Figure 6.10 Events in "cell suicide". The fragments of apoptotic cells are ingested by adjacent cells.

SUMMARY

- The life of a cell begins with its formation by cell division of a mother cell and ends either with the formation of daughter cells or with its death.

- A eukaryotic cell divides for one of the three reasons—reproduction, growth, and replacement.

- The stages through which a cell passes from one cell division to the next constitute the cell cycle.

- The cell cycle is divided into two major phases—M phase that ends with cytokinesis and interphase.

- M phase includes the process of mitosis in which duplicated chromosomes are separated into two nuclei, and cytokinesis, in which the entire cell is physically divided into two daughter cells.

- Interphase is typically a longer phase than M phase and is subdivided into distinct G_1, S and G_2 phases. G_1 is the period following mitosis and preceding replication, S is the period during which DNA and histone synthesis occur, and G_2 is the period following replication and preceding the onset of mitosis.

- It has been experimentally shown that entry of a cell into M phase is triggered by the activation of protein kinase called MPF (maturation-promoting factor) that includes a catalytic subunit called a cyclin-dependant kinase (Cdk) and a regulatory protein called cyclin.

- Mitosis begins with chromosome condensation and ends with decondensation. It is an equatorial division that is divided into prophase, metaphase, anaphase, and telophase.

- During prophase, chromosomes condense and become short; in pro-metaphase and metaphase, individual chromosomes are first attached to spindle fibres from both poles and then move to a plane at the centre of the spindle. During anaphase and telophase, sister chromatids are separated from one another into separate regions of the dividing cell, and all chromosomes return to interphase condition.

- Meiosis is a process that includes two sequential nuclear divisions, producing four haploid daughter nuclei containing only one member of each pair of homologous chromosomes as compared to the parent cell.

- Prophase first is the longest phase that is divided into five sub-stages: leptotene, zygotene, pachytene, diplotene, and diakinesis. The exchange of paternal and maternal chromosomes and thereby reshuffling of genes takes place in this phase.

SUMMARY

- ⊙ *Meiosis plays an indispensable role in gametogenesis in sexually reproducing animals. At the end of spermatogenesis, four haploid motile spermatozoa are formed, whereas in oogenesis, only one haploid non-motile and mature egg or ovum is formed.*

- ⊙ *Apoptosis is programmed cell death in a differentiated tissue. Examples of apoptosis are death of excess nerve cells, death of T lymphocytes, death of unwanted embryonic tissue, and the death of potential cancer cells.*

- ⊙ *Death of a cell by apoptosis is brought about by orderly fragmentation of its DNA by endonuclease and phagocytosis by neighbouring cells.*

REVIEW QUESTIONS

1. What is cell cycle? Describe the various stages of the cell cycle with the specific duration.

2. Define MPF. Explain the role of maturation-promoting factors in the control of cell cycle.

3. What is meant by a cell cycle checkpoint? What is its importance?

4. What is cell division? Explain mitosis with its significance.

5. How do the events of prophase of mitosis prepare the chromatids for later separation at anaphase?

6. Contrast the events that occur during cytokinesis in a typical plant cell and an animal cell.

7. Describe in detail both the sequential divisions that occur during meiosis.

8. Compare the events that take place during prophase I and prophase II of meiosis.

9. Explain the mechanism of genetic recombination that occurs during prophase I of meiosis.

10. Define apoptosis. Enlist the examples of programmed cell death.

11. Describe the events that take place during cell death.

12. Give the points of differences between mitosis and meiosis.

13. Compare spermatogenesis and oogenesis.

14. Write short notes on:

    i. G_0 stage

    ii. Cyclin dependent kinase

    iii. S phase in interphase

    iv. Longevity of interphase

    v.   Cell-division cycle (*cdc*) genes

    vi. Crossing over

EUKARYOTIC CHROMOSOMES AND VARIATION

CHROMOSOMES—THE CARRIERS OF GENES

Mendel studied inheritance of contrasting characters in the pea plant and proposed the laws of inheritance. Although he provided convincing evidences that inheritance of the traits were governed by discrete factors or elements (now referred to as genes), his studies were totally unconcerned with the physical nature of these elements, or their location within the organism. Mendel carried out his entire research without ever observing anything under a microscope.

By the 1880s a number of cell biologists were closely following the activities of cells and using rapidly improving light microscopes to observe newly formed cell structures. During cell division, the material of the nucleus became well organized into visible "threads" that in 1888 were named chromosomes, meaning coloured body by Waldeyer. In 1897, Walter Fleming observed that before cell division, each chromosome duplicates to form two identical chromosomes. During nuclear division, one of each pair of daughter chromosomes moves into each daughter nucleus. As a result of these events (now collectively called mitosis), the chromosomal complement of daughter cells is usually identical to that of the parental cell.

The German biologist Theodore Boveri studied the importance of chromosomes contributed by the male during the fertilization of the sea urchin eggs that had been fertilized by two sperms rather than just one, as is normally the case. This condition is called **polyspermy**, which is characterized by disruptive cell divisions and early death of the embryo. The presence of an extra set of chromosomes and extra centrioles, both introduced into an ovum by the second sperm, leads to abnormal cell divisions in the embryo during which daughter cells receive variable number of chromosomes. Boveri concluded that orderly process of normal development is dependant upon a particular combination of chromosomes; and this can only mean that the individual chromosomes must possess different qualities. This was the first evidence of a qualitative difference among chromosomes, which would be expected if they constituted a set of carriers of genetic information.

During the1880s, the process of meiosis was described, and in 1887 August Weismann proposed that meiosis included a "reduction division" during which the chromosome number was halved prior to formation of gametes. Sutton (1902) studied the formation of sperm cells in grasshopper and realized the presence of pairs of chromosomes, or homologous chromosomes that include one paternal (contributed by sperm) and other maternal (contributed by an ovum) chromosome and they perfectly correlate with the pairs of inheritable factors uncovered by Mendel. In 1933, T.H.Morgan discovered the role of chromosomes in the transmission of the hereditary characters.

NUMBER OF CHROMOSOMES

Each nucleus encloses a fixed number of chromosomes and for a given species the number of chromosomes remains constant. This number has been evolved during the course of evolution and characterization of the species. The number and morphology of chromosomes help in deciding the phylogeny and interrelationship of the group as well as the individual. During gamete formation, the chromosomes number is reduced to half the number of chromosomes in the somatic cell. The gametes with half the number of chromosomes are said to possess one genome that can be expressed as the genetic number or haploid or 'n' number. The number of haploid chromosomes in most animals and plants lies between 6 and 25. The nematode, *Ascaris megalocephala* has the smallest haploid number (n = 1). Among multicellular animals, the hermit crab has the greatest number of chromosomes (254), while in Protozoa, *Aulocantha* (a radiolarian) has a chromosome number as high as 1600. Each chromosome carries definite number of genes and that is species-specific. Table 7.1 shows the number of chromosomes and their genes present in some of the organisms.

Table 7.1   Number of chromosomes and genes in some of the organisms

Organism	Number of chromosomes	Number of genes
Bacteria (*Escherichia coli*)	1	4,406
Yeast (*Saccharomyces cerevisiae*)	16	5,885
Nematode (*Caenorhabditis elegans*)	12	19,099
Plant (*Arabidopsis thaliana*)	10	25,498
Fruit fly (*Drosophila melanogaster*)	18	13,601
Plant (*Oryza sativa*)	24	57,000
Mouse (*Mus musculus*)	40	30,000–35,000
Human (*Homo sapiens*)	46	34,000?

SIZE OF CHROMOSOMES

Chromosomes occur as a network of fibrils called chromonema in the interphase nucleus. Under the light microscope, at the time of cell division in eukaryotes, chromosomes appear as compact solid structures as a result of condensation of chromonema. There is a great variety in chromosome size and shape. Their size may range from 0.1μ to 30μ in length, the diameter varying between 0.2μ to 2μ. There are exceptionally giant (or polytene or salivary gland) chromosomes that may be as long as 2 mm. An individual organism may have all similar sized chromosomes (symmetrical karyotype) or they may have chromosomes with variable sizes (asymmetrical karyotype) in the cells at different regions. In human beings, there is gradation in the size of chromosomes with approximate length of 4 to 6μ. Generally plant cells have larger chromosomes than animal cells. *Trillium* (plant) contains the largest chromosomes that may attain 30–32 microns in length. Variation in size of the chromosome can be induced by environmental factors (like temperature), chemical treatment (like colchicine) or other mutagenic agents.

MORPHOLOGY OF CHROMOSOME

During interphase, with the help of electron microscope, the condensed portion of the chromosomal material (heterochromatin) can be distinguished from the less dense regions (euchromatin). The heterochromatin is the genetically inactive region while the euchromatin is the functional region, the other aspects of which are explained in chapter 2. The chromatin in the interphase nucleus represents an aggregation of chromosomes. Since the chromatin fibre is about 200–300 Å in diameter, it is not seen in the light microscope. In 1943, chromatin fibres were morphologically identified as chromosomes, and found to be Feulgen- positive.

The compact structures in dividing cells are metaphase chromosomes with definite morphological features. During cell division, the shape of the chromosomes is variable from phase to phase. They are very thin during interphase and become short and thick in prophase with a distinct and clear zone called centromere or kinetochore or primary constriction. The centromere occupies a constant position relative to the ends of chromosomes and on the basis of its position, four types of chromosomes (Figure 7.1) are distinguished as follows:

1. *Metacentric*   Here the centromere is located in the middle region of the chromosome and the two isochromatic arms of the chromosome have equal length. During anaphase, when chromosomes move towards the poles, metacentric chromosomes appear V-shaped.

2. *Submetacentric*   In such chromosomes, the centromere is located away from the middle of its arms. As a result, one arm of the chromosome is shorter than the other. During anaphasic movements, submetacentric chromosomes appear 'L' or 'J' shaped.

3. *Acrocentric*   In this case, the centromere is situated almost near one end of the chromosome. As a result, one arm of the chromosome is extremely short and other one very long.

4. *Telocentric* Here the centromere is located exactly at one end of the chromosome and hence it has only one long arm. Telocentric chromosomes are very rare and unstable.

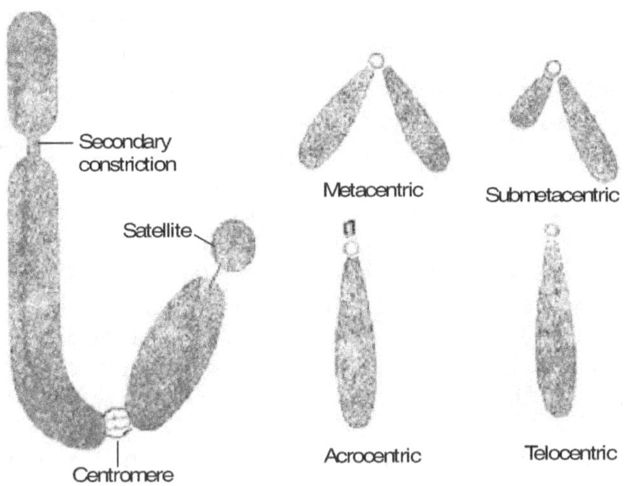

Figure 7.1 External morphology and variations in chromosomes

The metaphase chromosome (Figure 7.2) consists of 1) **two chromatids**, which are further composed of two fibrils called chromonemata or chromonemal fibrils. These fibrils remain coiled with each other either in paranemic type (fibrils are just lying parallel to each other) or plectonemic type (fibrils are twisted around each other). Each chromosome is chemically a continuous fibre of nucleoprotein [a complex of nucleic acid (DNA) and basic proteins (histones)] upon which the genes are arranged in a linear fashion. 2) **Chromomeres** in the form of thick and thin regions giving a bead-like appearance to a fibre, described by **Balbiani** (1876) and **Pfitzner** (1881). Chromomeres are clearly visible in the salivary gland chromosomes of Diptera or lampbrush chromosomes of amphibian oocytes, 3) **Centromere** or primary constriction or kinetochore as modified and thickened region of the chromosome where chromonemal fibrils of a chromosome are joined together. Its position is constant for a particular chromosome and it forms the basis for classification of chromosome into the above-mentioned four types of chromosomes. The microtubules of the mitotic spindle are found attached to the kinetochore that causes the daughter chromosomes to separate during cell division. Depending upon the presence or absence of centromere, the chromosomes are classified as acrocentric (chromosome without centromere), centric (chromosome with one or more centromeres), monocentric (chromosome with single centromere), bi- or dicentric (chromosomes with two centromeres) or polycentric (chromosome with more than two centrioles), 4) **Secondary constriction** I and II with definite position on the chromosome. Secondary constriction I is associated with formation of nucleolus and it is known as nucleolar organizer that contains the genes for synthesis of rRNA and proteins. Secondary constriction II is the satellite that remains attached to the rest of the body by a thread of chromatin. Satellite is found on the long arms of human chromosomes 1,10, 13, 16, and Y chromosome. Chromosomes with satellite are also called SAT chromosomes. The satellite DNA is a part of heterochromatin that also contains a type of repetitive DNA. The term satellite is a misleading term since it is not a satellite in any sense in the cell; when eukaryotic DNA is isolated by normal means, it breaks into pieces, because of its gigantic size. When the mixture of pieces is analysed by a density gradient method, most of the DNA runs in one band but repetitive DNA described here has a slightly different density and runs separate from the main band just as a satellite. It consists of short DNA sequences repeated over and over in tandem arrangement. Satellite sequences are thus clear examples of what might be called structural rather than genetic regions of DNA contained in eukaryotic genome. On the basis of the number of base pairs in repetitive sequence, there can be minisatellite DNAs (containing on an average 15 base pairs) or microsatellite DNAs (2 to 5 base pairs long). Minisatellites are key molecules in producing DNA fingerprinting that determines individuality, and 5) **Telomeres** in the terminal part of chromosomes. They have polarity that prevent other chromosomal segments from fusing with it.

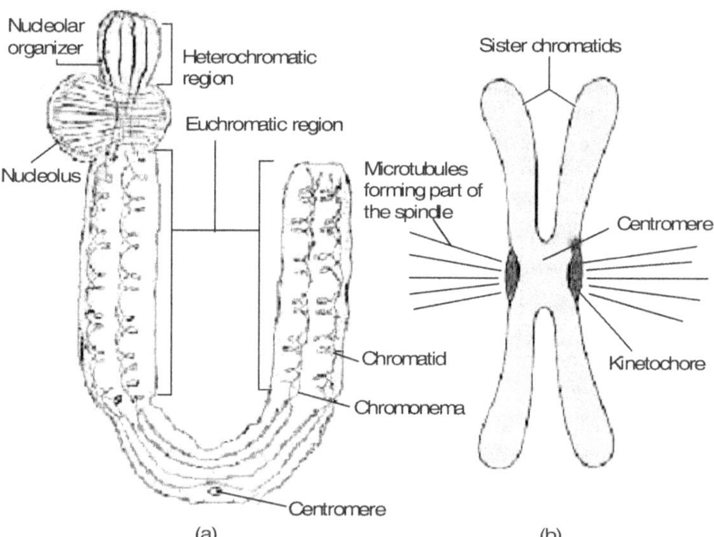

Figure 7.2   (a) Structure of a typical chromosome. (b) Metaphase chromosome consisting of two daughter chromatids.

CHROMOSOME BANDING

The chromosomes have a unique staining pattern allowing their specific identification. It is based on the differential staining techniques of the chromosomes that results in stained and unstained areas in the form of bands (Figure 7.3). These methods are called chromosome-banding techniques. Banding patterns can also be obtained by a diversity of treatment, including heating, digestion with acid, alkali or salt and treatment with a variety of chemical compounds. The main banding techniques are identified by letters, C, Q, G, R, T, and so on and are related to different staining methods used.

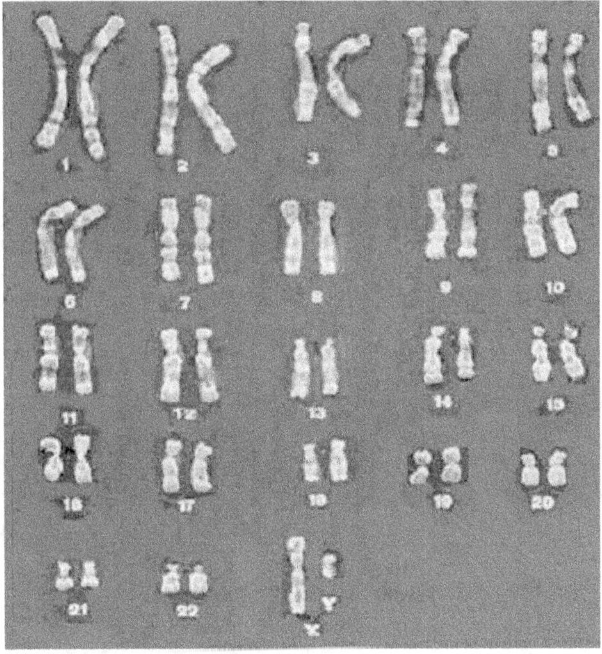

Figure 7.3 Chromosomes of normal human male showing differential banding patterns

Pardue and **Gall** discovered one of the first chromosome banding techniques by the development of the *in situ* hybridization procedure. They observed the unique staining pattern after heat denaturation of chromosomes followed by their staining with Giemsa staining method. In this, the centromeric regions of mitotic chromosomes take up the stain. This staining pattern is referred to as **C-banding**. These regions are composed of constitutive heterochromatin (the basis for naming it C-banding). The C bands apparently develop because much more stainable chromatin remains in this region after preliminary treatment to extract DNA and protein. Since there is only little DNA and protein remaining after extraction elsewhere along the chromosome, very little stain becomes absorbed and most of the chromosome looks pale.

A group of Swedish researchers, led by Tobjorn Caspersson developed another chromosome banding technique that developed even greater staining differentiation of metaphase chromosome. When the chromosomes are treated with the fluorochrome quinacrine mustard, **Q bands** with differential brightness are produced. These bands are rich in adenine and thymine bases.

Tau-Chiuh Hsue and Frances E. Arrighi have developed another banding technique. The bands produced by this method are called **G-bands**. These are stained with Giemsa and other non-fluorescent dyes. Another technique results in the reverse G-band staining pattern, called an **R-band** pattern. They are stained as green, brightly fluorescent bands with acridine orange. These are reciprocal G-bands. R-bands correspond to the regions on chromosomes having proteins lacking sulphur.

In addition to above-mentioned banding techniques, there is **T banding** technique in which staining of certain telomeric regions of chromosomes is involved.

HUMAN KARYOTYPE

In 1956, two cytologists, J.H. Tjio and A. Levan, working in Sweden, published their research article in relation to chromosome number in human beings, giving 46 as the 2n (diploid) number. Their counts were made from tissue culture preparation of lung tissue representing four different embryos. A group of cytogeneticists met at Denver, Colorado in 1960 and adopted a system of

Table 7.2 Classification of human chromosomes into seven groups

Group	Pairs	Description
A	1 to 3	Large, metacentric
B	4 to 5	Large, submetacentric
C	6 to 12 and X	Medium-sized, submetacentric
D	13 to 15	Large, acrocentric with satellite
E	16 to 18	16-metacentric, 17 and 18-small submetacentric
F	19 to 20	Small, metacentric
G	21 to 22 and Y	Short, acrocentric with satellite, chromosome Y is without satellite

classifying and identifying human chromosomes. The length of the chromosome and position of centromere serve as the major landmarks in the categorization of chromosomes. The 22 pairs of autosomes have been divided into 7 groups identified with letters A to G (Table 7.2). Numbers 1 to 22 are associated with autosomes in descending order of length while the 23rd pair is the sex chromosome pair as shown in Figure 7.4.

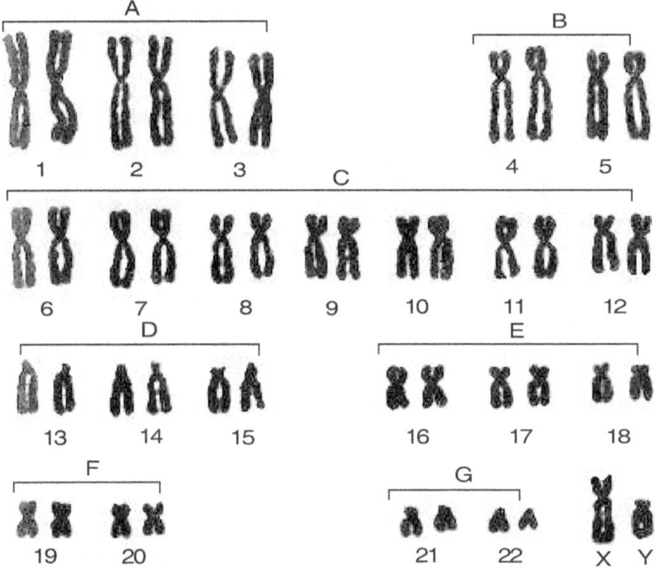

Figure 7.4   The karyotype of a normal human male

CHROMOSOMAL ABNORMALITIES IN HUMANS

Chromosomes are carriers of genes that are hereditary units. The constancy in the number and structure of chromosomes is very essential for the individual not only to maintain its phenotypic characters, but many a time even for its own survival. Occasionally however, there occur variations in both structure and number of chromosomes producing alterations in morphological, behavioural and other characters, which may be lethal for the individual. These changes are known as **chromosomal aberrations or chromosomal mutations**, since they bring about changes in genetic characters of the individual. It is different from gene mutation where a single gene is mutated. Structural and numerical alterations in human chromosomes result in abnormalities in man, some of which are familiar and interesting. These abnormalities are listed in the following flowchart.

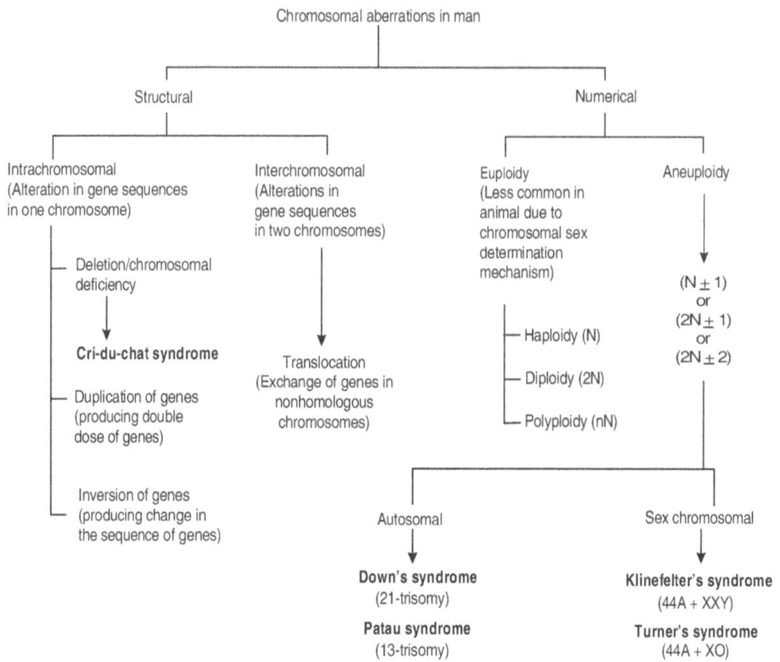

Cri-du-chat Syndrome

In 1963, Jerome Lejeune reported a case of partial monosomy in human beings in France. He described the clinical symptoms of this abnormality and referred it to as **Cri-du-chat syndrome** (Figure 7.5).

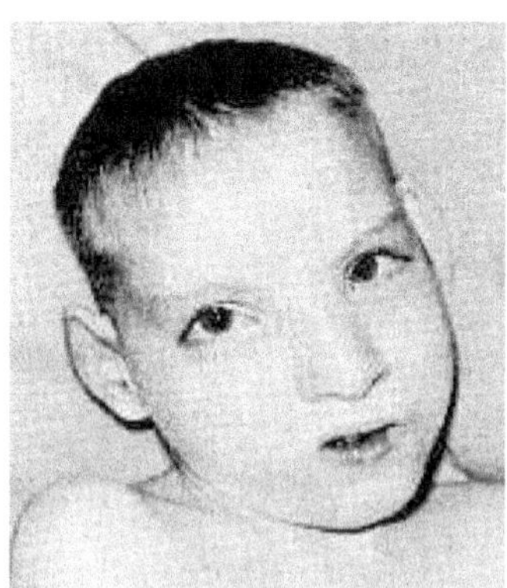

Figure 7.5   Patient with *Cri-du-chat syndrome* having microcephaly and catlike cry

This syndrome (a group of symptoms that characterizes the disease) is associated with the loss of about one half the short arm of chromosome 5. Infants with this syndrome exhibit multiple anatomical complications like microcephaly (small head), broad face and saddle nose, widely spaced eyes and physical and mental retardation. An infant afflicted with this syndrome has a cry similar to that of mewing of a cat, thus giving the syndrome its name. Cri-du-chat patients die in infancy or early childhood and do not transfer the chromosome deletion to the offspring.

Down's Syndrome

It is the best and most common chromosome-related disease syndrome that was first reported by Langdon Down in 1866. Formerly the syndrome was known as "Mongolism" since the patient with this disorder had an epicanthal fold in the eyelid, a phenotypic characteristic of the Mongolian race. The syndrome arises due to the trisomy of autosome number 21 arising from non-disjunction during meiosis in one of the parents; hence the condition is also referred to as **21-trisomy**. The karyotype of the afflicted patient thus shows 47 chromosomes. A patient with Down's syndrome (Figure 7.6), who illustrates typical facial features, that is, broad head with round face, slanting eyes, open mouth with large protruding tongue and are mentally retarded with an IQ range 25–50. The hand of a child with Down's syndrome shows the abnormal crease on the palm and little finger. An estimated frequency of the Down's syndrome is about 1 in 700 births.

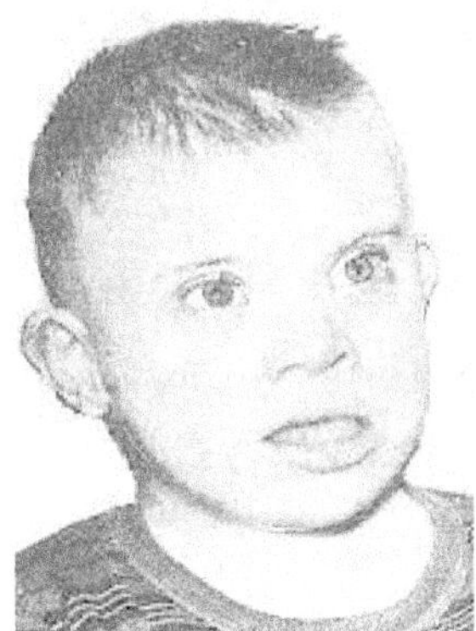

Figure 7.6 Facial features of a child with Down's syndrome

A special case should be cited in which an extra chromosome or more, probably a chromosome fragment of No. 21, is associated with a specific disease. This is the **Philadelphia chromosome** that is found in the leucocytes of patients with chronic granulocytic leukemia. It is apparently a deleted part of chromosome No. 21 and serves as an identification marker in the bone marrow cells of people who have this particular type of blood cancer.

Patau Syndrome

In 1960, Klaus Patau reported a case with autosomal trisomy of chromosome 13, described as Patau syndrome having 47 chromosomes. Affected infants are severely mentally retarded, thought to be deaf, and characteristically have a harelip, cleft palate, cardiac anomalies, posterior heel prominence and polydactyly. The trisomy-13 syndrome occurs in 1 in 20,000 newborns. The severe symptoms result in early death and so children or adults with this syndrome are almost non-existent.

Klinefelter's Syndrome

In 1942, H.F. Klinefelter reported an extra X chromosome in addition to the usual male (XY) chromosome complement (47, XXY) has been associated with the abnormal male syndrome known as Klinefelter's syndrome.

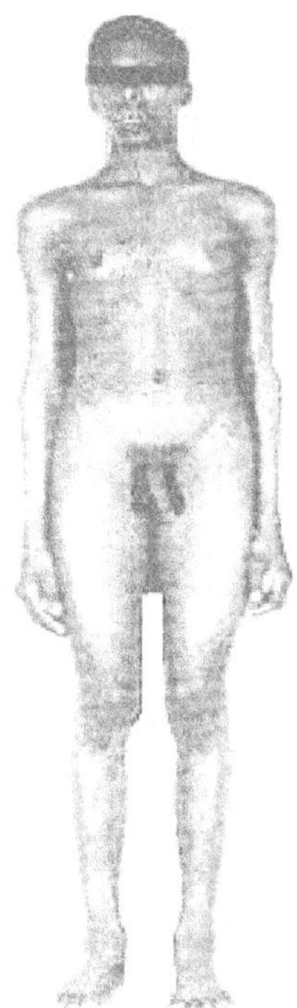

Figure 7.7   Male with Klinefelter's syndrome. Note enlargement of breast (gynecomastia), small testes, very little hair, and long legs.

Individuals with this syndrome are phenotypically males but with some tendency toward femaleness, particularly in secondary sexual characteristics. These features include enlarged breast (gynecomastia as in the Figure 7.7), underdeveloped body hair, small testes, and absence of spermatogenesis. The affected person shows longer legs and arms than normal individual and also feminine fat deposition. Most of the patients ate mentally retarded and have a variety of psychiatric problems. The buccal smears of these patients show Barr bodies (chapter 2). So such individuals are referred to as chromatin-positive males, since they have two X chromosomes in their genotype. Presumably, the XXY constitution originates either by fertilization of an exceptional XX egg by a Y sperm or an X egg by an exceptional XY sperm. The incidence of this syndrome is 1 in 500 live male births. Some of the more complex karyotypes associated with Klinefelter's syndrome are 48XXXY, 48XXYY, 49XXXYY, 49XXXXY, and 50XXXXXY. All of them show one or two Barr bodies in the epithelial cells of the buccal smear or in the neutrophils of the blood smear due to the presence of more than one X chromosome.

Turner's Syndrome

It is one of well-known examples of the aneuploidy in humans, which was reported by H. H. Turner in 1938. It occurs in about 1 per 2500 live female births.

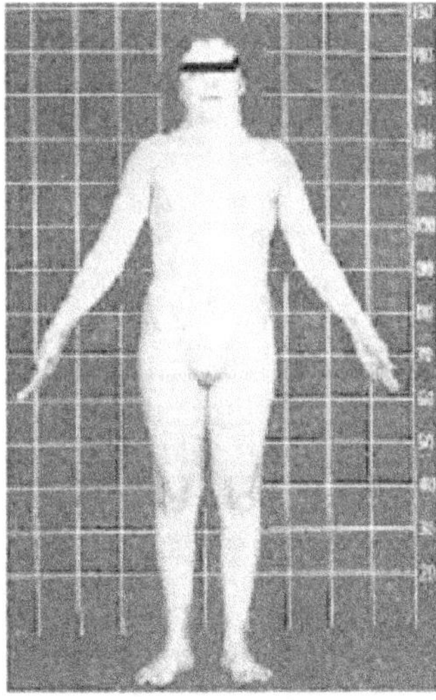

Figure 7.8 Female with Turner's syndrome (44A + XO). Note the underdeveloped breast with widely spaced nipples.

The patients with Turner's syndrome (Figure 7.8) are phenotypically females and genetically 44A + XO, i.e., only one X chromosome is present in their genotype. This monosomic chromosome complement has been with an abnormal female that is characterized by short stature, poorly developed or absence of ovaries, webbed neck, low set of ears, broad chest with underdeveloped breasts and widely spaced nipples. These adults have limited secondary sexual characters and are sterile. Epithelial cells of these patients are X-chromatin-negative since they have only one X chromosome.

Such abnormalities probably originate from exceptional eggs of sperms with no sex chromosome or from the loss of a sex chromosome in mitosis during early cleavage stages, after an XX or XY zygote has been formed. Mosaics with X/XX sex chromosomes show symptoms of Turner's syndrome but are usually taller than X and have fewer anomalies than non-mosaic 45X females. They show more femine characteristics, more normal menstruation, and may be fertile. Patients with partial deletion of one X chromosome are Barr positive and therefore may be misdiagnosed if a buccal smear is the only test used.

GIANT CHROMOSOMES

These chromosomes are strictly confined to certain types of cells of the organism. There are two types of giant chromosomes, namely polytene chromosomes and lampbrush chromosomes, which attain their largest size in the nuclei of their respective cells.

Polytene Chromosomes

Larvae of some dipteran insects like *Drosophila, Chironomus,* housefly and *Anopheles*, have specialized cells in the salivary gland that contain polytene chromosomes. Balbiani in 1881 was the first to observe salivary gland chromosomes in the salivary gland of chironomus larva and referred them as **salivary gland chromosomes**. The term **polytene chromosomes** is more preferable and was given by Koller due to the presence of many daughter chromatids formed due to repeated duplication of chromosomes that remain closely associated with parental strands. It makes the giant chromosome, a bundle of fibrils. This process is endoduplication or **endopolyploidy** or **endomitosis**. The giant chromosomes also exist frequently in other tissues, such as Malpighian tubules, muscle cells, and fat cells as well as in several species of protozoans and plants.

Cells from larval salivary gland of a common fruit fly *Drosophila* contain chromosomes that are about 100 times thicker than the chromosomes found in most other cells of the organism (Figure 7.9). During larval development, these cells stop dividing, but they keep growing in size. DNA duplication continues, providing the additional genetic material needed to support the high level of secretory activity of these enormous cells. The duplicated DNA strands remain attached to each other in perfect side-by-side alignment, producing giant chromosomes with as many as 1,024 times the number of DNA strands of normal chromosomes. The salivary gland chromosomes show somatic pairing at interphase because of their multistranded giant nature. They have another characteristic feature in relation to their number. The salivary gland chromosomes always appear to be half that of the normal cells. Along the length of the chromosome there is a series of dark bands alternating with other clear zones called interbands. The dark bands represent the heterochromatic (densely packed chromatin) region that stains deeply and are Feulgen-positive. In addition, they absorb ultraviolet light at 260 nm. The interbands have fibrillar structure that absorb very little ultraviolet light and are Feulgen-negative. Most important, the banding pattern is essentially constant from fly to fly and from tissue to tissue.

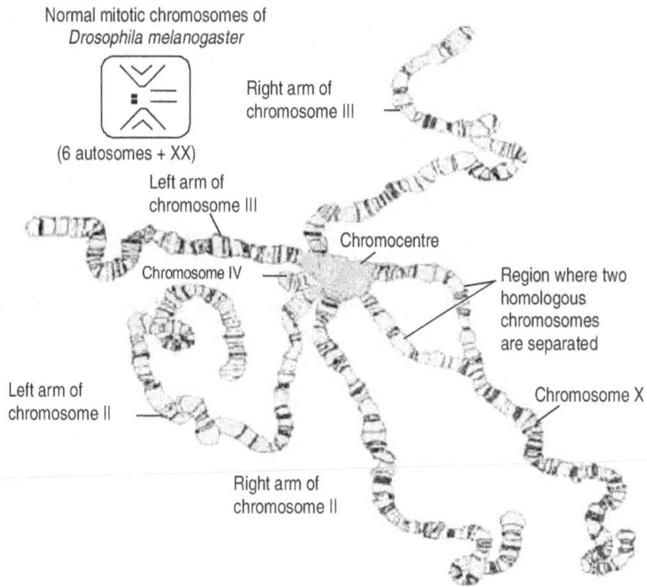

Figure 7.9   The polytene chromosomes in the cell of salivary gland of female *D. melanogaster*

The salivary gland chromosomes of *Drosophila melanogaster* appear as five long chromosomes and one quite short chromosome found attached to a central mass called **chromocentre** to which the single large nucleolus is attached. The unusual size of these chromosomes is a product of the type of growth that occurs in larval glandular tissues. As the larva develops, the synapsed chromosomes replicate usually nine times, giving rise to a polytene strand of about 1000 DNA double helix molecules. Thus, giant chromosomes are cable-like structures having many DNA strands (polytene). Of the six chromosomes in a single cell of the salivary gland of *D. melanogaster*, one short chromosome represents the fourth chromosome, and the other longer chromosome represents the X chromosome, while the remaining four are the right and left arms of the V shaped second and third chromosomes. In salivary gland nuclei from female larvae, the strand representing the 'X' is double like the others, while in nuclei from male fruit fly it is single. Y chromosome is also found attached to the chromocentre in the form of a separate strand.

Puffs on Polytene Chromosomes and Gene Expression

In salivary gland cells, some 5000 bands and interbands can be detected in the four polytene chromosomes that make up the *Drosophila* genome. The different staining properties of the bands and interbands are highly suggestive of different states of compactness or organization of the chromatin in these regions. Thus, it was initially postulated that each band would represent a single gene that encodes a single protein. Indeed, early exhaustive genetic mapping of selected regions of several *Drosophila* chromosomes revealed that the number of complementation groups corresponds very closely with the number of visible bands. However, the average band contains some 30,000 base pairs of DNA, more than enough to encode a single polypeptide chain (even if the gene's transcript contains several non-coding sequences called **introns**).

A particular useful property of the giant chromosomes is that obvious changes in morphology can be seen when the genes in a particular band become activated. The DNA decondenses into a much more open state, forming a distinctive **puff**. Several geneticists have found that at certain stages of larval development, some specific chromosomes show some enlargement. These enlarged bands are called puffs that are scattered here and there along the length of the polytene chromosome. Thus, puffs are expanded regions of chromatin that actively participate in the process of transcription. The larger and more diffuse a puff appears, usually the higher the rate of its transcription. The coherence of chromosome filament is loosened at puffed regions. The loose ring always starts at a single band. In small puffs, a particular band simply loses its sharp counter and becomes diffuse. At other loci or at other times a band may look as though it had "exposed" into large ring loops (Figure 7.10) around the chromosomes. Such nut like structures are called Balbiani rings, after E.G.Balbiani, who first described them in 1881.

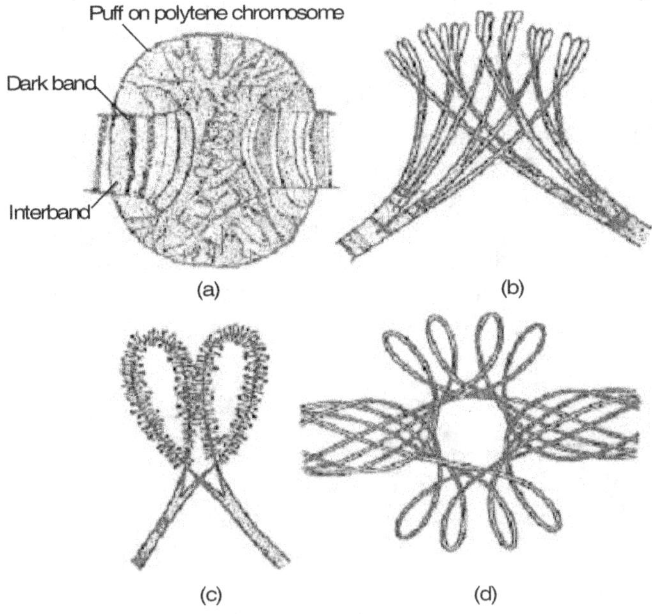

Figure 7.10 Balbiani rings in a polytene chromosome

As the organism progresses through development, puffs arise and recondense in a defined order as the synthesis of different mRNAs is turned on and off. Experimentally, puffing of different sets of bands can be induced by **heat shock** (the larva is heat-shocked at 37°C for 20 minutes prior to dissection) or by the addition of the insect hormone **ecdysone**, which stimulates genes to undergo transcription to synthesize proteins required for moulting and pupation. Electron microscopic examination of serial sections through large puffs can reveal the ribonucleoprotein (RNP) structure of the nascent RNA transcripts. Use of antibodies directed against various nuclear proteins can show that molecules such as RNA polymerases, hnRNP proteins, snRNPs, and topoisomerases become specifically concentrated in the puff regions.

Lampbrush Chromosomes

These are meiotic or "germ line" chromosomes having laterally projecting loops that are present during prophase first of oogenesis in many vertebrates, particularly amphibians. It was given this name because it is similar in appearance to the brush used to clean the chimney of the lamp in olden days. It was first discovered by Flemming (1882) in amphibian oocytes. They were studied in great detail by Ruckert (1892) who coined the term "lampbrush" for for the extra large chromosomes present in the oocytes of shark. In addition to fishes and amphibian oocytes, the lampbrush chromosomes are also found in other vertebrates like reptiles and birds as well as in some invertebrates, for example, *Sagitta, Sepia, Echinaster* and in several species of insects. Grun (1958) noted the presence of lampbrush chromosomes in certain plants.

Lampbrush chromosomes are up to $800\,\mu$m long, and are larger than salivary gland chromosomes; they thus provide very favourable material for cytological studies. During oogenesis, at the stage of female meiotic prophase, the homologous chromosomes are paired, and each has duplicated to produce two chromatids at the lampbrush stage. Each lampbrush chromosome contains a central axis region, where both chromatids are highly condensed with a series of lateral loops (Figure 7.11).

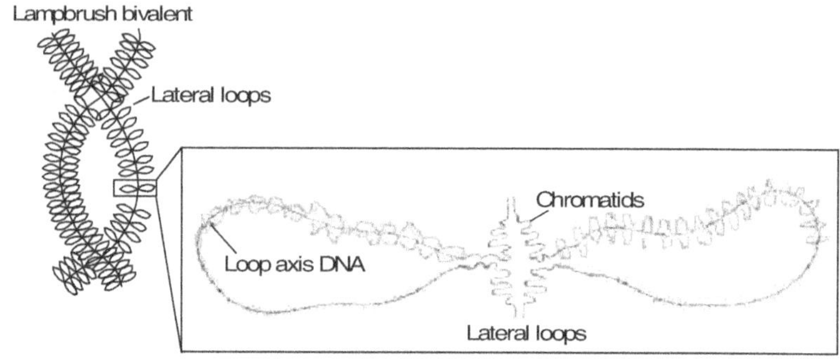

Figure 7.11 Schematic representation of giant lampbrush chromosome of amphibian oocyte

The chromosomal axis is a highly flexible structure whose diameter varies from 30–50 Å and it contains two bivalent chromosomes each with two chromatids, thus four chromatids in all are present; these chromatids further give rise to lateral loops. They are extensible and elastic, and as a chromosome is stretched, pairs of loops become widely spaced along the axis but the bases of the loops do not open out considerably. Biochemically, the chromosomal axis is made up of DNA and proteins. The lateral loops projecting from central axis are Feulgen-positive. Each lateral loop has an axial fibre covered with matrix. The axial fibre is composed of DNA while the matrix is made up of RNA and proteins (Figure 7.12).

The loops of lampbrush chromosomes are transcriptionally active regions of single chromatids. The integrity of both the central axis and the lateral loops is dependant on DNA. When these chromosomes are treated with DNase, both the axis and the loops are fragmented, while treatment with RNase or proteases removes surrounding matrix material but does not destroy the continuity of either the axis or the loops. Electron microscopy of RNase and protease-treated lampbrush chromosomes reveals a central filament of just over 20 Å in diameter in the lateral loops. Since each loop is a segment of one chromatid, and since the diameter of the DNA double helix is 20 Å, these lampbrush chromosomes must be unineme structures. It is also concluded that the axial region contains two DNA molecules, one from each of the two tightly paired chromatids.

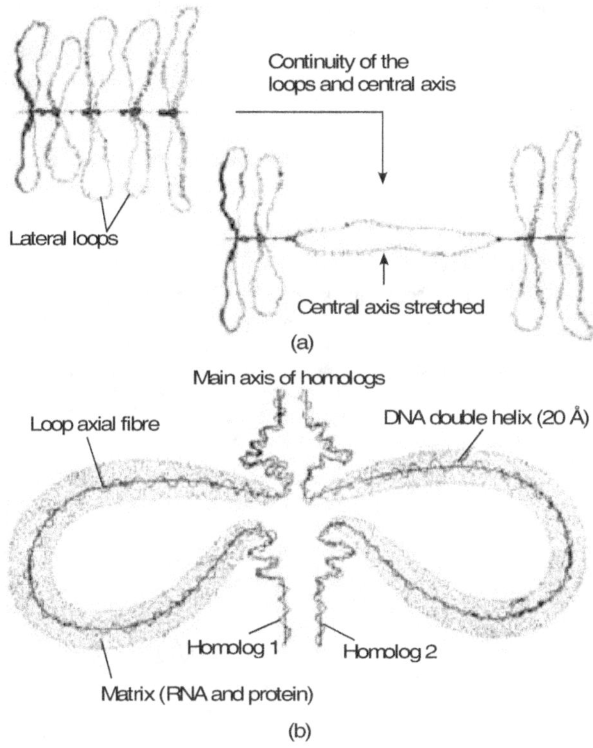

Figure 7.12 (a) The lampbrush chromosome stretched to show the continuity of the loops and central axis, and (b) Single pair of loops showing axial fibre and matrix.

Functionally, the lampbrush chromosomes are involved in the synthesis of RNA and proteins. Each loop is believed to be representing one long operon, which consists of repetitive cistrons (structural genes). Each gene locus codes for RNA. The presence of RNA and proteins in the loops, continuous production of "nucleoli", all suggest that they are the sites of intensive nucleoprotein metabolism. The loop is supposed to synthesize at high rate because of repetitive gene sequence. There are also reports that the lampbrush chromosomes help in the formation of yolk present in the egg. Thus they might play a role in the formation of deutoplasmic reserves.

SUMMARY

- Waldeyer in 1897 coined the term chromosomes for the material of the nucleus that became well organized into visible "threads" during cell division.

- Walter Fleming observed that each chromosome duplicates to form two identical chromosomes before cell division.

- Boveri provided the first evidence of a qualitative difference among chromosomes and Morgan discovered the role of chromosomes in the transmission of the hereditary characters.

- Each nucleus encloses a fixed number of chromosomes and for a given species, the number of chromosomes remains constant. The number of haploid chromosomes in most animals and plants lies between 6 and 25.

- During interphase, the condensed portion of the chromosomal material (heterochromatin—inactive region) can be distinguished from the less dense regions (euchromatin—active region).

- On the basis of the position of the centromere, four types of metaphase chromosomes are distinguished, viz. metacentric, submetacentric, acrocentric, and telocentric.

- Each metaphase chromosome consists of two chromatids, chromomeres, secondary constrictions I and II, centromere, and telomeres.

- Chromosomes have a unique staining pattern that allows their specific identification with stained and unstained areas in the form of bands. Different banding techniques involve C-banding, Q-banding, G-banding, T-banding and R-banding patterns.

- In 1956, two cytologists, Tjio and Levan, confirmed 46 as the 2n (diploid) chromosome number in human beings.

- A group of cytogeneticists adopted a system of classifying and identifying human chromosomes on the basis of the length of the chromosome and position of the centromere. The 22 pairs of autosomes and one pair of sex chromosomes have been divided into 7 groups identified with letters A to G.

- Lejeune reported partial monosomy in human beings in France in the patients with Cri-du-chat syndrome that is associated with the loss of the short arm of chromosome 5. Down discovered "Mongolism" in the patient due to the trisomy of autosome number 21. Klinefelter reported an abnormal male with extra X chromosome with genotype 47XXY. Turner reported an example of aneuploidy in humans who were phenotypically female and genetically 44A +X0, i.e., with only one X chromosome.

- Chromosomal aberrations or chromosomal mutations are the variations in structure and number of the chromosomes that produce alterations in morphological, behavioural and other characters, which may be lethal for the individual.

- Certain types of animal cells have giant chromosomes such as polytene chromosomes and lampbrush chromosomes.

- Larvae of dipteran insects like Drosophila, Chironomus, *housefly* and Anopheles, have specialized cells in the salivary gland that contain polytene chromosomes that are about 100 times thicker than the normal chromosomes.

- Polytene chromosomes have alternate bands, interbands and distinctive puffs or Balbiani rings, which are the sites of activated genes that undergo transcription. Puffs can be induced by heat or ecdysone.

- Lampbrush chromosomes are meiotic or "germ line" chromosomes having laterally projecting loops and are present during prophase first of oogenesis in amphibians and other vertebrates.

- Each lampbrush chromosome contains a central axis region, where both chromatids are highly condensed with a series of lateral loops that synthesize RNA and proteins.

REVIEW QUESTIONS

1. Explain the contribution of different cytologists in the discovery of chromosomes.

2. Describe the morphological features of a metaphase chromosome.

3. What is chromosomal banding technique? Enumerate various banding patterns.

4. Explain the karyotype of normal human beings with the complete classification key.

5. Define chromosomal aberrations. Describe various syndromes associated with structural and numerical alterations in chromosomes of human beings.

6. What are giant chromosomes? Explain its types with morphological features and significance.

7.   Write short notes on:

    i.   Chromocentre

    ii.   Satellite chromosome

    iii.   Chromosomal deficiency

    iv.   Down's syndrome

    v.   Endomitosis

    vi.   Germ line chromosomes

    vii.   Puff induction

    viii.   Polyspermy

DNA—CHEMICAL NATURE, STRUCTURE AND REPLICATION

DNA—HISTORICAL DEVELOPMENT

By the early twentieth century, geneticists had associated the presence of genes with chromosomes. Research began to focus on exactly which chemical component of the chromosome comprised this genetic material. The sophisticated biochemical procedures have provided a rather extensive picture of the molecules that make up the chromosome. In general, all chromosomes contain four major classes of molecules: DNA, RNA, histone proteins, and non-histone proteins. In non-dividing eukaryotic cells (in G_0) and those in interphase (G_1, S, and G_2), the chromosomal material or chromatin appears randomly dispersed showing "beads-on-a-string" configuration when observed under the electron microscope (as explained in chapter 2). Chromatin consists of fibres containing DNA and protein in approximately equal masses, along with a small amount of RNA.

As early as 1871, Miescher, who was working with pus cells, spermatozoa, and nucleated red blood cells of chickens, isolated nuclei and demonstrated that the major component was a material he called "nuclein". Further analysis and determination of the major component of nuclear material, specifically chromosomes, was a particular type of nucleic acid, DNA. With the help of a very specific staining procedure called Feulgen method, the qualitative and quantitative differences in DNA was determined. In 1942, using cytochemical procedures, Brachet demonstrated the presence of another nucleic acid, RNA. In 1944, Avery *et al.*, first implicated DNA as the genetic material. Mirsky and Pollister (1946) showed that there were proteins associated with chromosomal material. On the basis of Chargaff's chemical data, Wilkins and Franklin's X-ray diffraction data, Watson and Crick proposed the DNA model in 1953.

CHEMICAL STRUCTURE OF DNA

It is the deoxyribonucleic acid (DNA) that carries genetic information of all living forms except RNA viruses. The basic building block of this macromolecule is nucleotide. Each nucleotide is composed of 1) a five-carbon (pentose) sugar, deoxyribose, 2) a phosphoric acid (biologically a phosphate), and 3) a cyclic nitrogen-containing compound called a base, either a pyrimidine or purine, as shown in Figure 8.1.

The pentose sugar in DNA lacks the oxygen on the carbon 2′(pronounced as two prime) position, hence called 2-deoxy-D-ribose or simply deoxyribose. The four nitrogen bases found in DNA are adenine (A), guanine (G), thymine (T), and cytosine (C). The A and G are purines and T and C are pyrimidines. The nucleoside in DNA is called deoxynucleoside that is complex of a deoxyribose sugar and a nitrogen base, where the purines with their N-9 position always attach to carbon atom 1′ of deoxyribose sugar and pyrimidines with their N-1 attached to C-1′ of deoxyribose sugar.

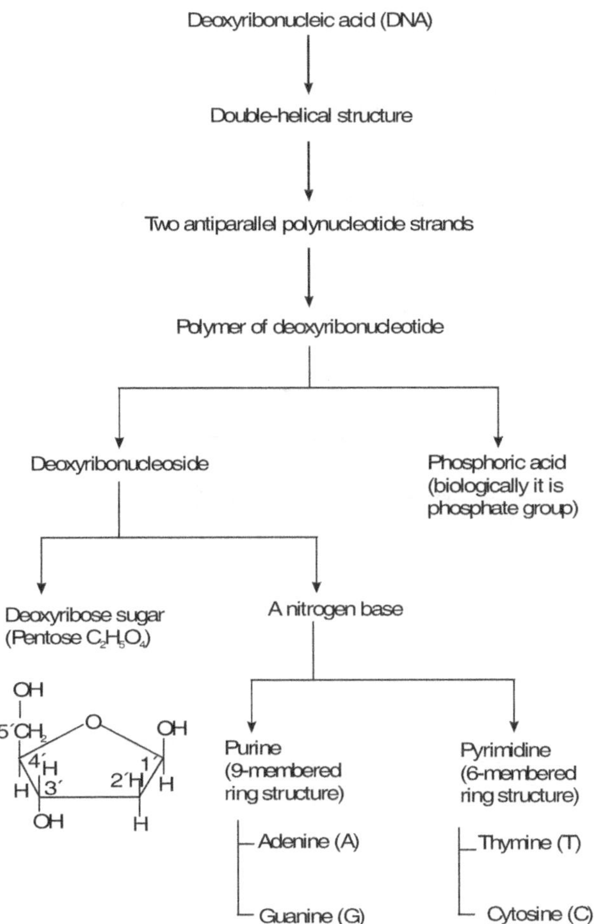

Figure 8.1   The building blocks of deoxyribonucleic acid (DNA)

On the other hand, the deoxyribonucleotide consists of a deoxyribose sugar, a nitrogen base, and a phosphate group, where a phosphoric acid or phosphate group is attached to carbon 5′ of the deoxyribose sugar that is already joined with any one of nitrogen bases. Four types of nitrogen bases, deoxynucleosides, and deoxyribonucleotides with their terminologies and abbreviations are mentioned in Figure 8.2. Each deoxyribonucleotide is a polar structure, where the edge where the phosphate is located is called 5' end, while the other edge is called 3′ end.

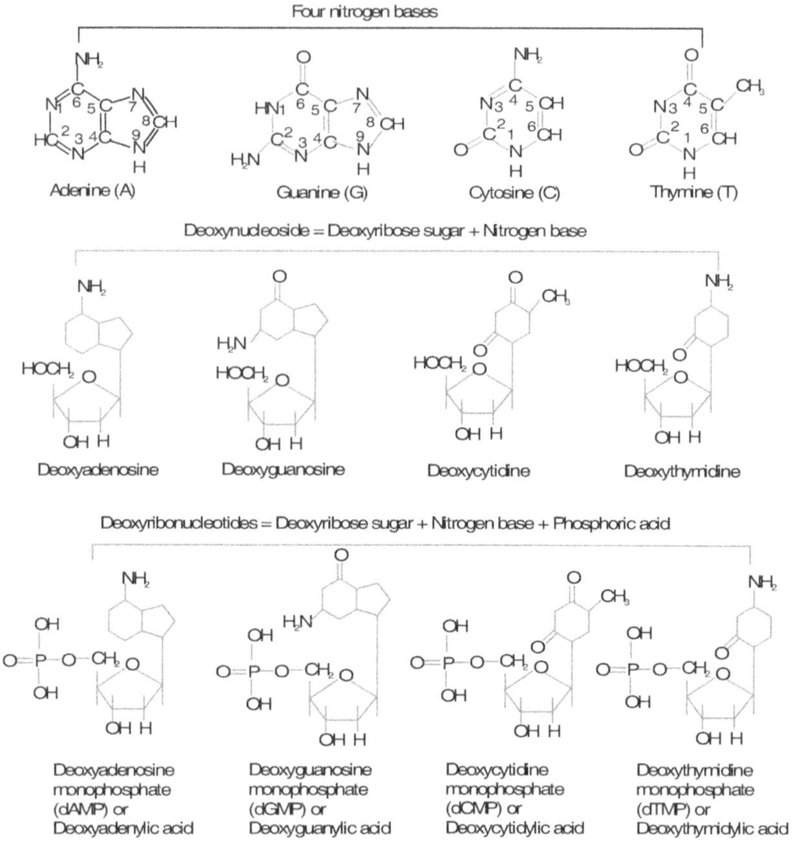

Figure 8.2 Chemical structures of nitrogen bases, deoxyribonucleosides, and deoxyribonucleotides of DNA

The groups of phosphoric acid –OH are both ionized at physiological pH since one of the –OH groups has a pKa of around 2 and a second of around 7. This means that they are highly hydrophilic. But the nitrogen bases are almost water-insoluble and DNA is strongly negatively charged.

The deoxyribonucleotides were known to be covalently linked to one another to form a linear polymer or strand, with a backbone composed of alternating sugar and phosphate groups joined by 3′, 5′-phosphodiester bonds (Figure 8.3). The nitrogen bases were thought to project from the backbone like a column of stacked shelves. Since each of the stacked nucleotides in a strand has polarity, the same direction is attributed to the entire strand. One end is 3′ end, and the other is the 5′ end.

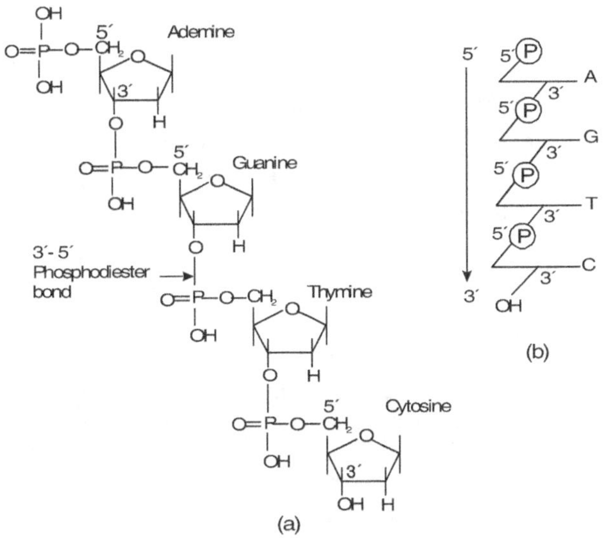

Figure 8.3 (a) Extended structure of a polynucleotide strand of DNA, and (b) Shorthand method for representation of a polynucleotide strand showing polarity

CHARGAFF'S RULE FOR BASE COMPOSITION

As per earlier concept, DNA was considered as a simple repeating tetranucleotide (e.g. —ATGCATGCATGC—). In 1950, Erwin Chargaff discovered the equivalence rule that suggested that despite wide compositional variation exhibited by different types of DNA, the total number of purines is equal to the total amount of pyrimidines (A+G = T+C); the amount of adenine equalled the amount of thymine (A=T) and the amount of guanine equalled the amount of cytosine (G=C). Chargaff analysed the **base composition** that was accomplished by hydrolysing the bases from their attached sugars, separating the bases in hydrolysate by paper chromatography. His equivalence rule has been found to apply to almost universally all organisms.

However, he found that the ratios of the four bases were quite variable from one type of organism to the other. For example, DNA isolated from higher plants and animals was rich in adenine and thymine (A : T) and relatively poor in guanine and cytosine (G : C). In the case of humans, AT/GC ratio was 1.40 : 1, indicating higher percentage of A and T, whereas DNA isolated from microorganisms (viruses, bacteria, and lower animals and plants) was in general rich in guanine and cytosine and relatively poor in adenine and thymine. For example AT/GC ratio of DNA of *Mycobacterium tuberculosis* was 0.60 : 1. Based on these observations, Chargaff discovered the following base composition rules in DNA:

(A) = (T), (G) = (C), and (A) + (T) $\neq$ (G) + (C)

These differences in base compositions give the specificity and individuality of one organism from another. It reflects the phylogenic, evolutionary and taxonomical relationship between different organisms.

X-RAY CRYSTALLOGRAPHY OF DNA

The structure of DNA was deciphered only after many types of experimental evidence and theoretical considerations were combined. The crucial evidence was obtained by X-ray crystallography, the process that requires tremendous skill. Some chemical substances, when they are isolated and purified, can be made to form crystals. The positions of the atoms in a crystalline substance can be inferred from the pattern of diffraction of X-rays passed through it (Figure 8.4). The English chemist Rosalind Franklin and biophysicist Maurice Wilkins succeeded in obtaining the X-ray diffraction photographs of samples containing very uniformly oriented DNA fibres. Such studies demonstrated that DNA was a helical structure with a diameter of 20 Å and one complete turn of helix is about 34 Å long. These findings provided significant information for Watson and Crick to propose the model of DNA.

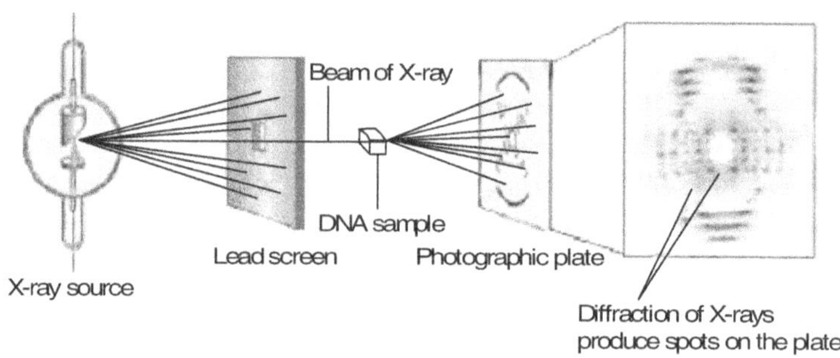

Figure 8.4   X-ray crystallography technique revealing the helical nature of DNA

WATSON AND CRICK PROPOSAL FOR DNA MODEL

Taking into consideration Chargaff's chemical data, Wilkins and Franklin's X-ray diffraction data, and inference drawn from model building, in 1953, J.D.Watson and F.H.C.Crick proposed that DNA exists as a double helix in which the two polynucleotide strands are coiled about one another in a spiral (Figure 8.5a) with a uniform diameter of 20 Å. Each polynucleotide chain consists of a sequence of deoxyribonucleotides linked together by 3′–5′ phosphodiester bonds. Both the polynucelotide strands are held together in their helical configuration by hydrogen bonding between bases in opposite strands, the resulting base-pairing being stacked between the two chains perpendicular to the axis of the molecule like the steps in a spiral staircase. The double-helical structure of DNA is right-handed (that is, it twists to right, as do the threads on the most screws).

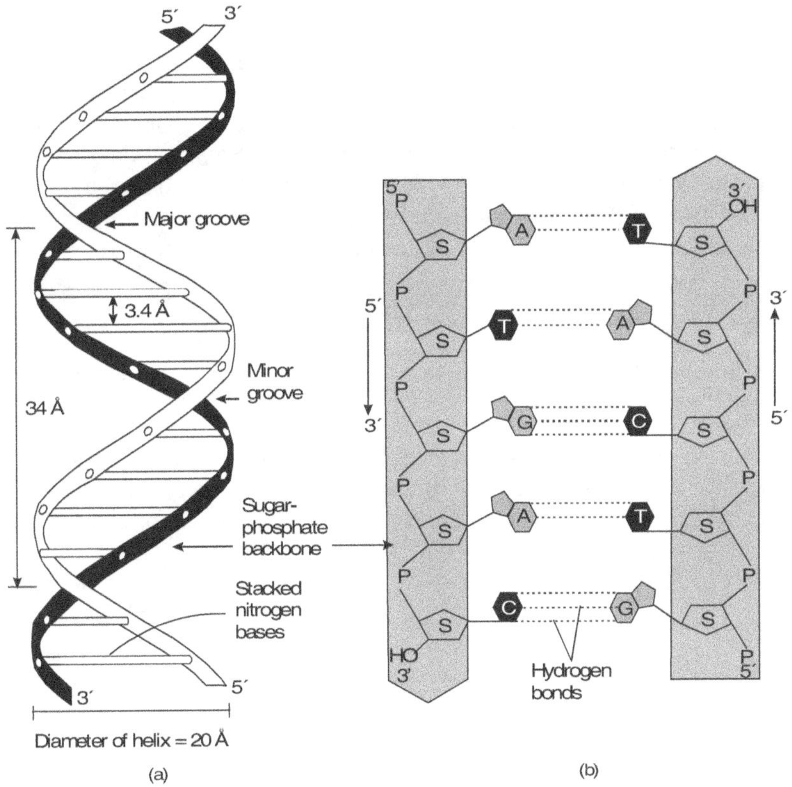

Figure 8.5   (a) Watson–Crick double-helical model of the DNA molecule, and (b) Molecular structure of DNA showing sugar–phosphate backbones of polynucleotide strands with their antiparallel nature.

The base-pairing is specific; adenine is always paired with thymine and guanine is always paired with cytosine. Thus, all base pairs consist of one purine and one pyrimidine. In their most common structural configurations, adenine and thymine form two hydrogen bonds, and guanine and cytosine form three hydrogen bonds. Since both the strands of DNA are complementary to each other, if the sequence of bases in one strand of a DNA double helix is known, the sequence of bases in the other strand is also known because of the specific base-pairing. This property makes the DNA uniquely suited to store and transmit genetic information. The base pairs in DNA are stacked 3.4 Å apart with 10 base pairs per turn (each base pair rotates 36° with respect to the adjacent pair) of the double helix. Thus, one complete turn of the DNA helix has the length of 34 Å. The DNA double helix shows two external grooves, a deep wide **major groove** and a shallow narrow **minor groove**. Proteins that bind to the DNA often fit into these grooves.

The sugar–phosphate backbone of two polynucleotide strands run antiparallel to each other, i.e., they have opposite chemical polarity (Figure 8.5b). As one moves unidirectionally along a DNA double helix, the phosphodiester bonds in one strand go from a 3′ carbon of one nucleotide to a 5′ carbon of the adjacent nucleotide, whereas those in the complementary strand go from a 5′ carbon to a 3′ carbon. This opposite polarity of the complementary strands is very significant in considering the mechanism of replication of DNA.

ALTERNATE FORMS OF DNA HELIX

The structure of DNA is not monotonous. The DNA model proposed by Watson and Crick was based on the X-ray diffraction data in which the fibres of DNA had been prepared when "wet" or fully hydrated. It is thought to represent the majority of DNA molecules present in living cells with the **B form** or **B-DNA** (right-handed double helix). Further experiments revealed that DNA is a much more polymorphic (indicating the presence of alternate forms of DNA) than expected.

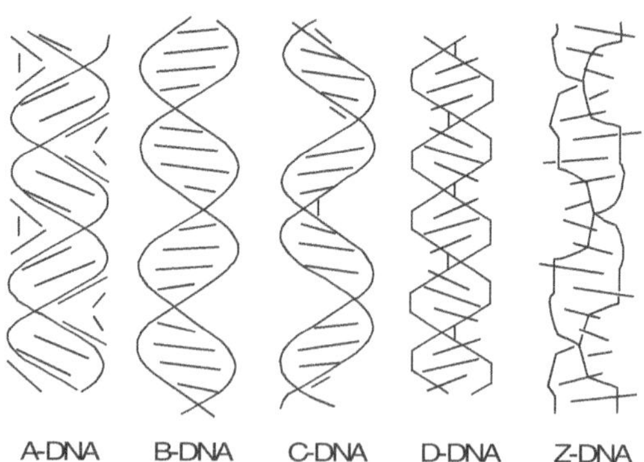

Figure 8.6   Five different molecular forms of DNA double helix. The A-, B-, C- and D-DNA are alternate DNA conformations for right-handed DNA helix and Z-DNA is polymorphic form for left-handed DNA.

When the DNA fibres were prepared under lower humidity conditions and processed for X-ray crystallography, they had a more crystalline structure and generated more sharply defined X-ray diffracted patterns. The molecular arrangement of DNA in the dried fibres is somewhat different from the hydrated DNA fibres and it is referred to as the **A form** of DNA. In 1978, Alexander Rich studied the X-ray diffraction pattern of G–C–rich DNA and found that the strands of this synthetic molecule twisted in a counterclockwise spiral indicating its left-handed nature. Because of the zig-zag conformation of the sugar–phosphate backbone, this new DNA structure is referred to as Z-DNA. The comparative account of A, B, and Z-DNAs are mentioned in Table 8.1. In addition to the above-mentioned types, C-DNA and D-DNA are also alternate forms of right-handed DNA helix (Figure 8.6).

Table 8.1   A comparison of some morphological parameters of three major alternate forms of DNA double helix

Parameter	A-DNA	B-DNA	Z-DNA
Overall proportion	Short and broad	Longer and thin	Elongated and slim
Helix sense	Right-handed	Right-handed	Left-handed
Residue per turn	11	10.5	22.6
Distance between adjacent bases	2.4 Å	3.4 Å	3.7 Å
Rotation per residue	$32.7°$	$36°$	$-9°, -15°$
Diameter of helix	25.5 Å	23.7 Å	18.4 Å
Major groove	Narrow, deep	Wide, deep	Flattened
Minor groove	Wide, shallow	Narrow, deep	Narrow, deep

SUPERCOILED DNA IN PROKARYOTES

In 1963, Jerome Vinograd and co-workers discovered in the prokaryotic cell that two closed, circular DNA molecules of identical molecular weight could exhibit very different rates of sedimentation when centrifuged through a density gradient. The folded DNA becomes more compact than the relaxed counterpart and hence occupies less volume and moves rapidly in response to a centrifugal force.

The chromosome in prokaryotes is actually "a naked DNA double helix". The length of the circular DNA molecule of *E. coli* is about $1100\,\mu$m while the entire cell has a diameter of only 1–$2\,\mu$m. It clearly indicates that the chromosome must exist in a highly folded or coiled configuration within the cell. When the chromosomes of *E. coli* are isolated by very gentle procedure in the absence of ionic detergents (commonly used to break cells) and is kept in the presence of 1M salt to neutralize the negative charged phosphate groups of DNA, the chromosomes remain in a highly condensed state and are referred to as "folded genome" that is apparently the functional state of the *E. coli*. In its folded genome state, the single DNA molecule of *E.coli* is arranged into about 50 loops or domains. DNA in this state is said to be **supercoiled** (Figure 8.7), much like a rubber band in which two ends are twisted in opposite directions or like a tightly coiled telephone cord. The folded chromosome can be relaxed by treatment with either DNase or RNase enzymes that digest DNA or RNA respectively. DNA is referred to as **negatively supercoiled** when supercoiling results from being underwound (DNA with greater number of base pairs per turn of helix), and positively supercoiled when overwound. Many biologically functional chromosomes (e.g. mitochondrial, viral, bacterial) are invariably negatively supercoiled.

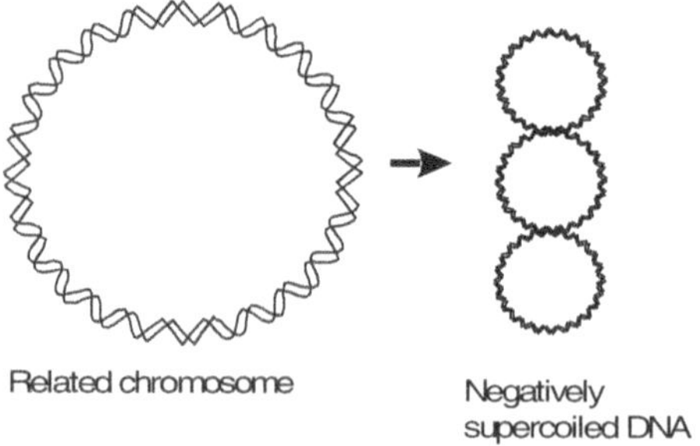

Figure 8.7   Diagrammatic representation of relaxed chromosome and supercoiled DNA of *E. coli*

Supercoiling is not only restricted to small, circular prokaryotic genomes but it also occurs in linear, eukaryotic DNA where stretches of the molecules are wrapped around histone octamers to form nucleosomes. Supercoiling plays a significant role in allowing the DNA of chromosomes to be compacted so as to fit inside the microscopic nucleus of the cell. Negatively supercoiled DNA also helps in separation of strands during replication of DNA and in the process of transcription to synthesize RNA. In prokaryotic and eukaryotic cells, there are enzymes called **topoisomerases** that are able to change the supercoiled state of a DNA duplex. Topoisomerases are divided into two classes according to the mechanism of action, viz. type I topoisomerases act by creating a nick in one strand of the DNA duplex, allowing that strand to rotate around the other strand and then resealing the strand. Type II topoisomerases (include DNA gyrase) introduce a double-strand break in DNA, then pass a DNA segment through the break, and reseal the molecule. Topoisomerases are essential in DNA replication and transcription. They also play an indispensable role in the separation of duplicated pairs of chromosomes during mitosis.

DENATURATION AND RENATURATION OF GENOME

DNA plays the key role in all the biosynthetic and hereditary functions of living organisms. It carries genetic information from one generation to the next. Gene is the unit of heredity that corresponds to a particular segment of DNA. The sum of the genetic information that the offsprings inherit from their parents is equivalent to the sum of the entire DNA segments present in the fertilized egg at the start of life. All the individuals of a species population contribute the same set of gene. Thus, each species of organism has a unique complement of genetic information that is referred to as its **genome**. In the case of humans, the genome is defined as all of the genetic information that is present in a single haploid set of 23 chromosomes (22 autosomes + X in ovum and 22 autosomes + Y in sperm).

The complexity of the genome can be understood by the ability to denature and renature itself. One of the first properties of DNA to be investigated was the denaturation of the double helix by heat. When the temperature of a solution of DNA is slowly raised and at a specific temperature (referred to as melting temperature = T_m), separation of strands begins. This is known as **thermal denaturation** or **DNA melting**.

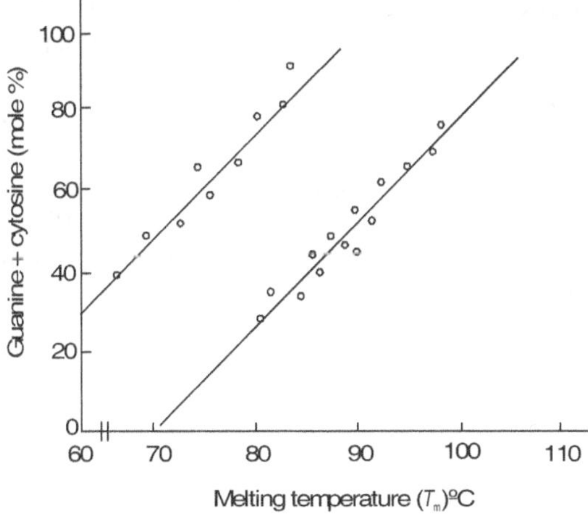

Figure 8.8   Dependence of thermal denaturation of DNA on its GC content. Higher the GC content, higher was the T_m. DNA from different sources was dissolved in two different solutions at pH 7.0.

When double-helical DNA molecules are heated above physiological temperatures (to near 100°C), their hydrogen bonds break and complementary strands often separate from each other, a process called DNA denaturation. DNA molecules with high GC content (%G + %C) are more resistant to thermal melting than AT-rich molecules (Figure 8.8). This is because each GC base pair is held together by the three hydrogen bonds rather than the two hydrogen bonds that hold each AT base pair, higher temperatures are necessary to separate GC-rich strand than to break apart AT-rich molecules.

When heated solutions of totally denatured DNA are slowly cooled, single strands often meet their complementary strands and reform regular double helices. The ability of DNA to reassociate into double helix when the temperature of the solution rapidly drops approximately 25°C below the T_m, is called **renaturation** or **reannealing**. This property of the genome helped the molecular biologists to develop a methodology called nucleic acid hybridization in which artificial hybrid DNA molecules can be formed by slowly cooling mixtures of denatured DNA from two different species. For example, hybrid DNA molecule can be formed containing one strand from a man and one from a mouse.

DNA REPLICATION

Every living organism has the ability to duplicate itself either by asexual or sexual reproduction, every living cell duplicates itself by cell division, similarly the genetic material duplicates by replication. As a carrier of genetic information, DNA performs heterocatalytic or autocatalytic functions. In a heterocatalytic role, DNA directs the synthesis of chemical molecules other than itself, e.g. the synthesis of RNA, proteins, etc., whereas as an autocatalytic activity, DNA directs the synthesis of DNA itself. The formulation of the structure of DNA by Watson and Crick in 1953 was accompanied by a proposal for its "self-duplication". Each strand of DNA double helix can serve as **template** for the synthesis of a new strand to form duplicated DNA. Watson and Crick proposed that replication of DNA involves breakdown of weak hydrogen bonds that hold the duplex together, and the process is followed by a rotation and separation of both the polynucleotide strands (Figure 8.9), much like the separation of the two halves of a zipper. Because of the complementary nature of both strands of DNA, each separated strand acts as a template as it contains the information required for the synthesis of the other strand. Each base of template strand attracts a complementary nucleotide available within the cell in the presence of a polymerizing enzyme called DNA polymerase that holds the nucleotide in a position to continue the synthesis of the new strand. Finally, replicas (genetically identical with each other) of double helical molecules are formed.

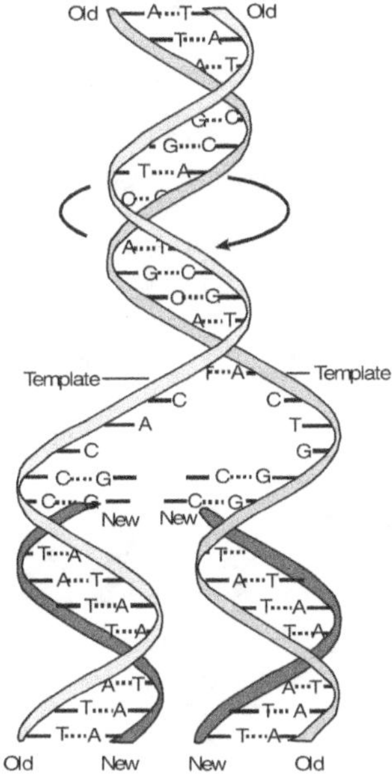

Figure 8.9   Watson and Crick proposal for replication of DNA double helix

Semiconservative DNA Replication

According to the proposal made by Watson and Crick for DNA duplication, once the DNA replication is initiated, both the old strands of the duplex serve as templates that direct the synthesis of new complementary strand. As a result, each daughter duplex should be hybrid in nature since it consists of one complete strand from parental duplex and one complete strand that has been newly synthesized. Thus, each daughter DNA duplex retains half of the parental DNA and the replication of this type is said to be **semiconservative** (Figure 8.10a). When these two hybrid duplexes replicate themselves, there would be formation of four duplexes, two of which would contain a single strand derived from the original chromosome and two of which contain totally new DNA strands. In addition to semiconservative type of replication of DNA, the double helical DNA molecule can replicate by conservative or dispersive methods. In **conservative** replication, both strands of parent DNA would be conserved and the new DNA molecule would consist of two newly synthesized strands (Figure 8.10b). When two of such DNA molecules replicate themselves to form the second generation, four daughter DNA molecules would result, one of which would contain only the fully conserved duplex, while the other daughter DNA molecules would contain only newly synthesized DNA. The third of its kind, that is **dispersive** replication (Figure 8.10c), involves fragmentation of parent

DNA molecule and intermixing of pieces of parent strands with newly synthesized pieces, thereby each strand of daughter DNA molecules would contain a mixture of old and new DNA; that is neither the strands nor the duplex is conserved.

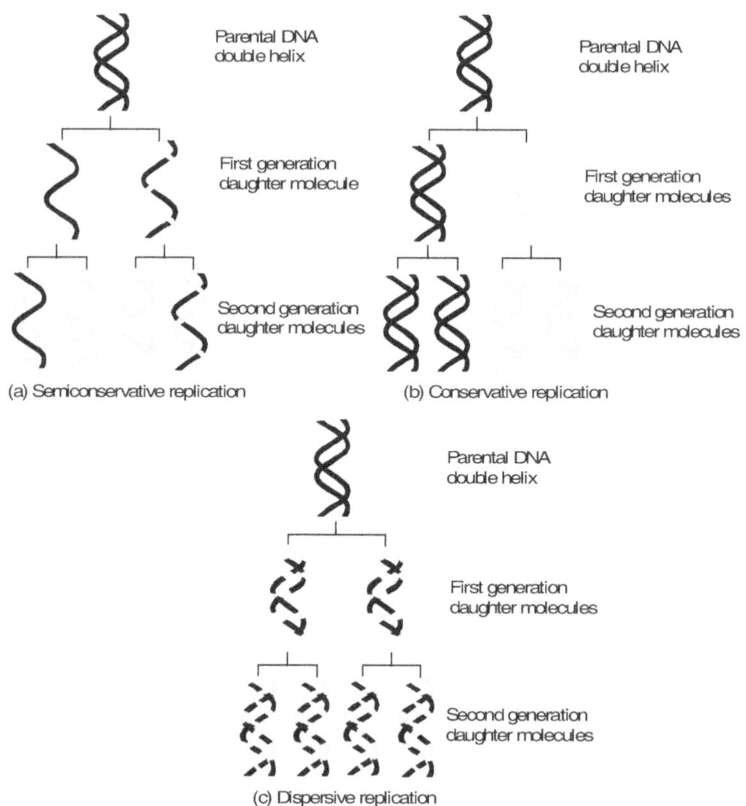

Figure 8.10   Three alternate methods of DNA replication

Meselson and Stahl's Experiment

In 1958, Matthew Meselson and Franklin Stahl of California Institute of Technology confirmed the semiconservative nature of DNA replication in bacteria using radioisotope through a series of experiments. They grew *E. coli* cells for many generations in media containing ^{15}N-ammonium chloride (^{15}N is a "heavy" isotope of nitrogen but not a radioisotope) as the sole nitrogen source. As a result, the nitrogen bases of the DNA of these bacteria contained only the heavy nitrogen isotope and when such bacteria were isolated from the medium, their DNA was extracted and subjected to density gradient centrifugation, only denser DNA with heavy nitrogen can be visualized in the centrifuge tube (Figure 8.11).

Depending on the nature of the nitrogen source (either ^{15}N or ^{14}N) utilized by *E. coli* to synthesize its genetic material, the heavier DNA containing ^{15}N, being denser, settle further down the tube than the lighter DNA containing ^{14}N.

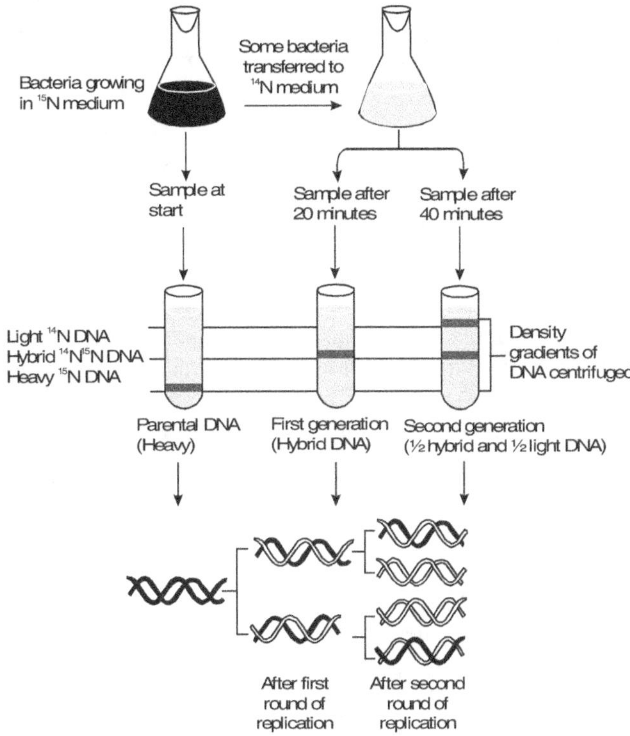

Figure 8.11 Meselson–Stahl's experiments demonstrating that DNA replication in bacteria is semiconservative

Bacteria cultured in a medium containing ^{15}N were washed free of the medium and transferred to ^{14}N-containing medium and incubated to grow for a specific length of time. Samples were removed and processed at increasing intervals over a period of several generations. The position of a band of particular DNA is directly related to the percentage of ^{15}N atoms in DNA as opposed to ^{14}N atoms. Since the process of replication is semiconservative, the density of DNA gradually decreased up to the point where one generation time had reached. After one generation, all the DNA molecules were ^{15}N–^{14}N hybrids, and their density were found halfway between that expected for totally heavy and totally light DNA. As replication continues after the first generation, the newly synthesized DNA continues to contain only ^{14}N atoms, and two types of DNA duplexes, those containing ^{15}N–^{14}N hybrids and those containing only ^{14}N appear in the gradient.

As the time of bacterial growth in the ¹⁴N-containing medium continues, a greater and greater percentage of lighter DNA molecules increased and the condition was reflected by increased thickness in the band of lighter DNA molecules in the centrifuge tube. At the same time, the original heavy parental strands remained intact and present in hybrid DNA molecules that occupied a smaller and smaller percentage of the total DNA present in the daughter molecules. This experiment clearly demonstrated that one DNA strand served as a template for the other and one of the parent strands retained in the daughter DNA molecules confirming the semiconservative type of replication.

REPLICATION OF DNA IN BACTERIA

As described earlier, the bacterial chromosome is a circular double-stranded DNA molecule. Its replication is also a semiconservative type where two original strands, called parental strands, are separated and each acts as a template for synthesizing a new strand; each new double helix has one old and one new strand. Before cell division occurs in *E. coli,* there must be a complete duplication of chromosome, and cell division follows about 20 minutes later.

In 1963, an autoradiographic technique was developed by John Cairns to provide striking visual evidences of the process of bacterial replication. He confirmed the circular nature of the bacterial chromosome and revealed the overall pattern of its replication. In many cases, circular chromosomes were caught in the act of replication that gave rise to what are called theta configurations because they have appearance of the Greek letter **theta** (θ) as shown in the Figure 8.12.

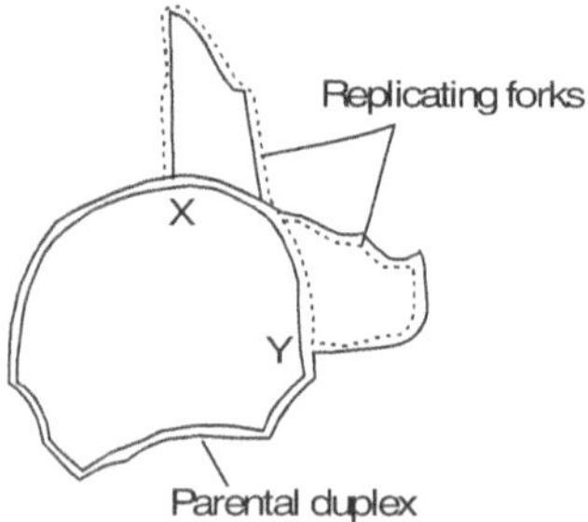

Figure 8.12 Visualization of theta (θ) configuration of replicating *E. coli* chromosome obtained by Cairns. X and Y are replicating forks.

Each theta structure is composed of three distinct lengths of DNA, one of which represents the unreplicated portion of the chromosome and the other two the pair of daughter molecules in the process of formation. The points at which the pair of replicated segments come together and join the non-replicated segments are called the replicating fork. Replication of bacterial DNA begins at a specific site on the bacterial chromosome called the **origin**. The origin of replication on the *E. coli* chromosome is called *oriC* which consists of approximately 245 specific base sequences very rich in A–T pairs, presumably to facilitate separation of strands.

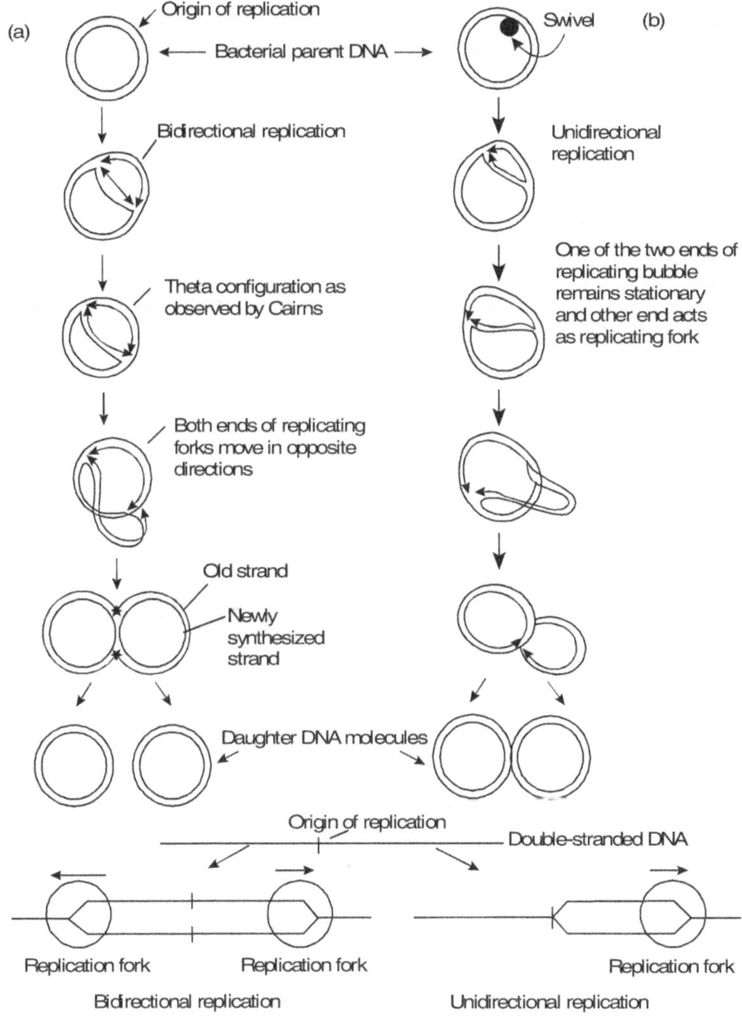

Figure 8.13   (a) Bidirectional replication of *E. coli* DNA duplex showing movement of two replication forks in opposite direction as compared with (b) Unidirectional method of replication, where only one replication fork moves from the point of origin.

At the time of initiation of replication, a protein called DNA-A binds in multiple copies to this region and causes separation of strands. This permits the main unwinding enzyme (**helicase**) that works at each replicating fork, to attach and begin progressive unwinding of strands in both directions, that is, **bidirectionally** (Figure 8.13a). The two replicating forks travel in opposite directions until they meet at a point across the circle from the origin, where replication is terminated. The two newly replicated duplexes detach from one another and are ultimately directed into two different cells.

Bidirectional replication from a fixed origin has also been demonstrated in several organisms with chromosomes that replicate as linear structures. Replication of DNA molecule in chromosomes of eukaryotes is also bidirectional. However, bidirectional replication is not universal. The chromosome of coliphage P2, which like the lambda chromosome is circular during replication, replicates unidirectionally from a unique origin, where one of the two branched points (either X or Y as shown in Figure 8.12 and 8.13b) was believed to be a replicating fork and other a terminus containing a "swivel" that served as an axis of rotation for unwinding the double helix.

DNA Polymerases in Prokaryotes

DNA polymerases are the enzymes responsible for the synthesis of new DNA strands during replication of DNA duplex. In 1957, Arthur Kornberg and his co-workers isolated an enzyme from bacterial extracts that had the ability to add nucleotides to the pre-existing DNA chain. The enzyme was called **DNA polymerase** or **Kornberg enzyme** but after the discovery of additional DNA polymerizing enzymes, it was named **DNA polymerase I**. For the reaction to proceed, the enzyme required the presence of DNA and all four deoxyribonucleoside triphosphates (dATP, dTTP, dCTP, and dGTP). The enzyme is active only in the presence of Mg^{2+} ions and pre-existing DNA. This DNA must provide two essential components, one serving a **primer** function and the other a **template** function. DNA polymerase is only capable of adding nucleotides to the 3′ hydroxyl terminus of the existing strand (Figure 8.14). The DNA strand that provides the enzyme with necessary 3′ OH terminus is called primer. The α-phosphate of a nucleoside triphosphate molecule forms a 3′, 5′ phosphodiester bond with a free 3′ OH in the growing polynucleotide chain, and a molecule of pyrophosphate (P–P) is simultaneously released. All biologically important DNA polymerases have critical properties in that they are only capable of synthesizing DNA in the 5′-to-3′ (written as 5′ → 3′) direction.

In 1969, further studies revealed that in addition to DNA polymerase I, there are other two distinct DNA polymerases present in bacterial cells and these are called **DNA polymerase II** and **DNA polymerase III**. A typical bacterial cell contains about 300 to 400 molecules of DNA polymerase I per cell,

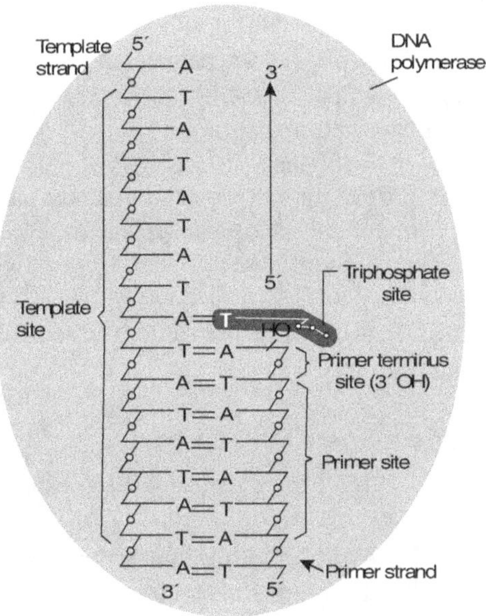

Figure 8.14 Role of DNA polymerase in the synthesis of new strand of DNA by interacting with template strand, primer, and incoming nucleoside triphosphate

only about 40 copies of DNA polymerase II, and 10 copies of DNA polymerase III. Some of the other differences in all three polymerizing enzymes are given in Table 8.2.

Table 8.2 Some of the differences in the properties of prokaryotic DNA polymerases

Parameter	DNA polymerase I	DNA polymerase II	DNA polymerase III
Chemical nature	Polypeptide	Polypeptide	Polypeptide
Number of copies per cell	Approx. 300–400	About 40	10 copies
Molecular weight	109,000	120,000	>250,000
Function	Principal repair polymerase, primer excision	Error-prone repair polymerase	Principle replication polymerase

DNA polymerase II has similar activity to that of DNA polymerase I, in addition it is repair enzyme and brings about chain elongation in the $5' \rightarrow 3'$ direction, using 3'OH groups. **DNA polymerase** III is a multimeric complex enzyme that plays a significant role in DNA replication. Its holoenzyme form contains ten subunits, viz. alpha (α), beta (β), epsilon (ε), theta (θ), tau (τ), gamma (γ), delta (δ), delta dash (δ'), chi (χ), and psi (φ). All subunits have different functions. For example, alpha subunit has to perform editing or proof-reading activity, beta subunit is a repair polymerase, gamma is an organelle polymerase, and delta subunit is the principal replicative polymerase. All the ten subunits are essential for DNA replication *in vitro*. The core enzyme contains three subunits α, β, and θ, while the rest of the seven subunits increase the speed and efficiency of the DNA polymerase.

Semidiscontinuous DNA Synthesis

None of the three prokaryotic DNA polymerases are able to construct daughter DNA strands in the $3' \rightarrow 5'$ direction. The opposing chain directions ($5' \rightarrow 3'$ and $3' \rightarrow 5'$) of the parental duplex mean that the two daughter strands being synthesized at each replicating fork must also run in opposite directions. Therefore, the overall direction must be $5' \rightarrow 3'$ for one daughter strand and $3' \rightarrow 5'$ for the other daughter strand. As mentioned earlier, all known forms of DNA polymerase can add nucleotide precursors to DNA only in the $5' \rightarrow 3'$ direction, since the chemical reaction catalysed by the three enzymes allows a nucleotide triphosphate to react only with the free 3'OH end of the growing polynucleotide strand. Consequently, one of the newly synthesized strands grows towards the replication fork where the DNA strands are being separated, while the other strand grows away from the fork (Figure 8.15). Apparently, the strand that assembled in a continuous fashion is known as the **leading strand**. In contrast, the strand that grows away from the replication fork is synthesized discontinuously in the form of small pieces of DNA (approximately 1000–2000 nucleotides long in *E. coli* and about 100–200 nucleotides long in eukaryotic cells). These small pieces or segments of DNA are often called "Okazaki fragments" after R.Okazaki of Nagoya University, Japan, who first identified them. The strand being extended in overall $3' \rightarrow 5'$ direction grows by the synthesis of Okazaki fragments (synthesized $5' \rightarrow 3'$) and is referred to as **lagging strand** because the initiation of each fragment must wait for parental strands to separate and expose additional template. Subsequently, the Okazaki fragments are joined to form a continuous strand by the enzyme called DNA ligase. Because one strand is synthesized continuously and other discontinuously, replication is said to be semidiscontinuous.

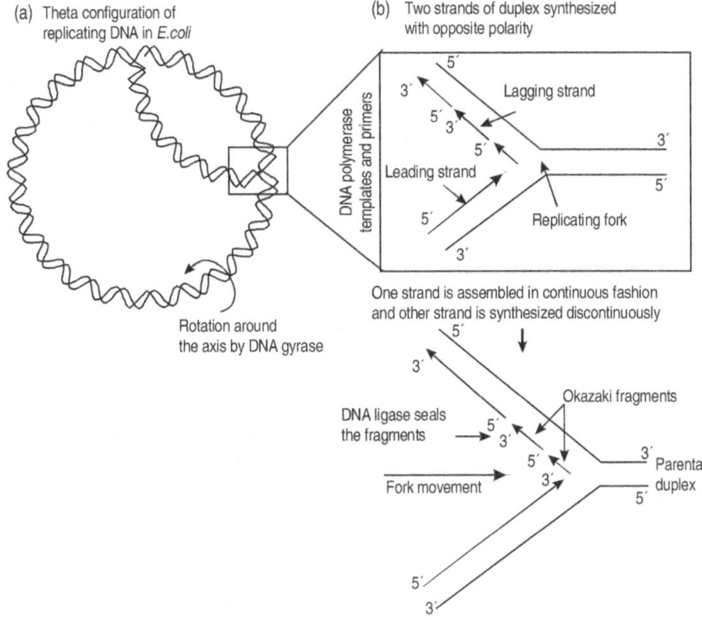

Figure 8.15 (a) DNA replication at macromolecular level, and (b) Semidiscontinuous synthesis of daughter DNA duplexes at molecular level

Replication Complex and Associated Proteins

DNA replication is a complex process. It is carried out by a dynamic complex of enzymes and proteins, often called the replication complex or the **replisome** (Figure 8.16). DNA replication in *E. coli* requires at least two dozens of different gene products. Some of them are listed in Table 8.3 with their functions. Unwinding and movement of the replication fork occurs progressively in the presence of enzymes and proteins. Three different types of proteins appear to help in unwinding the strands of DNA duplex.

1. *DNA unwinding proteins* or *DNA helicases*   There are two types of helicases involved in the unwinding of the double helices. One helicase, the product of the *rep* gene, binds to and stimulates separation of the strand that has 3′ to 5′ polarity in the direction of replication fork movement. The other helicase binds to and helps unwinding of the strand that has 5′ to 3′ polarity in the direction that the fork is moving.

2. *Single-stranded DNA-binding proteins (SSBPs)*   These are tetrameric proteins that b ind tightly to single-stranded regions of DNA produced by the action of helicases and prevent the formation of double-stranded hairpin

loops by rewinding of single-stranded DNA. Thus, SSBPs assist in stabilizing the extended single-stranded templates required for polymerization.

3. *DNA gyrase* This essentially relaxes the supercoiled DNA and releases torsion strain that builds up during replication by travelling along the DNA and acting like a "swivel".

Table 8.3 Some of the enzymes and proteins associated with replisome

Component of replication complex	Role in replication
DNA helicase	Catalyses unwinding of DNA ahead the replication fork
DNA polymerases	Polymerizes daughter DNA strands and repair the gap in lagging strand
DNA gyrase	Relaxes supercoiled DNA and release tension caused by helicase action on DNA duplex
DNA primase	Synthesizes primers to initiate Okazaki fragment synthesis
RNase H	Removes primers from lagging strand
Single-stranded DNA-binding proteins (SSBPs)	Stabilize extended single-stranded templates for polymerization

The replisome involves the enzymes, viz. DNA polymerases, DNA primase, nucleases, and DNA ligases. As mentioned earlier, out of the three types of polymerizing enzymes, DNA polymerase I and II are meant for DNA repair while DNA polymerase III is the principal replication enzyme. DNA primase is an enzyme that synthesizes RNA primers for DNA synthesis. RNA primers begin the synthesis of the Okazaki fragments of the lagging strand which are then extended by DNA polymerases. The primase depends on the DNA-B helicase for efficient priming activity. Primase and helicase form a functional complex called **primosome** as a mobile complex. It moves along the DNA in the same direction as the fork movement. The nucleases, exonuclease or endonuclease, break down the polynucleotide chain into its precursor nucleotides. They either attack the 3′ or the 5′ end of the phosphodiester bonds that held the nucleotide to form polynucleotide strand. All bacterial polymerases have exonuclease activity. An exonuclease degrades nucleic acids by removal of the terminal nucleotides one at a time. In contrast, DNA ligases seal the ends of Okazaki fragments and establish a continuous strand by catalysing phosphodiester bond formation.

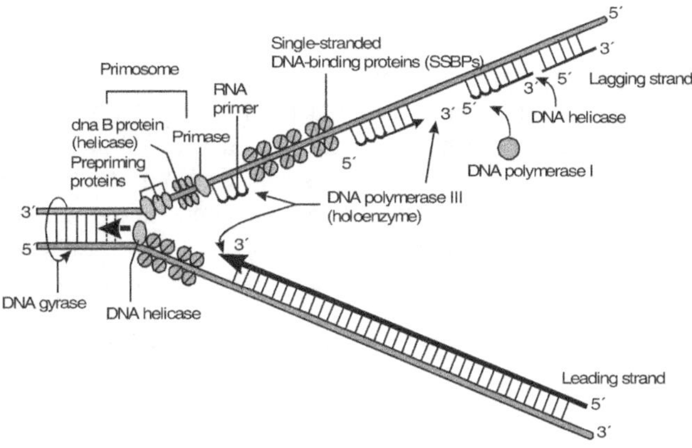

Figure 8.16   Replication complex of *E. coli*

Alternative Means for Replication

An alternative way to replicate circular DNA is the rolling circle mechanism. By this method, many viral DNAs, the fertility (F) factors in bacteria during conjugation and the DNA in certain genes are replicated.

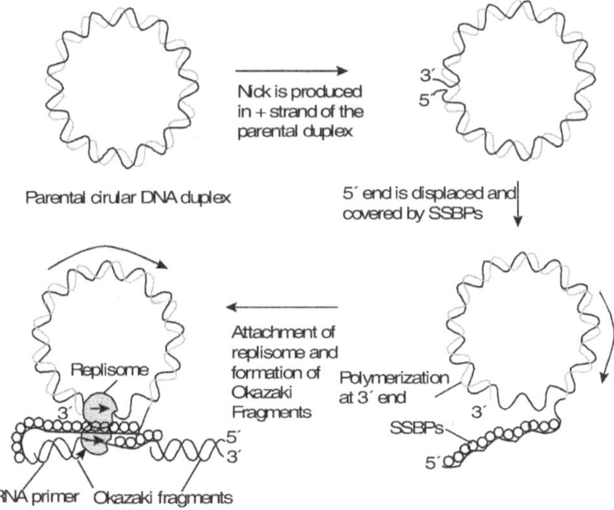

Figure 8.17   Rolling circle mechanism for replication of circular DNA duplex

Rolling circular mechanism (Figure 8.17) is facilitated by nicking one strand (+) of the parental circular double-stranded genome and extending the free 3′ end.

The 5′ end then becomes displaced from the duplex, thereby permitting DNA polymerase III to add deoxyribonucleotides to the free 3′ OH. As replication proceeds, 5′ end of the cut strand is rolled out as a free tail of increasing length, which soon becomes covered by the single-stranded DNA- binding proteins. Such a replicating structure is called a rolling circle because the unravelling of the SSBPs-bound single-stranded DNA is accompanied by rotation of the duplex template about its axis. Subsequently, the 5′-ended tails quickly become converted to double-stranded DNA through formation of Okazaki fragments that grow from RNA primers started by primase molecules contained in the locomotory primosomes. The movement of DNA polymerase–primosome complex (replisome) propelled by its helicase component then polymerizes the daughter DNA strand.

DNA REPLICATION IN EUKARYOTES

The chromosomes in eukaryotes have much complex structure than that of prokaryotic chromosomes. The duplication of the chromosomes of eukaryotes involves not only the replication of their giant DNA molecules, but also the synthesis of the associated histones and non-histone chromosomal proteins. However, at the molecular level, the replication of DNA in eukaryotes is quite similar to that of prokaryotes regardless of the complexity of its genome. It appears to involve the same enzymes and mechanism as in prokaryotes. The eukaryotic DNA polymerases have the same absolute requirements for template and primer as prokaryotic polymerases. Eukaryotic DNA replication is semiconservative and semidiscontinuous.

In 1957, J.H. Taylor and P. Wood provided experimental evidence in support of semiconservative mode of replication in eukaryotes by using autoradiography technique and light microscopy in dividing root tip cells of the bean, *Vicia faba*. They labelled *V. faba* chromosomes by growing root tips for 8 hours (less than one cell cycle) in medium containing radioactive ^{3}H-thymidine (tritiated). The root tips were then removed from the medium, washed and transferred to non-radioactive medium containing colchicine that arrests the metaphase chromosomes. The distribution of radioactive DNA at the first and second metaphase were determined by autoradiography. It was found that in the first generation of duplication, both chromatids of chromosome were labelled. However, at the second metaphase (c-metaphase as it contained colchicine), only one of the chromatids of each pair was radioactive (Figure 8.18). These results indicated that the replication of DNA in eukaryotes is semiconservative.

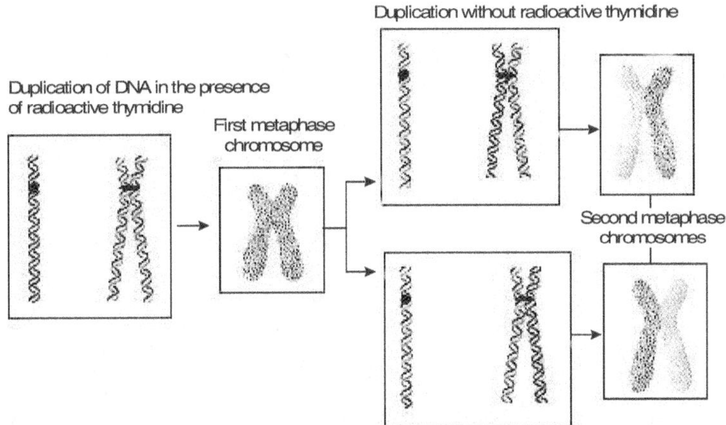

Figure 8.18 Taylor and Wood's experiment to demonstrate semiconservative mode of replication of DNA in root tips of the bean

Like prokaryotic DNA polymerases, there are five different types of DNA polymerases which have been isolated from eukaryotic cells and they are designated as alpha (α), beta (β), epsilon (ε). Polymerase α is tightly associated with the primase, which initiates synthesis of primers at the 5′ end of each Okazaki fragment as the polymerase–primase complex moves along the lagging strand template. Polymerase β plays a role in DNA repair mechanism. Polymerase γ replicates mitochondrial DNA. Most of the fragments on the lagging strand are assembled by polymerase δ. Polymerase ε appears to play some role in replication of nuclear DNA. All the eukaryotic DNA polymerases elongate DNA strands in the 5′ → 3′ direction by the addition of nucleotides to a 3′OH group, and none of them is able to initiate the synthesis of a daughter DNA strand without a primer.

The number of nucleotides in Okazaki fragments synthesized in the lagging strand during eukaryotic DNA replication is only 100–200 nucleotides, whereas they are 1000 to 2000 nucleotides long in prokaryotes. This is due to the fact that cells of higher organisms have much more DNA than bacteria, and eukaryotic polymerases incorporate nucleotides into DNA at much lower rates (about 1 μm of DNA per minute) than in prokaryotes (about 30 μm per minute). To replicate an entire chromosome in a reasonable amount of time, eukaryotic DNA is replicated in smaller units called **replicons**. The sizes of the replicons in eukaryotic chromosomes are generally between 15 and 100 μm in length (50 to 300 kilobases), and each replication has its own origin of replication from where the replicating forks move bidirectionally. As indicated in

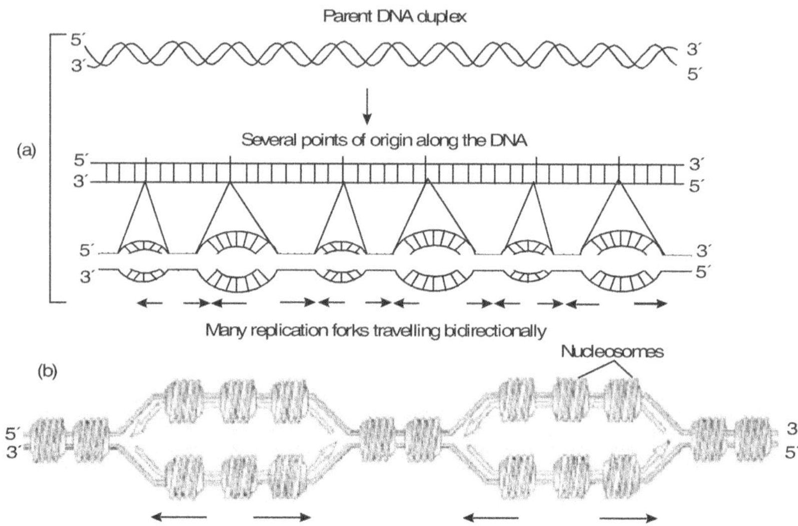

Figure 8.19 (a) Replication of eukaryotic DNA initiated at many points along the DNA. (b) Diagrammatic representation of arrangement of nucleosomes after DNA replication.

Figure 8.19a, replication of DNA occurs at separate sites along the same DNA molecule by formation of replication forks at several points. The replication forks travel away from each other until they meet a neighbouring fork. The chromosomes of a yeast cell contain approximately 400 origins of replication scattered throughout their DNA. In the higher organisms, DNA is tightly complexed with histone octamers to form nucleosomes. Electron microscopic observations (Figure 8.19b) of replicating DNA reveals that nucleosome-like particles are present on both daughter DNA duplexes very near to the replication forks.

SUMMARY

- *Deoxyribonucleic acid (DNA) is a double helix made of polynucleotides that carry genetic information of all living organisms, except RNA viruses. Each nucleotide of DNA is composed of a deoxyribose, a phosphoric acid, and one of the nitrogen bases.*

- *Deoxyribonucleotide is a polar structure with 5′ end and 3′ end. As a result, the DNA duplex has polarity.*

- *In 1950, Chargaff analysed base composition of DNA and discovered the equivalence rule and suggested that the total number of purines is equal to the total amount of pyrimidines (A+G = T+C).*

- *Franklin and Wilkins obtained the X-ray diffraction photographs of DNA fibres and demonstrated that DNA was a helical structure with a diameter of 20 Å and one complete turn of helix is about 34 Å long.*

- *In 1953, Watson and Crick proposed the DNA model that exists as a double helix with two polynucleotide strands coiled about one another to form a spiral with uniform diameter of 20 Å showing alternate major and minor grooves.*

- *There are five different molecular forms of DNA double helix referred to as A-, B-, C- and D-DNA in the form of right-handed DNA helix and Z-DNA is polymorphic form for left-handed DNA.*

- *The DNA molecule in E. coli is supercoiled as is arranged into about 50 loops or domains that can be relaxed under the influence of topoisomerases.*

- *Thermal denaturation or DNA melting is the separation of strands of DNA when the temperature of a solution of DNA is slowly raised and reaches a specific temperature (T_m). DNA molecules with high GC content are more resistant to thermal melting than AT-rich molecules.*

- *Renaturation or reannealing of DNA is the ability of DNA to reassociate into double helix when the temperature of the solution rapidly drops approximately 25°C below the T_m.*

- *In semiconservative type of replication, each daughter DNA duplex retains half of the parental DNA. The second type of replication is conservative in which both strands of parent DNA would be conserved and the new DNA molecule would consist of two newly synthesized strands. In the third type called dispersive replication, fragments of parent DNA molecule are mixed with newly synthesized pieces, thereby each strand of daughter DNA molecules would contain a mixture of old and new DNA.*

SUMMARY

- According to Watson and Crick DNA duplication model, once the DNA replication is initiated, both the old strands of the duplex serve as templates that direct the synthesis of new complementary strand.

- Meselson and Stahl confirmed the semiconservative nature of DNA replication in bacteria using radioisotope through a series of experiments.

- In 1963, John Cairns developed an autoradiographic technique and provided striking visual evidences for bacterial replication. He caught the bacterial chromosome in the act of replication that gave rise to theta configurations, which was further processed under the influence of three types of DNA polymerases bidirectionally using templates and primers.

- DNA polymerase can add nucleotide precursors to DNA only in the $5' \rightarrow 3'$ direction. Consequently, one of the newly synthesized strands grows towards the replication fork called leading strand, while the other strand called lagging strand grows away from the fork and is synthesized in the form of small pieces of DNA known as Okazaki fragments. Because one strand is synthesized continuously and other discontinuously, replication is said to be semidiscontinuous.

- DNA replication in E. coli requires at least two dozens of different gene products that together form the replication complex or the replisome.

- Rolling circle mechanism is an alternative way to replicate circular DNA and is a method by which viral DNAs, the fertility factors in bacteria during conjugation and the DNA in certain genes are replicated.

- Eukaryotic DNA is replicated in many smaller units called replicons that involve five different types of DNA polymerases and each replicon has its own origin from which replication forks move bidirectionally.

REVIEW QUESTIONS

1. Explain the molecular structure of DNA.

2. Give an account to support the statement that the DNA molecule is polar and negatively charged.

3. Describe Chargaff's rule for base composition of DNA.

4. Explain the technique of X-ray crystallography of DNA. How does it help in proposing DNA model?

5. Describe the properties of the DNA model proposed by Watson and Crick.

6. Why does GC-rich DNA melt at higher temperature than AT-rich DNA?

7. How can you distinguish different forms of DNA?

8. What does it mean when we say that replication is semiconservative? How was this feature of replication demonstrated in bacterial cells?

9. Explain the contribution of John Cairns to demonstrate bacterial replication and how is it interpreted?

10. Contrast the role of DNA polymerases in bacterial replication.

11. What are the reasons that DNA molecule is replicated as lagging and leading strands?

12. Why is the replication of DNA referred to as semidiscontinuous?

13. Explain the replisome in bacteria with the role of each of its component.

14. Describe the rolling circle mechanism for DNA replication.

15. What are the differences between bacterial and eukaryotic DNA replication.

16. Write short notes on:

    i. Deoxyribonucleotide

    ii. Phosphodiester bond

    iii. Thermal denaturation

    iv. Reannealing of DNA

    v. Negatively supercoiled DNA

    vi. Grooves in DNA model

    vii. Topoisomerases

    viii. DNA gyrases and helicases

    xi. Primosomes

    x. Replicons

    xi. Single-stranded DNA-binding proteins

    xii. Replisome

    xiii. Okazaki fragments

DNA MUTABILITY AND ITS REPAIR MECHANISM

MUTATION

Genes are made up of DNA that transmits the developmental potential from parents to offspring during reproduction. DNA is a purely informational molecule and the information in DNA is encoded in the sequence of nitrogen bases present in its strands. The stability of DNA is maintained and accurately replicated over generations and hence the normal functioning and even the life of a cell or multicellular organism is continued. Nevertheless, the replication of DNA is not perfectly accurate, sometimes the enzymatic machinery that replicates DNA does some "mistake", which changes the base sequence in DNA. Any change in DNA sequence is reflected in the change of sequence of corresponding RNA or protein molecules. The sudden and inheritable changes in DNA, or in gene/s, or in chromosomes, or in genotype of an organism are called **mutations**. Some of the chromosomal mutations were described in chapter 7. Mutational changes are away from Mendelian line and recombination of genes. Mutations may occur both in somatic as well as germinal cells. Somatic mutations are merely acquired and insignificant from evolutionary point, whereas those of germinal cells are heritable and have evolutionary significance. The idea of mutations first originated from the observations of Hugo de Vries (1880) on the plant species *Oenothera lamarckiana* (an evening primrose) showing different somatic variations.

DNA is a complex organic molecule of finite chemical stability; yet it can suffer from spontaneous damage due to change in single base pair, the substitution of one base pair for another, or duplication or deletion of single base pair. Such mutations that occur at molecular level are referred to as **point mutations**. Living cells require correct functioning of thousands of proteins that are encoded by normal base sequence of DNA; however, damaged DNA causes alteration in the structure and properties of proteins or enzymes that adversely affect the trait or phenotype. A gene, genome, cell or individual carrying a given mutation is called **mutant**. Mutations may occur spontaneously by environmental effect or can be induced in the laboratory either by radiation, physical factors, or chemicals called **mutagens**.

NATURE OF MUTATIONS

In a particular gene, there is a specific number and order of sequence of base pairs that contain the hereditary information required for the functioning of a particular gene. In turn, genes play an essential role in the development of the phenotypic character of the individual. However, many natural mutations destroy the function of a gene completely. The following are the various ways in which changes are brought about in the base sequence in DNA.

Deletion Mutation

It is the loss or deletion of one or few bases from the sequence. The deletion of even single base will result in the alteration of genetic code and hence the function of the gene. For example, GCA GGA CAT ATC is the normal sequence of nitrogen bases on DNA, the sequence of amino acids in a protein to be synthesized will be Ala-Gly-His-Ile and if the base 'C' is removed from the first sequence, the message will be changed to GAG GAC ATA TC and this leads to synthesis of abnormal protein containing the sequence of amino acids Glu-Asp-Ile. Deletion mutations have been frequently reported in some bacteriophages.

Insertion or Addition Mutation

It is the type of mutation that can be brought about if one or few bases are added to the normal base sequence. For example, the base 'A' is inserted into above-mentioned normal sequence between first and second genetic code in DNA, thereby the sequence will get stretched and altered as GCA AGG ACA TAT C that will get subsequently synthesize aberrant protein with the sequence of amino acids Ala-Arg-Thr-Tyr. This incorrect and inactive protein may even cause the death of a cell.

Mutations that arise from the insertion or deletion of the individual nucleotide in the normal sequence of base pairs in DNA and that consequently cause the alteration in the reading frame are collectively referred to as **frameshift mutations**. These mutations invariably affect the normal functioning of a gene and ultimately the phenotype they control.

Substitution Mutation

These are most common among the types of gene mutations where a particular base is replaced by another base (Figure 9.1). Substitution mutations alter the phenotype of an organism and are of great genetical importance. On the basis of the type of nitrogen base substituted, they may be of following types:

Transition   When a purine (e.g. adenine) base of a triplet codon of a cistron is substituted by another purine (e.g. guanine) or when a pyrimidine (e.g. thymine) is replaced by another pyrimidine (e.g. cytosine) then such type of substitution is known as transition. Such kind of mutations occur due to tautomerization.

Transversion   It is the base pair substitution where a purine is replaced by a pyrimidine or a pyrimidine is replaced by a purine. For example, adenine is replaced by thymine and vice versa or guanine is replaced by cytosine and vice versa.

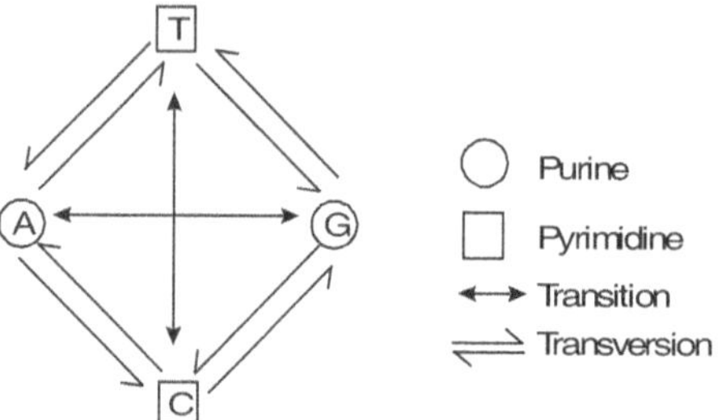

Figure 9.1   Schematic representation of possible base substitution in DNA

In DNA, the bases are not static, the hydrogen atoms can move from one position in a purine or pyrimidine to another position, for example, from amino group to a ring nitrogen. Such chemical fluctuations are called **tautomeric shift**. Although such phenomenon is rare, they may be of considerable importance in the metabolism of DNA since they can alter the base-pairing potential. The structures of bases shown in Figure 9.2 are the common, more

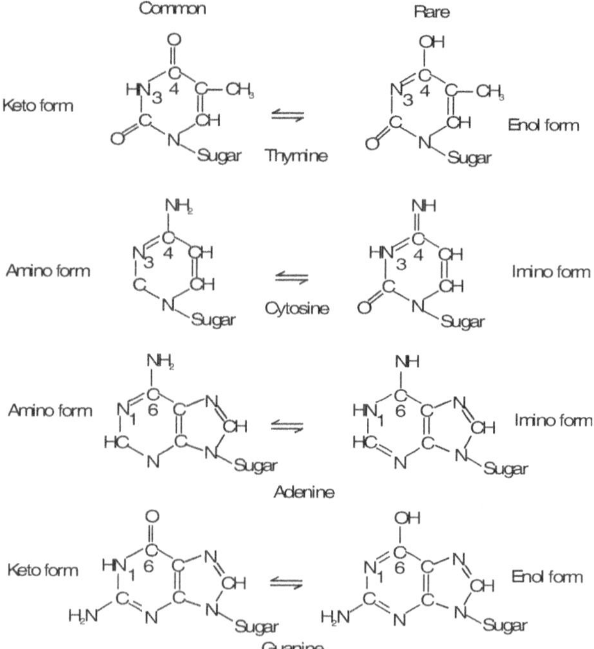

Figure 9.2   Tautomeric forms of four common bases in DNA

stable forms, in which adenine always pairs with thymine and guanine always pairs with cytosine. The more stable keto forms of thymine and guanine and amino forms of adenine and cytosine may infrequently undergo tautomeric shift to less stable enol and amino forms respectively. Such less stable tautomeric forms exist only for a very short period of time. However, if any one of such rare forms is incorporated into nascent DNA strand, there can be a formation of adenine–cytosine or guanine–thymine base pairs.

Sickle cell anaemia is a molecular disease in which the normal shapes of red blood cells become distorted to become sickle shaped and their oxygen-carrying capacity is drastically decreased. Haemoglobin (Hb) packed in the red corpuscles is a conjugated protein consisting of globin as a protein part and haeme as a prosthetic group. Globin in Hb contains two identical alpha polypeptide chains (each chain with 141 amino acids) and two identical beta chains (each chain with 146 amino acids). Each polypeptide is encoded by specific gene containing a particular sequence of nitrogen bases. The amino acid sequences of the "normal" beta chain and the sickle-cell beta chain have been determined by direct chemical analysis. It was found that, in the "normal" beta chain, the amino acid at 6th position was glutamic acid, while in sickle cell beta chain the 6th amino acid was replaced by valine. Further analysis indicated that the sixth code in DNA for normal glutamic acid to be placed in a beta chain was CTC whereas the sixth code in the mutated DNA was found to be CAC. The substitution of a base T (thymine) by the base A (adenine) in a template strand of DNA that is to be transcribed to form mRNA caused this drastic effect.

Inversion

The base sequence in DNA can also be altered by inversion where a segment of DNA strand is removed and reinserted in a reverse manner so that the base-pairing is reversed in its direction. For example, GCA GGA <u>CAT</u> ATC is the normal sequence and if there is reversion at a point in the above-mentioned sequence, then it changes to GCA GGA <u>TAC</u> ATC. As a result the sequence of amino acids in a polypeptide chain during protein synthesis will be Ala-Gly-<u>Tyr</u>-Ile instead of normal Ala-Gly-<u>His</u>-Ile. Of course this protein will be inactive and cannot perform its normal attributed function.

DELETION AND INSERTION

In such mutations a base in a sequence is removed and added simultaneously in another sequence so that deletion and addition occur at the same time. For example, the normal sequence of DNA can be altered by deletion of 'C' from the second position of the first code and in turn the base 'A' is added, the sequence would become GAA GGA CAT ATC and as a consequence the amino acid sequence changes to Glu-Gly-His-Ile. This protein will also be aberrant and that may lead to phenotypic change.

All the examples mentioned in our present discussion are the mutations that alter the code (the sequence of nitrogen bases in DNA) so that one amino acid is replaced by another amino acid in a polypeptide chain during protein synthesis. Such mutations are called **missense** mutations. In contrast, there is another type of mutation called **nonsense** mutation, where the code is mutated to termination code (TAC, TAA or TAG on the DNA) or stop codon (UAG, UAA or UGA, the triplet bases on mRNA) so that there is synthesis of shorter polypeptide chain. For example, the normal base sequence on DNA is ATG TTT GTG TAT GAC GGG AAA CAT and it is destined to synthesize octapeptide with the sequence of amino acids in a polypeptide chain as Met-Phe-Val-Tyr-Asp-Gly-Lys-His. If the base 'T' in fourth code at third position is substituted by a nitrogen base 'A', the original sequence will become ATG TTT GTG TAC GAC GGG AAA CAT and since TAC is one of the termination codes it will terminate the protein synthesis after joining only three amino acids in a polypeptide chain. Thus it will form an abnormal protein that cannot be utilized by the cell.

BACK MUTATIONS AND SUPPRESSOR MUTATIONS

The mutation of a "wild-type" gene to a form that results in a mutant phenotype is usually called "**forward mutation**". Sometimes, however, the terms "wild-type" and "mutant are quite arbitrary. They may simply represent two different phenotypes. For example, consider two alleles (alternate form of genes) controlling brown and blue eye colour in humans both to be the "wild type". However, in a population consisting almost entirely of brown-eyed individuals, the allele for blue eyes might be thought of as a mutant allele. In any case, the mutation events are often reversible. That is, a subsequent mutation may occur that restore the original "wild type" phenotype. This is referred to as **back mutation, reverse mutation** or **reversion**. Restoration of original phenotype may occur due to 1) a true back mutation at the same site in the gene as the original mutation, restoring the wild-type nucleotide sequence in DNA, or by 2) the occurrence of a second mutation at a different location in the genome which in one or other way compensates for the first mutation. Mutations of this kind are called suppressor mutations since they suppress the effect of the original mutations.

INDUCED MUTATION AND MUTAGENS

The mutation that occurs without a known cause is called spontaneous mutation. It may result from a mistake during DNA replication, or may be caused by mutagenic agents present in the environment. In contrast, induced mutations are those which can be induced artificially in the living organisms by exposing to abnormal environment such as radiation, atomic explosion, ultraviolet light, certain physical conditions (i.e., temperature), and chemicals that react with

DNA (or RNA, in RNA viruses). The substances or the agents that induce mutations are called **mutagens** or **mutagenic agents**. Various classes of mutagens are as follows:

Physical Mutagens

They may be of different types such as the following.

Ionization radiation The portion of the electromagnetic spectrum containing wavelengths that are shorter and of higher energy than visible spectrum (wavelength below about 0.1 μm). It can be further subdivided into ionization radiation (X-rays, gamma rays, and cosmic rays) and non-ionizing radiation (UV-ultraviolet light). X-rays have high energy and they penetrate the tissues deeper than UV and visible light. This results in release of energy and makes the atoms to lose electrons, leaving positively charged free radicals or ions. This process is referred to as **ionization**. Many such ions collide with other molecules causing the release of further electrons. The net effect is formation of an ion tract of high energy that passes through matter or living tissues. As a result, the free radicals or ions react with DNA producing a variety of DNA damages including breakages and rearrangements in base sequence, damaged bases and sugar rings, prevention of DNA replication by depolymerization thereby stopping cell division. The effect of ionization radiation may be direct causing ionization of atoms within DNA molecule, or indirect causing production of free radicals, which then react with DNA.

Ultraviolet (UV) radiation Ultraviolet light is a non-ionization radiation. Atenberg for the first time demonstrated that UV radiation could cause mutation. In the upper atmosphere, there is an ozone layer that absorbs most of the UV light coming from the sun and thus protects the living organisms on the earth. However, UV radiation falls on the skin and causes damage in DNA mainly in the form of **dimerization**. The most effective wavelength of ultraviolet for inducing mutation is about 2,600 Å. UV-induced damages are called **photoproducts**. UV rays are mainly absorbed by the pyrimidine bases especially thymine. The principal effect of UV radiation is the generation of thymine dimer (T=T) involving adjacent bases (Figure 9.3).

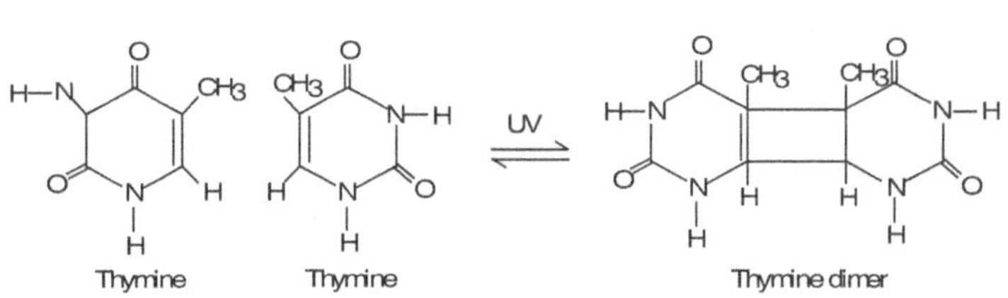

Figure 9.3 Effect of UV radiation on adjacent thymine monomers in DNA to form thymine dimer

In addition to dimerization, UV radiation may also induce damage to single bases, often by hydrating them. Thus another UV-induced photoproduct is pyrimidine hydrates. Several lines of evidence indicate that thymine dimerization is probably the major mutagenic effect of UV. Dimers apparently affect the DNA double helix and interfere with accurate DNA replication.

Temperature It affects the rate of all chemical reactions. For every 10°C rise in the temperature there is a two-fold increase in the rate of the chemical reaction and at the same time it increases the mutation rate two or three-fold. Probably, temperature affects the thermal stability of DNA and the rate of interactions of other substances with DNA.

Chemical Mutagens

The classes of chemical mutagens are as follows.

Deaminating agents These mutagens bring about the deamination of nitrogen bases in DNA. The examples of such mutagens are **nitrous acid** that is non-specific and **sodium bisulphate**, which specifically deaminates cytosine and converts it to uracil. In a similar reaction, adenine is deaminated to hypoxanthine by nitrous acid. The effect of deamination reflects in transitional mutation.

Alkylating agents These are strong mutagens that carry one, two, or more alkyl groups in reactive form. These agents add ethyl or methyl groups to guanine. Guanine with methyl or ethyl group becomes analogue to adenine. Consequently they remove the alkylating guanine producing **apurinic gaps**. The overall process is referred to as **depurination**. The deletion of base from the normal sequence makes DNA mutated. Further, the apurinic gap can be filled by an incorrect base thereby changing the codon in mRNA and amino acid in the polypeptide chain to be synthesized.

The examples of alkylating agents are dimethyl sulphate (DMS), diethyl sulphate (DES), methyl methane sulphonate (MMS), ethyl ethane sulphonate (EES), ethyl methane sulphonate (EMS), mustard gas, mustard compounds, phosphoric esters and ethylene amines.

Base analogues These chemical agents have structural similarities with bases of DNA. If they are available, they may get incorporated into a newly synthesizing strand of DNA during replication. 5-bromouracil (5-BU) or its nucleoside 5-bromodeoxyuridine (5-BrdU) resembles thymine in its keto form resembles thymine and hence it substitutes thymine in the replicating strand of DNA. As a result, A–T pair becomes A–BU pair that ultimately causes damage to DNA. Similarly 2-aminopurine (2-AP) is another base analogue that can incorporate opposite to thymine in the first round of replication and then pair with a cytosine at the next round to produce AT–GC transition.

Intercalating agents These chemical mutagens are polycyclic molecules that insert themselves between the stacked base pairs of the DNA duplex, thereby doubling the distance between adjacent base pairs. Chemicals like ethidium bromide, acridine dyes such as proflavin and acridine orange are intercalating agents, which induce frameshift mutations and blocks the process of replication.

In addition to the above-mentioned chemical mutagens. There are certain other chemicals of non-specific type such as aflatoxins B_1 produced by certain fungi, benzopyrene, purrolizidine alkaloids, certain plant-derived flavonoids, malonaldehyde, an oxidized product of polyunsaturated fatty acids, tobacco tar, certain food additives, hair dyes, some of the pesticides and the planar molecules such as psoralens that induce cross-linking between DNA bases when exposed to UV light. All these chemicals induce mutations in DNA by one or other way and are responsible for transformation of normal cell into cancer cell.

Biological Mutagens

These are mutation-inducing agents that are derived from biological sources, e.g. **restriction enzymes** and **transposons**. The restriction enzymes are molecular scissors that are isolated from bacteria. They have site-specific activity and are used by the bacteria to destroy various viral DNAs that might enter the cell, whereas transposons are **mobile genetic elements** made up of specific DNA sequences that can move around the genome. Examples include viruses (e.g. bacteriophage Mu, retroviruses), episomal plasmids (e.g. F factor), and transposable genetic elements (e.g. P-element, Ty-elements). These agents may have mutagenic ability since they can insert themselves in the gene sequence and can interrupt the normal functioning of the gene or they may carry gene/ regulatory elements affecting endogenous gene expression.

DNA REPAIR MECHANISM

Every living cell has to continue life in spite of the impact of various physical, chemical, biological or environmental mutagens and for that the cell has its own DNA repair mechanisms. DNA polymerases initially make a significant number of mistakes in assembling polynucleotide strands. The frequency with which genes mutate spontaneously is called **mutation rate**. Mutations occur much more frequently in certain regions of the gene than others, and such favoured regions are called hot spots (first discovered by S. Benzer in his classic genetic analysis of the phage T4 *rII* gene). The observed mutation rate of one for every 10^6 bases replicated would result in about 1,000 mutations every time a human cell divided. Fortunately, human cells have at least three DNA repair mechanisms at their disposal:

1. A *proofreading mechanism* corrects replication as DNA polymerase makes them. DNA polymerase performs proofreading function. It recognizes mispairing of bases and removes incorrectly introduced nucleotide and tries to link it in the replicating strand again. In addition to DNA polymerase, other proteins of the replication complex also help in proofreading. The error rate for this process is only about 1 in 10,000 base pairs and lowers the overall error rate for replication to about one base in every 10^{10} bases added to replicating strand.

2. A *mismatch repair mechanism* scans DNA immediately after it has been replicated and corrects any base-pairing mismatches. This system involves a set of proteins that surveys newly replicated molecule, looks for remaining mismatched base pairs and introduces correct base pairs. When mismatch repair fails, DNA sequences are mutated. One form of colon cancer arises in part from a failure of mismatch repair.

3. An *excision repair mechanism* removes abnormal bases and replaces them with functional bases. The group of certain enzymes that includes endonuclease, DNA polymerase and DNA ligase constantly inspects the cell's DNA to detect the occasional errors such as thymine dimers to correct them.

Mechanisms for the repair of damaged DNA are probably universal from simple bacterial viruses to humans. *Escherichia coli*, for example, possesses three different mechanisms for the repair of DNA containing thymine dimers: 1) Photorepair mechanism, 2) Dark excision repair mechanism, and 3) Post-replication recombination repair.

Photorepair Mechanism

It was discovered for the first time in bacteria with damaged DNA due to ultraviolet radiation. UV radiations cause thymine dimers to appear in DNA to become mutated. In prokaryotes and eukaryotes, there are photoreactivating enzymes that form the major line of repair for thymine dimers. This enzyme splits thymine dimers directly without the removal of any nucleotides. The photoreactivating enzyme derives energy from visible light (specifically blue light) and binds to thymine dimer and cleaves the bond between thymine dimer to convert them into thymine monomers (Figure 9.4).

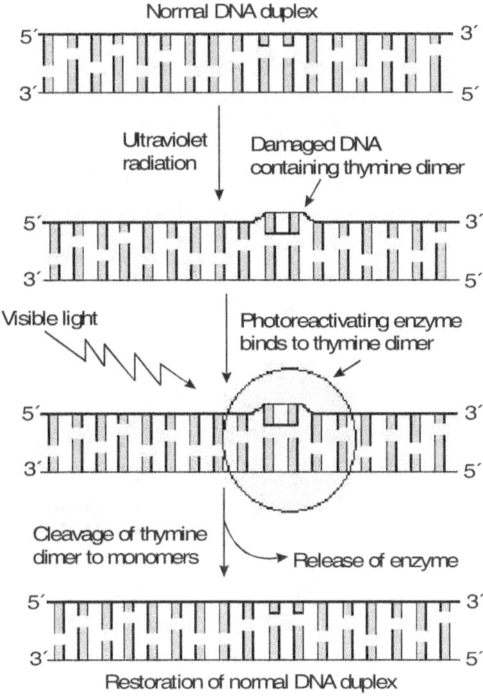

Figure 9.4   Role of photoreactivating enzyme in repair of thymine dimer

Dark Excision Repair Mechanism

It involves a sequence of enzyme-catalysed steps so that thymine dimers are removed from the DNA molecule and a new segment of DNA is synthesized (Figure 9.5). The repair process occurs in mutant strains of *E. coli* stored in the dark. The step in excision repair is catalysed by an enzyme endonuclease that identifies thymine dimers, or the distortion in the DNA duplex that they cause, and breaks the phopshodiester backbone of the DNA strand containing thymine dimer producing nicks on either side of the damage. Then probably the 5′ → 3′ endonuclease activity of DNA polymerase I removes a segment of a strand adjacent to endonuclease cut including thymine dimer. DNA polymerase I then synthesizes the segment of DNA using complimentary strand as template. Finally, DNA ligase catalyses phosphodiester linkages between adjacent nucleotides restoring normal sequences of bases in DNA.

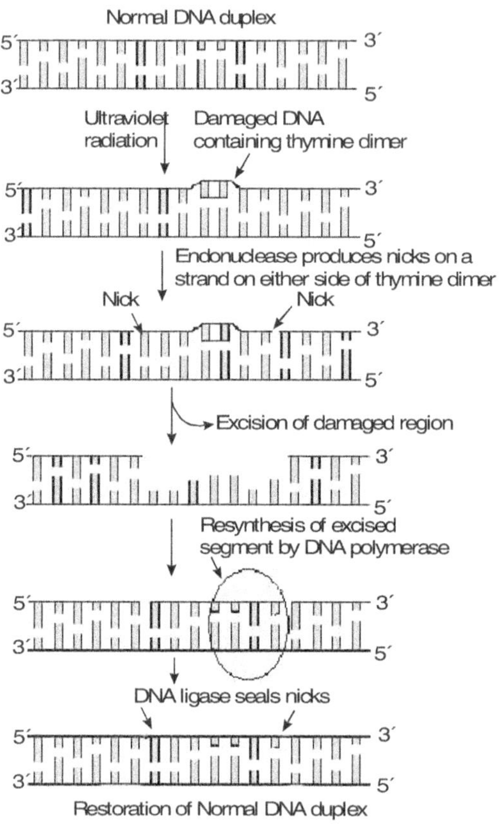

Figure 9.5 Pathway for removal of thymine dimer from damaged DNA by dark excision repair mechanism

Post-replication Recombination Repair

It is a type of dark excision repair mechanism that occurs in *E.coli* involving replication of damaged chromosomes followed by recombination. The simplified mechanism for post-replication recombination repair method is represented in Figure 9.6. When UV-induced mutated DNA molecule is replicated, gaps are formed in the nascent complementary strands opposite the thymine dimers because DNA polymerase cannot use the distorted strand of DNA duplex as a template. As a result of semiconservative type replication, progeny with double helices containing thymine dimers in one strand and gaps in the complementary strand is produced. If the sister chromatid exchange (SCE) takes place between them, the dimers and gaps may combine to form one chromosome (that will be nonviable) and the intact, undamaged segments combine to form the other chromosome (that will be functional and produce a viable cell). SCE represents the interchange of DNA replication products at homologous chromosomal loci and it is a highly sensitive method for detecting DNA damages also in eukaryotes.

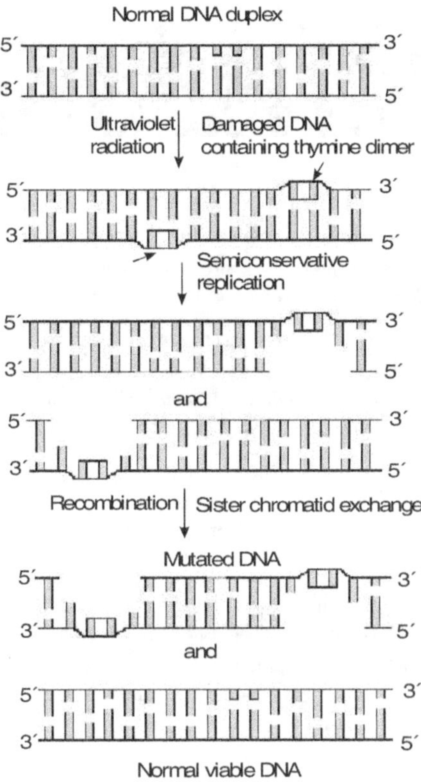

Figure 9.6   Post-replication recombination repair mechanism for UV-induced damaged DNA

XERODERMA PIGMENTOSUM (XP)

It is a rare recessive genetic disorder in humans. The victims of XP possess a deficient repair system that is unable to remove segments of DNA damaged by ultraviolet radiation. It is an autosomal disease characterized by extreme sensitivity to sunlight; even very limited exposure to direct rays of the sun can produce large number of dark-pigmented spots on exposed areas of the body such as the face. The patient has elevated risk of developing disfiguring and fatal skin cancers. About 20 per cent of patients with this disorder also exhibit neurological abnormalities due to premature death of nerve cells and mental retardation. Genetic analyses of cells from individuals with xeroderma pigmentosum suggest that mutations in as many as six different genes that are related to replication of DNA and repair mechanism can cause this disease.

SUMMARY

- *The sudden and inheritable changes in DNA are called mutations. A gene, genome, cell or individual carrying a mutation is called* mutant *that may occur spontaneously by environmental effect or they can be induced in the laboratory either by radiation, physical factors, or chemicals called* mutagens.

- *Genes have a specific number and order of sequence of base pairs, which is essential for the development of the phenotypic character of the individual. Many natural mutations can destroy the function of a gene completely by deletion, insertion, substitution (transition or transversion), inversion, or by deletion and insertion.*

- *Forward mutation is mutation of a "wild-type" gene. The process called back or reverse mutation or reversion may sometimes restore the original phenotype, whereas suppressor mutation occurs as a second mutation at a different location in the genome which in one or other way compensates for the first mutation.*

- *Spontaneous mutation occurs without a known cause. It may result from mistake during DNA replication, or may be caused by mutagenic agents present in the environment. In contrast, induced mutations are induced artificially in the living organisms by exposing to abnormal environment such as radiation, atomic explosion, ultraviolet light, certain physical conditions and chemicals.*

- *Cell survival depends on continual repair of DNA. In human cells, there are three DNA repair mechanisms namely proofreading mechanism, mismatch repair mechanism and excision repair mechanism that remove abnormal bases and replaces them with functional bases.*

- *Escherichia coli has three different mechanisms for the repair of UV-induced DNA damage and these are photorepair mechanism, dark excision repair mechanism, and post-replication recombination repair.*

- *Xeroderma pigmentosum is the recessive autosomal disorder in humans in which victims have a deficient repair system and the disease is characterized by extreme sensitivity to sunlight, fatal skin cancers, neurological abnormalities and mental retardation.*

REVIEW QUESTIONS

1. What is mutation? Explain different classes of mutagens with their effects on DNA.

2. Describe various kinds of mutation with suitable examples.

3. Distinguish between transition and transversion.

4. Evaluate the effect of ionization radiation and alkylating agents on base sequence of DNA.

5. How does the action and mutagenic effect of 5-bromouracil differ from that of nitrous acid?

6. Why is sickle cell anaemia called a molecular disease?

7. What are different mechanisms for repair of damaged DNA in humans?

8. How does a bacterium repair its UV-induced damage in DNA?

9. What is tautomeric shift? Describe different tautomers that can be found in mutated DNA.

10. Define frameshift mutation. Explain the types with suitable examples.

11. How does the photorepair mechanism differ from dark excision repair mechanism?

12. Write short notes on:

    i. Spontaneous mutation

    ii. Forward mutation

    iii. Mutant

    iv. Xeroderma pigmentosum

    v. Sister chromatid exchange

    vi. Transposons

    vii. Base analogues

    viii. Dimerization

    ix. Apurinic gap

    x. Reversion

    xi. Point mutations

TRANSCRIPTION -THE SYNTHESIS OF RNA

From genotype to phenotype

The Central Dogma

Modification of Central Dogma due to RNA viruses

Chemical nature of RNA

Transcription—the basic process of DNA directed RNA synthesis

Mechanism of transcription in prokaryotes

Transcription in eukaryotes

Three major classes of RNAs

Ribosomal RNA (rRNA)

Transfer RNA (tRNA)

Messenger RNA (mRNA)

Post-transcriptional modifications of mRNA

Gene silencing by RNA interference (RNAi)

Ribozyme (Catalytic RNA)

FROM GENOTYPE TO PHENOTYPE

It was a well-established fact that chromosomes contain genes, which possessed a unique molecular constituent, deoxyribonucleic acid (DNA), but there was no way to show that this molecule carried the genetic information. So it made sense to approach the nature of gene by asking how genes function within cells. From the very beginning of this speculation, the simplest hypothesis was that the genetic information within genes determines the order of 20 different amino acids within the polypeptide chains of proteins to produce a particular phenotype or in other words how does the flow of information take place within the cell. There are many steps in the establishment of the relation between the genotype and phenotype. Genes cannot, all by themselves, be directly responsible for production of brown colour in the eye, dimpled cheeks, particular height of the plant, or a specific colour of the seed. The molecular basis of the phenotype was identified before the proof that DNA was genetic material. In 1940s, Beadle and Tatum showed with the help of a series of experiments that when there was an alteration in the gene structure (change in the normal sequence of nitrogen bases in DNA), it always resulted in alteration in phenotype. They suggested that each mutation causes a defect in only one enzyme in a metabolic pathway and described the **one gene-one enzyme** hypothesis indicating that genes always encode an enzyme. If there is mutation in the gene, it resulted in the production of an aberrant enzyme. This hypothesis has undergone several modifications in the light of our present-day knowledge of molecular biology. Since several enzymes are made up of more than one polypeptide chain or subunits, and each chain is encoded by its own separate gene, the hypothesis is correctly modified as **one gene-one polypeptide** relationship indicating that a gene controls the production of a single, specific protein. Later on, it was discovered that all the genes do not always code for proteins, some of the genes code for different forms of RNA that do not become translated into polypeptides. However, in 1956, Francis Crick proposed the central dogma of molecular biology for the flow of genetic information from DNA to the protein.

THE CENTRAL DOGMA

Crick and his co-workers proposed the central dogma of molecular biology for the transfer of genetic information from genes to protein that occur in two steps. The first step is known as **transcription** in which the messenger RNA (ribonucleic acid) is assembled as a complementary copy of one of the two strands of DNA. The mRNA retains the same information as the gene itself. In this process, there is not much difference between the coded language (script) of DNA and RNA, hence it is a transcription due to the fact that genetic code on DNA is in the form of ATGC and codons on the mRNA are in the form of AUGC, thus T in DNA is replaced by U (for uracil) in RNA. This process is catalysed by the enzyme called DNA-dependant RNA polymerase or simply RNA polymerase.

The second process is referred to as translation, which is a highly complex process that requires the participation of dozens of different components, including ribosomes containing ribosomal RNA (rRNA) and a third major class of RNA known as transfer RNA (tRNA). This step involves the change of codons of mRNA to the particular sequence of amino acids in a polypeptide chain to form a protein. The process is called translation because it involves a complete translation of the language of codons on mRNA to amino acids in protein that form a phenotype. Thus, messenger RNA (mRNA) is intermediate between a gene and its polypeptide (Figure 10.1). The central dogma in simple words is a one-way transfer of genetic information from DNA to RNA, from RNA to protein and protein does not code for protein, RNA or DNA. Thus once the information is passed to protein, it cannot get out again.

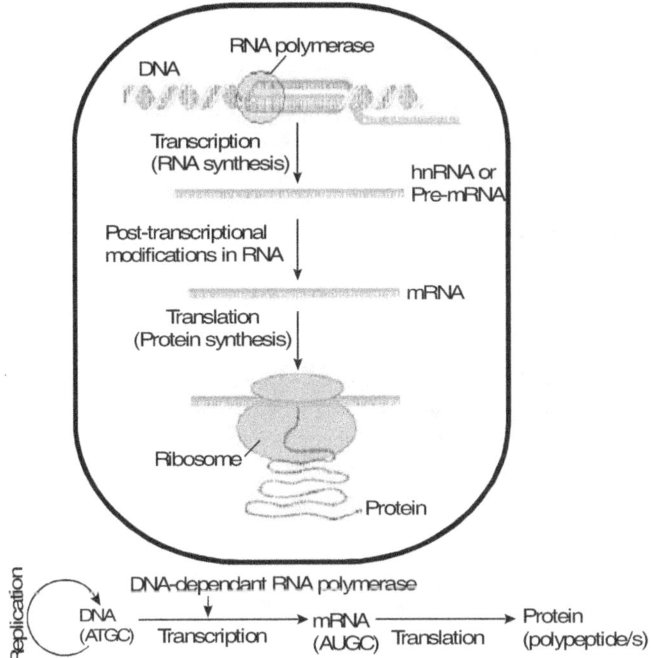

Figure 10.1   Central dogma of molecular biology showing one-way transfer of genetic information from DNA to protein

MODIFICATION OF CENTRAL DOGMA DUE TO RNA VIRUSES

In 1962, Temin reported the presence of an enzyme called **RNA-dependant DNA polymerase** or **reverse transcriptase** in RNA viruses such as tobacco mosaic virus, influenza virus, poliovirus (see chapter 4) that have RNA as genetic material instead of DNA. This enzyme can synthesize DNA duplex from a single- stranded RNA template. Baltimore (1970) also independently reported the activity of this enzyme in certain RNA tumour viruses. Their findings gave rise to the new concept of **"reverse transcription"** or **Teminism** suggesting that flow of information is always not unidirectional from DNA to RNA but it can be from RNA to DNA (Figure 10.2).

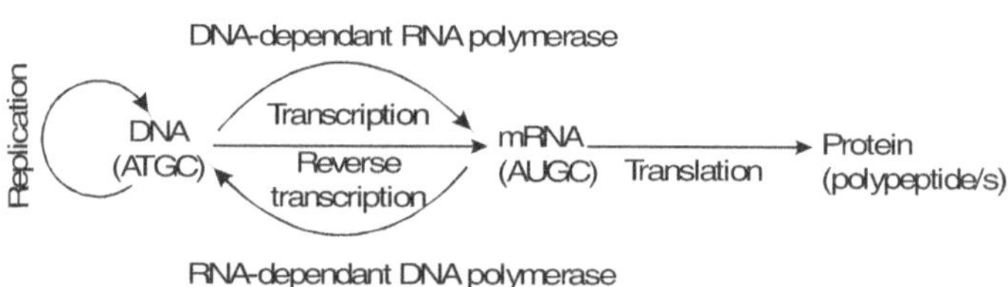

Figure 10.2   Modified central dogma showing reverse transcription

CHEMICAL NATURE OF RNA

Ribonucleic acid (RNA) is a key intermediate between the genotype and its expression in the synthesis of protein. Chemically, the non-genetic RNA is a polynucleotide similar to DNA, but it differs from DNA in the following ways:

- ❖ RNA is usually single-stranded, i.e., it has only one polynucleotide strand where adjacent ribonucleotides are linked by phosphodiester bonds. At some places the exposed nitrogen bases are complementary hence this single-stranded molecule may become folded to form complex shapes due to internal base-pairing.

- ❖ Each ribonucleotide consists of ribose (a pentose sugar with the molecular formula $C_5H_{10}O_5$), rather than deoxyribose found in DNA, a nitrogen base and phosphoric acid.

- ❖ The nitrogen bases in RNA are purines (adenine and guanine) and pyrimidines (cytosine and uracil). The base uracil (U) is similar to thymine but lacks the methyl ($-CH_3$) group. The complementary base-pairing rules are same as in DNA, except that adenine pairs with uracil.

- ❖ RNA molecules do not usually have complementary base ratios (Table 10.1). The amount of adenine does not often equal the amount of uracil, and the amount of guanine also usually differ from each other.

Table 10.1 The base composition of RNA from various sources

RNA source	Proportion of the four main bases			
	Adenine	Guanine	Cytosine	Uracil
Poliomyelitis virus	0.30	0.25	0.25	0.21
E.coli	0.24	0.32	0.24	0.21
Yeast	0.25	0.24	0.22	0.27
Allium cepa (onion seed)	0.24	0.29	0.24	0.20
Cucurbita pepa (pumpkin seed)	0.25	0.30	0.24	0.19
Rat kidney	0.19	0.30	0.31	0.20

TRANSCRIPTION—THE BASIC PROCESS OF DNA-DIRECTED RNA SYNTHESIS

In eukaryotes, most of the cell's DNA is confined to the nucleus, and during transcription, there is flow of this genetic information from nucleus to the cytoplasm where actual synthesis of protein takes place. To understand this mechanism, Crick and his colleagues proposed the **messenger hypothesis** suggesting that messenger RNA (mRNA) forms as a complementary copy of one strand of DNA of a particular gene. This mRNA molecule is then responsible for transfer of genetic information from nucleus to the cytoplasm of the cell and acts as template for the synthesis of protein. Each gene consists of a definite number of nucleotides arranged in a specific order in DNA, and codes for the particular protein is expressed as a sequence in mRNA. This hypothesis was found correct since it has been tested for several times for genes that code for proteins. Similarly to understand the relationship between a specific nucleotide sequence in DNA and a specific amino acid sequence in a protein, Crick proposed the **adapter hypothesis** indicating that there must be an adapter molecule that can bind a specific amino acid with one region and recognize the sequence of nucleotides with another region. Later on, these adapter molecules called transfer RNA (tRNA) were identified. They identify the codons on mRNA and accordingly carry the specific amino acid to translate the coded language of DNA into the language of proteins. Thus, tRNA adapters line up themselves for the short span of time on mRNA and insert the amino acids in a proper sequence to elongate the polypeptide chain, constituting the process called **translation**. This hypothesis has been confirmed on the basis of actual observations of the expression of thousands of genes during which tRNA molecule acts as the intermediary between the sequence of nucleotides on mRNA and the sequence of amino acids in a protein.

Transcription is much simpler than translation since it involves synthesis of one nucleic acid from the other. And it is accomplished by a single enzyme that works in conjunction with a variety of accessory proteins. The enzymes responsible for transcription in both prokaryotic and eukaryotic cells are called **DNA-dependant RNA polymerases**, or **simply RNA polymerases**. In addition to enzymes, the process requires a DNA template for complementary base-pairing, the appropriate **ribonucleoside triphosphates** (ATP, GTP, CTP, and UTP) to act as substrate and the **promoter** as the site on the DNA to which RNA polymerase binds prior to initiating transcription. The synthesis of RNA from DNA takes place in three steps:

1. *Initiation*   It is the step in which RNA polymerase binds the promoter region on DNA and   begins the synthesis of RNA.

2. *Elongation*   In this step the RNA strand is extended using appropriate ribonucleotides and the majority of RNA synthesis takes place.

3. *Termination*   During this step, elongation of RNA ceases and nascent RNA (transcript) dissociates from the template.

MECHANISM OF TRANSCRIPTION IN PROKARYOTES

In *E. coli* and other prokaryotic cells, a single RNA polymerase is responsible for the synthesis of all kinds of RNAs (mRNA, tRNA, and rRNA) from the template strand of DNA. *E. coli* RNA polymerase is a complex, multimeric protein and the most studied of all RNA polymerases. It has a molecular weight of about 490,000 and consists of six subunits (polypeptides chains)—two identical alpha (α) subunits, one chain of each beta (β) and beta dash (β'), omega (ω), and sigma (σ) subunits (Figure 10.3). Some of the characteristics of each subunit of RNA polymerase are given in Table 10.2.

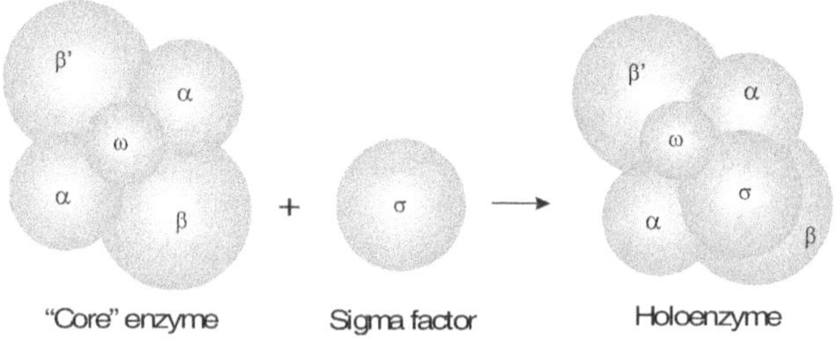

Figure 10.3   The complex structure of *E. coli* RNA polymerase

The completely assembled molecule of RNA polymerase is called **holoenzyme** that can be represented as $\alpha_2\beta\beta'\omega\sigma$ in which attachment of sigma (σ) subunit or factor is not firm. On its dissociation from holoenzyme, the resultant enzyme called **core enzyme** ($\alpha_2\beta\beta'\omega$) does not lose its catalytic activity. The sigma factor helps in the recognition of start signals on the DNA molecule and directs RNA polymerase in selecting the initiation site (promoter). Once the synthesis of RNA is initiated and RNA molecule becomes 8–9 nucleotides long, the sigma factor dissociates from the holoenzyme and then the core enzyme brings about the elongation of mRNA or any other RNA.

Table 10.2   Some of the characteristics of components of *E. coli* RNA polymerase

Subunit	Gene responsible	Molecular weight	Location	Function
α_2	*rpo* A	41,000 each	Core enzyme	Promoter binding
β	*rpo* B	155,000	Core enzyme	Nucleotide binding
β'	*rpo* C	165,000	Core enzyme	DNA template binding
ω	---	12,000	Core enzyme	---
σ	*rpo* D	95,000	Sigma factor	Initiation

As mentioned earlier, transcription involves three steps—initiation, elongation and termination (Figure 10.4). Unlike replication, transcription does not progress along the entire length of a DNA. When genes have to express themselves as per the need of the cell, only part of the segment of the DNA constituting the gene is transcribed and only one of the strands of a DNA duplex is transcribed. From this strand RNA is synthesized and hence it is known as **sense strand**, while the other strand, which is not transcribed, is called **anti-sense strand**. The region of the sense strand of DNA that is actually transcribed into RNA is referred to as the **coding region**.

The first step of transcription is the binding of RNA polymerase to a DNA molecule. The attachment of the enzyme occurs at the particular location called **promoter**. However, the accessory sigma factor is added to RNA polymerase before it attaches to the promoter. The attachment of the sigma factor to the core enzyme increases the affinity of the enzyme promoter sites in DNA. Once transcription has begun, the sigma subunit dissociates from the sense strand, while the core enzyme continues to elongate the RNA strand. Bacterial promoters are located in the region of the DNA strand just upstream from the site of initiation of RNA synthesis. Hundreds of promoters have been sequenced in *E.coli*, and these sequences have suprisingly little in common. There are two short stretches of 5 to 10 nucleotide sequences within these promoters. One of these stretches is approximately 35 bases upstream from the initiation site and is known as the **–35 sequence** or the **recognition sequence**. It occurs as a minor variation of the sequence (consensus sequence) TTGACA that is recognized by the sigma factor. The second consensus sequence is present approximately 10 bases upstream (**–10 sequence**) from the initiation site and occurs as a variation of the sequence TATAAT. This site is known as **Pribnow box** in *E. coli* and in eukaryotes the same region is called the **Hogness box**, both named after their discoverers. Usually this region is referred to as **TATA box** and it is believed to orient the RNA polymerase, as a consequence RNA synthesis proceeds from left to right. It is also the region at which the DNA duplex opens to form the **open promoter complex** (Figure 10.4a).

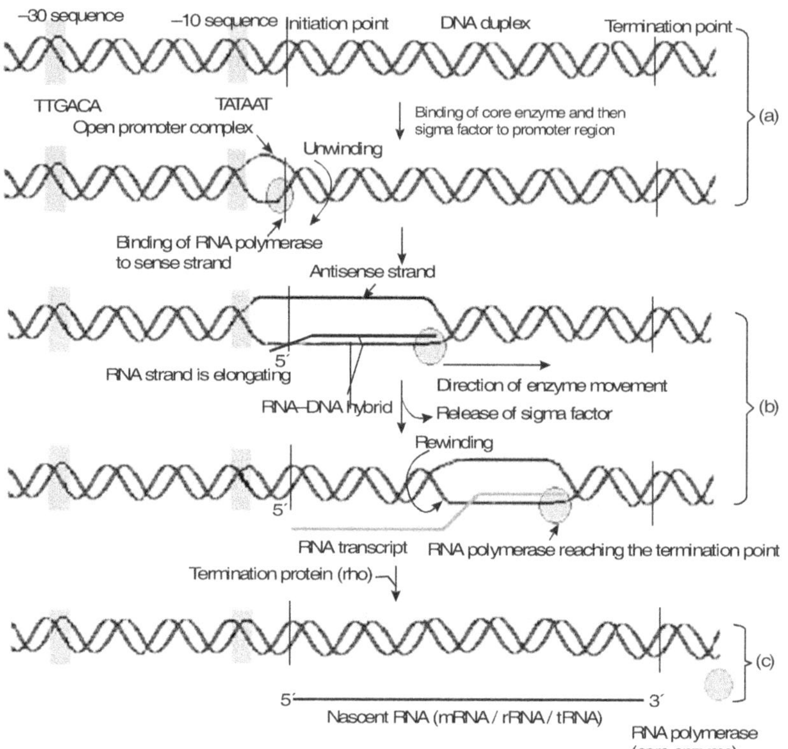

Figure 10.4   Three distinct processes of transcription (a) Initiation—binding of RNA polymerase to partially unwound DNA, (b) Elongation of RNA transcript, and (c) Termination of process in the presence of rho-termination protein and release of nascent RNA.

Once the open promoter complex is formed, the holoenzyme (core enzyme associated with the sigma factor) initiates RNA synthesis using the coded language present on the sense strand of DNA. The initiating nucleoside triphosphate (usually pppA or pppG) binds to the enzyme and forms a hydrogen bond with the complementary base of the sense DNA strand. Then the incoming nucleoside triphosphate is selected by its ability to form a hydrogen bond with the next base in DNA strand. Two nucleotides are then joined by a phosphodiester bond. After addition of about 8 nucleotides in a growing RNA strand, RNA polymerase loses the sigma factor becoming the core enzyme that carries most of the elongation process (Figure 10.4b). The core enzyme moves along the DNA template and extends the growing RNA chain in the 5′ to 3′ direction by adding complementary nucleotides one by one. As the RNA polymerase moves along the sense strand and as the RNA strand is being extended, there is a formation of unstable RNA–DNA hybrid, which soon dissociates and nascent RNA is displaced.

Just as transcription is initiated at a specific point on the DNA molecule, it also terminates when a specific nucleotide sequence is reached. In some cases, a termination protein called **rho factor** is required to cease the process (Figure 10.4c), whereas in most cases, the polymerase can stop the transcription and carry out the release of nascent RNA without addition of factor. Finally the core enzyme is dissociated and nascent RNA (it may be mRNA, rRNA, or tRNA) is released. Following the release of core enzyme from termination site, it interacts with sigma factor to reform holoenzyme that becomes available for the next round of RNA synthesis.

TRANSCRIPTION IN EUKARYOTES

The synthesis of RNA in bacteria is accomplished entirely by single RNA polymerase whereas to achieve the same objective there are three RNA polymerases in eukaryotic cells: RNA polymerase I, II, and III. Each has different properties and different specific roles. They act at different promoters to transcribe the various kinds of RNA in the cell:

1. RNA polymerase I (called pol I) occurs at the nucleolus and synthesizes the large (28S, 18S and 5.8S) ribosomal RNAs that play an indispensable role in the formation of ribosomes.

2. RNA polymerase II is located in nucleoplasm and makes hnRNA (heterogeneous nuclear RNA) that undergoes modifications to form messenger RNA, which carries genetic information from nucleus to the cytoplasm where protein synthesis takes place.

3. RNA polymerase III also occurs in the nucleoplasm and synthesizes various transfer RNAs that pick up activated amino acids and bring them to the site of protein synthesis and also helps them in the synthesis of the 5S ribosomal RNA.

Like bacterial RNA polymerase, these enzymes are large and complex proteins with a molecular weight of about half a million. However, they differ from bacterial RNA polymerase in that they contain about 8 to 14 polypeptides (subunits) rather than five. A further complexity of eukaryotic cells is that the genomes of mitochondria and chloroplasts are transcribed by their own organelle-specific RNA polymerases. In addition to the subunits that make up the enzymes, each of them is assisted in its operations by a variety of auxiliary proteins referred to as **transcription factors** (TFs). These factors determine the rate at which a particular gene or group of genes is transcribed.

Eukaryotic RNA polymerase I, II, and III can be distinguished on the basis of their sensitivity to α-amanitin, a highly toxic protein made up of eight amino acids, isolated from the poisonous mushroom. RNA polymerase II activity is very sensitive to α-amanitin, whereas RNA polymerase I is not affected by this compound. The activity of RNA polymerase III is inhibited in animals at higher doses but to a lesser degree than RNA polymerase II. A person suffering from food poisoning due to ingestion of α-amanitin-producing mushrooms, shows deterioration in liver function over the following days due to lack of production of new mRNAs that are necessary for continuous synthesis of proteins.

Eukaryotic Transcriptional Promoters, Enhancers and Silencers

In sharp contrast to RNA polymerase I which synthesizes only a single type of RNA (pre-rRNA), RNA polymerase III is responsible for regulated synthesis of different transfer RNA transcripts in all eukaryotic cells. RNA polymerase II acts on genes specifying certain smaller nuclear RNAs (snRNAs usually rich in uracil) as well as on every protein-coding gene to form heterogeneous nuclear RNAs (hnRNAs that modify to form mRNAs), whether the gene is always expressed or is expressed only at a particular stage of development in one tissue of a multicellular organism. The promoters of the genes of RNA polymerase II contain three distinct regions that are centred at sites lying between–25 bp and –100 bp (Figure 10.5).

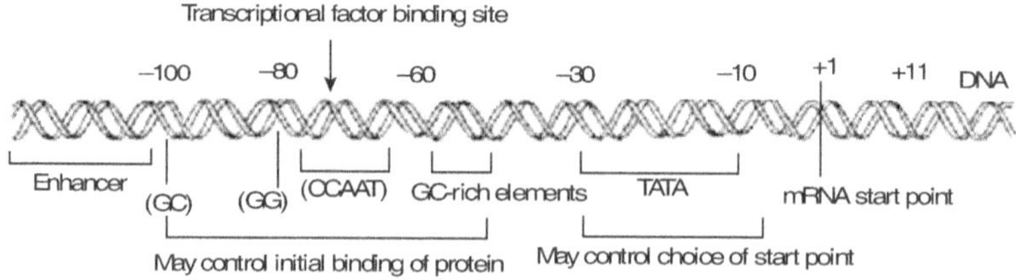

Figure 10.5   Essential sequences in mammalian RNA polymerase II promoter sites

Like the pribnow box in prokaryotes, the promoter for eukaryotic gene located 20 bp upstream to the starting point is called TATA box. It contains a sequence of bases that is either identical to or very similar to the oligonucleotides 5'-TATAAA-3'. Other sequence homologies found for RNA polymerase promoters are present farther upstream from the start site. For instance, a CCAAT homology (or **CAT box**) occurs between –70 and –80 base pairs. Another sequence called **GC box** includes GC-rich elements (GGGCGG) found in one or more copies at –60 or –100 bp upstream in many genes. It has become clear that some transcriptional factor (TF) binds the CCAAT sequence that resides between the GC-rich elements in the thymidine kinase promoter. In eukaryotes, transcription factors are proteins that are needed to initiate transcription. In fact, three of them (TFIID, TFIIA, TFIIB) bind sequentially to the promoter region prior to the polymerase, and a fourth (TFIIE) must bind later for transcription to begin. Another highly purified mammalian factor called SP1 is required for the transcription of genes whose promoters include multiple copies of the GC-rich region.

In addition to the sequences that constitute a promoter itself, there are outside elements that drastically alter the efficiency of transcription by RNA polymerase II. One of these activating regions is known as **enhancers**. In eukaryotic DNA sequences, for example in human β-globin promoter, enhancers are located 100 to 200 base pairs upstream and interact with proteins other than RNA polymerase and may increase the rate of transcription 200 fold. Enhancer sequences were first discovered in DNA tumour viruses such as SV40 and polyoma, where they are located near the origin of DNA replication, which is also the region that initiates RNA synthesis both early and late after infection. Deletion of enhancer sequences from the genome of SV40 reduces early gene expression by a factor of over 100. Many enhancers have now been characterized that play a key role in the regulation of gene expression. A particular enhancer element can act over considerable distance and function preferentially or exclusively in certain cell types. This tissue-specific function manifests itself as species-specific and that may also in part explain the host range of some animal viruses. Likewise, tissue-specific enhancers could provide the basis for differential expression of genes during development or in the various cell types of a mature organism.

There are other regulatory sites referred to as silencers that repress the expression of genes. Transcription in eukaryotes is often regulated by enhancers and silencers that can influence gene expression from distances of greater than 1000 nucleotide-pairs. Enhancers and silencers are orientation- and position-independent; they work equally when positioned on either side of a gene. Considerable evidence shows that methylation of bases in DNA can alter the expression. Steroid hormones (for example glucocorticoids, oestrogen, testosterone, ecdysone, etc.) are known to be major regulators of transcription in higher animals. These hormones are known, in some cases, to become bound to receptor proteins that then activate transcription via sequence-specific binding to enhancer elements.

THREE MAJOR CLASSES OF RNA

All three major types of RNAs, viz. messenger RNA (mRNA), transfer RNA (tRNA), and ribosomal RNA (rRNA) are derived from precursor RNA that are considerably longer than the "mature" RNA product. The initial RNA molecule (called **primary transcript or pre-RNA**) copied from the template strand of DNA undergoes a series of modifications to form a **mature transcript** that may no longer remain as an exact copy of the DNA. However, bacterial mRNAs are rarely processed. Usually in bacteria mRNA is transcribed on one end and its translation is initiated from the other end before the transcription is completed. Eukaryotic pre-mRNA is also called **heterogeneous nuclear RNA** (hnRNA) because it represents a diverse nucleotide sequence with large molecular weight and it is found in the nucleus of eukaryotic cell. RNA processing requires a variety of small RNAs (90 to 300 nucleotides long) and their associated proteins. These RNAs are called **small nuclear RNAs** (snRNAs) due to their small size and location in the nucleoplasm. There are more than a dozen different low molecular weight snRNAs that are rich in uracil residues.

Ribosomal RNA (rRNA)

It was explained in chapter 2, that the eukaryotic genes located in nucleolar organizer (rDNA genes) that encodes ribosomal RNA is part of the moderately repeated fraction of the genome. Eukaryotic cells contain millions of ribosomes, each of which contains several molecules of rRNA associated with dozens of ribosomal proteins. Ribosomal RNA (rRNA), which is stable or insoluble RNA, constitutes the largest part (up to 80%) of the total cellular RNA. To furnish the cell with such a large number of rRNA, the DNA sequences encoding rRNA called **rDNA sequences** are normally repeated hundreds of times. In interphase, the clusters of rDNA are gathered together as a part of one or more irregular shaped nucleoli, which serve as ribosome-producing organelles. Nucleoli disappear during mitosis and then reappear in the nuclei of the daughter cells. The regions of a chromosome that contain rDNA are known as **nucleolar organizers**.

Eukaryotic cells consist of four types of rRNA molecules namely, 28S, 18S, 5.8S and 5S. The 28S, 5.8S and 5S rRNAs occur in 60S ribosomal (large) subunit, while 18S rRNA is present in 40S ribosomal (small) subunit. The large and small subunits together form a eukaryotic ribosome (80S). The prokaryotic cells consist of three kinds of rRNA molecules viz. 23S, 16S, and 5S rRNA. The 23S and 5S rRNAs occur in 50S large ribosomal subunit and

16S rRNA occurs in 30S small ribosomal subunit, together they form 70S ribosome in prokaryotes. The most remarkable property of rRNA is a high degree of internal base-pairing (the single-stranded RNA folds back on itself) to give an elaborate three-dimensional structure. There appears to be remarkable homology in the general shape of analogous rRNAs in prokaryotes and eukaryotes. For example, the small subunit rRNA from the prokaryotes, eukaryotic cytoplasm, and eukaryotic organelles have very different nucleotide sequences, yet their three dimensional base-paired structures are quite similar (Figure 10.6).

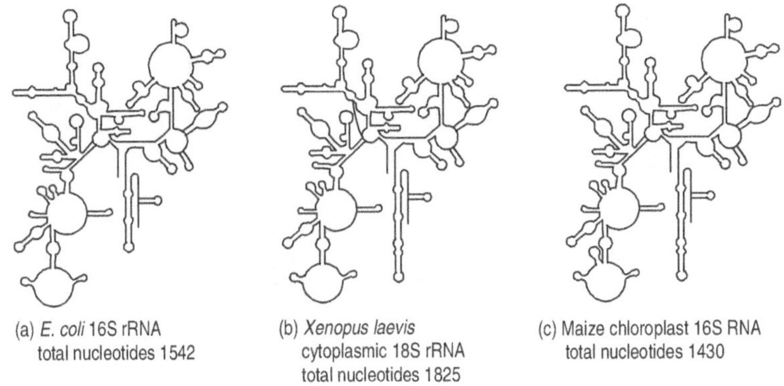

Figure 10.6 Three-dimensional structures of small subunit rRNAs from (a) *E. coli*, (b) *Xenopus laevis*, and (c) Maize chloroplast. These rRNAs have different sequences, yet their structures are quite similar.

In bacteria, the genes responsible for synthesis of all three types of rRNA are clustered in one region to form a single operon working as a functional unit. The synthesis of rRNA is initiated at a promoter and completed at a terminator sequence, whereas in eukaryotes, the rDNA genes appear to be clustered together as repeated sequence in the regions of the genome and often in the secondary constriction (**nucleolar organizer**) of the chromosome. The number of these genes per cell has been estimated to be 450 per haploid genome of African clawed toad *Xenopus*. Furthermore, it has been calculated that the average ribosomal gene has a mass of 8.7×10^6 daltons. As a result, there are approximately 900 copies of ribosomal genes per cell in the normal diploid *Xenopus*. Using a special technique, it is possible to spread and visualize the rDNA genes in action. **Non-transcribed spacer** separates each tandem repeat of transcriptional unit of ribosomal genes. Each repeat unit has a **coding region** with the genes for 18S, 5.8S and 28S rRNA molecules that exist next to each other in the order mentioned and **internal transcribed spacers** whose corresponding RNAs are degraded during processing, one each between 18S and 5.8S and another between 5.8S and 28S gene (Figure 10.7).

The 5S rRNA is synthesized from a separate RNA precursor present outside the nucleolus. The *Xenopus* genome has two different sets of genes encoding 5S rRNA. The oocyte-specific genes are expressed exclusively during the long period of oogenesis, when the maturing oocyte stores various macromolecules for use after fertilization. The somatic 5S genes are expressed primarily after fertilization and throughout subsequent development of the frog. Transcription of oocyte-specific 5S genes requires three factors in addition to RNA polymerase III. Two of these called TFIIIB and TFIIIC are general factors. In contrast, the third factor TFIIIA is an oocyte-specific protein that binds tightly to the internal control region. The 5S rRNA is stored in the cytoplasm of the egg as a ribonucleoprotein (RNP) particle consisting of 5S rRNA combined with a specific protein molecule. After fertilization, the 5S rRNA is used to build the large subunit of ribosome in association with 28S rRNA, which further associates with its small counterpart to actively participate in protein synthesis to fulfil the demand of proteins in the developing zygote.

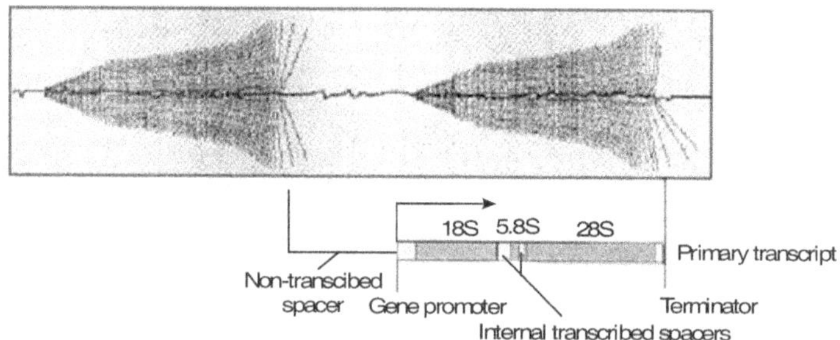

Figure 10.7 The portion of the rDNA gene in a nucleolus transcribing into rRNA in *Xenopus*. Three of the rRNAs (28S, 18S and 5.8S) are derived from a single primary transcript (called pre-rRNA).

Transfer RNA (tRNA)

These are small RNA molecules containing about 70 to 90 ribonucleotides, and are also known as soluble RNA (sRNA) playing a significant role in transfer of amino acids from the "amino acid pool" to the site of protein synthesis. Transfer RNA (tRNA) or soluble RNA constitute about 15% of the total cellular RNA. Eukaryotic cells are estimated to have approximately 60 different species of transfer RNA, each encoded by a DNA sequence that is repeated a number of times within the genome. The degree of repetition is species-specific. For example, there are about 300 tRNA genes in yeast cell, about 850 in fruit flies, and humans have about 1,300 genes. Genes responsible for tRNA synthesis are located within small clusters scattered around the genome. A single cluster typically contains multiple copies of different tRNA genes. According to Crick's adapter hypothesis, the translation of mRNA into proteins requires a molecule that links the information in mRNA codons with specific amino acids in proteins and that molecule is a tRNA. For each of the 20 amino acids, there is at least one specific type ("species") of tRNA molecule. Its molecular structure helps itself to relate three functions such as (i) it carries activated amino acid, (ii) it associates with mRNA molecules, and (iii) it interacts with ribosomes.

Robert Holley (1965) and his colleagues worked out the complete sequence of alanine tRNA molecule in yeast. Later on about 100 different "species" of tRNA were sequenced. All tRNA molecules have a generalized two-dimensional cloverleaf structure (Figure 10.8a) that has several unique properties including the following.

- presence of DNA-like double helical segments due to internal pairing of complementary bases,

- presence of a number of unusual nucleotides like pseudouridine ψ or psi, inosine, dihydroxyuridine (DHU),

- presence of guanine residues at the 5′ terminal end and unpaired CCA sequence at 3′ end that is called amino acid attachment site,

- the amino acid stem consists of seven paired bases,

- the T-stem or T-loop or TψC loop contain seven unpaired bases and five paired bases. This loop helps in binding of tRNA molecule to the ribosome,

- the D-stem or DHU loop or simply D-loop is composed of three or four base pairs. It assists in binding of aminoacyl synthetase,

- the anticodon stem or loop consists of seven unpaired bases, the 3rd, 4th and 5th of which from 3′end of the molecule constitute the anticodon that is complementary to the codon of mRNA,

- an extra small arm with variable number of nucleotides that may be absent in some tRNA molecules.

In 1973, S.H. Kim proposed the most acceptable three-dimensional structure of phenylalanine tRNA molecule found in yeast (Figure 10.8b). According to Kim, each tRNA molecule has a conformation in the shape of the letter 'L' with a thickness of 20 Å that is maintained by complementary base-pairing within its own sequence. This conformation of tRNA molecule allows it to combine specifically with binding sites on the ribosomes. The L-shaped model can be easily derived from the two-dimensional cloverleaf model. At the 3′end of each tRNA molecule is a site to which its specific activated amino acid binds covalently. The anticodon loop consists of three nucleotides that are complementary to the triplet of nucleotides on the mRNA and specify the amino acid. For example, a codon on mRNA is AUG (coding for methionine), the corresponding anticodon on tRNA will be UAC. The codons and anticodons are antiparallel to each other.

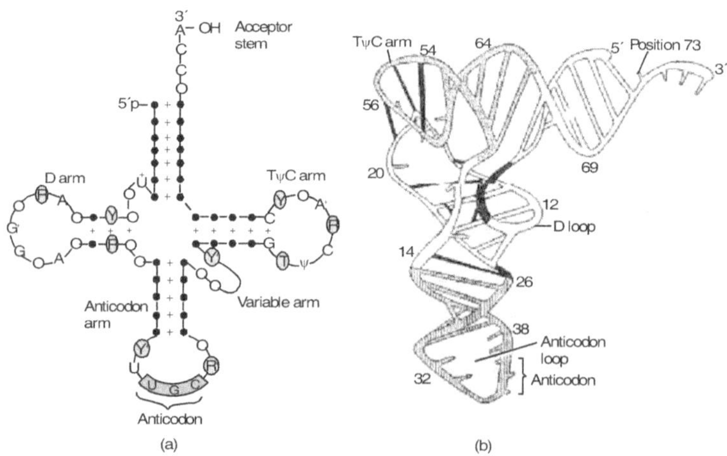

Figure 10.8 (a) Generalized two-dimensional cloverleaf structure of tRNA, (b) Diagrammatic representation of three-dimensional model of yeast tRNA^{Phe}

Messenger RNA (mRNA)

The eukaryotic cell consists largely of rRNA, tRNA, and snRNA. In addition, there is a group represented by RNAs of diverse (heterogeneous) nucleotide sequence with molecular weights up to 80S, or 50,000 nucleotides that are found only in the nucleus. These RNAs are referred to as **heterogeneous nuclear RNAs** (hnRNAs). Based on the experimental observations, the relationship between hnRNA in the nucleus and messenger RNA (mRNA) in the cytoplasm was interpreted. These studies revealed that the much larger hnRNA molecules is processed to form small mRNA molecule and even though mRNA and their hnRNA precursors constitute only a small percentage (about 5%) of total RNAs of most eukaryotic cells, they constitute a high percentage of RNA that is being synthesized by that cell at any given moment.

All mRNA precursors are synthesized by RNA polymerase II from the template strand of DNA in association with a number of auxiliary proteins, called general transcription factors. Generally, each of the RNA polymerase is believed to have its own set of transcriptional factors. The promoter for RNA polymerase lies to the 5' side of the transcription unit. In most of the cases, the critical portion of the promoter lies between 24 and 32 bases upstream from the site at which transcription is initiated. This region contains a sequence of bases that is similar to 5′-TATAAA-3′, which is known as TATA box where the pre-initiation complex forms between RNA polymerase II and general transcription factors. Further in the process transcription factors play a key role in separating the DNA strands of the duplex, allowing the RNA polymerase II to proceed along the sense strand to synthesize RNA molecule. *In vitro* studies show that about 15 to 100 nucleotides are added to the growing mRNA chain per second.

POST-TRANSCRIPTIONAL MODIFICATIONS OF mRNA

In eukaryotes, the mature mRNA is several times smaller than the immediate product of transcription, the hnRNA (or pre-mRNA or primary transcript). This is due to fact that once hnRNA is synthesized, it undergoes a series of post-transcriptional modifications, which include both the addition of nucleotides and loss of some nucleotides. The nuclear proteins catalyse all of these reactions. The following three remarkable changes take place in hnRNA to form mRNA.

1. Addition of "cap" (7-methyl guanosine) at 5′end,

2. Addition of "poly(A) (adenylic acid) tail" at 3′end, and

3. RNA splicing (removal of intervening sequences or introns)

Addition of "Cap" (7-methyl Guanosine) at 5′end

At the 5′end (the first end made by RNA polymerase) of the primary transcript, a special nucleotide, 7-methyl guanosine (7-MeG) is added as a "**cap**" in a 5′ – 5′ linkage. The addition of cap at 5′end makes the resultant mRNA resistant to the attack made by ribonucleases. Thus the cap protects the mRNA from

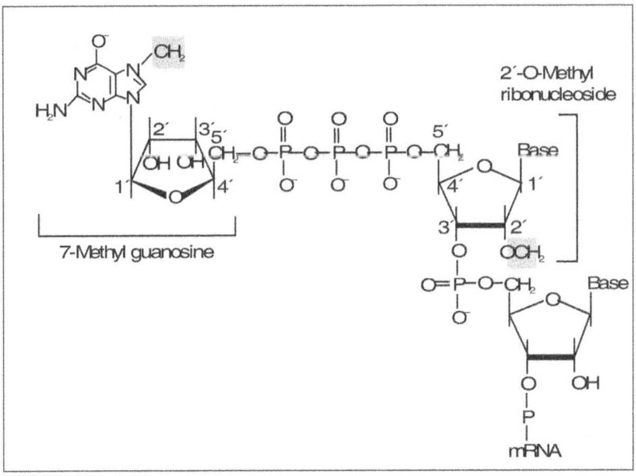

Figure 10.9   Molecular structure of the "cap" attached to 5′ end of mRNA

degradation by nucleases. During capping, a cap of methylated guanosine is added. Sometimes, the cap also includes methylation of additional sugar of both nucleotides: (7MeG)-5'PPP-5'(G or A with possibly methylated ribose)-3'-P—linked to mRNA as shown in Figure 10.9. It is apparent that the methyl group helps in the export of the mRNA from nucleus to the cytoplasm, since non-methylated mRNA is confined to the nucleus. Finally, the cap also assists in binding mRNA to the ribosome, thereby facilitating translation of mRNA.

Addition of "Poly(A)" (Adenylic Acid) Tail at 3'end

At this 3'end of the pre-mRNA, a sequence of RNA nucleotides is recognized by a nuclease that cuts the RNA to make a new 3' end. Then a second enzyme adds a string of about 200 adenosine residues that form a **poly(A)tail** to the newly generated end by an enzyme, called poly(A) polymerase. This energy-dependent process is called polyadenylation. The poly(A) tail is invariably found approximately 15 nucleotides downstream from a sequence AAUAAA that serves as a recognition site for a nuclease that cleaves the pre-mRNA just downstream from the site. The poly(A) tail is necessary for the transport of most mRNAs through the nuclear pore and seems to be important for the longevity of the mRNA in the cytoplasm. Yet mRNAs of most of the histone proteins are not polyadenylated at their 3'ends, they last for the normal period of time.

RNA Splicing (Removal of Intervening Sequences or Introns)

It is the most remarkable post-transcriptional change in mRNA, in which the intervening sequences or introns are removed and subsequently exons (coding sequences) are joined by DNA ligase to form mature mRNA. The length of hnRNA is much more than mRNA due to the presence of intervening sequences or introns that do not code for amino acids. The number of introns per gene is variable and for a given protein, it is not the same in all organisms. For example, there are two introns in the globin gene, while the gene responsible for coding of mitochondrial cytochrome b in yeast is interrupted by six introns. In the globin gene, the junctions between exons and introns have specific sequences that are recognized by an RNA–protein complex (small nuclear ribonucleoproteins bounded with specific proteins) appropriately called the **spliceosome**. Once the splicing complex lands at the 5' splice site, the proteins of spliceosome take over and catalyse the splicing reactions. Each intron forms hairpin loops in such a way that exons are brought nearer to each other. Endonuclease produces nicks at each junction of the exons and introns. As a consequence, looped introns are cleaved from the mRNA. DNA ligase then seals the cut ends of exons bringing them together to form mature mRNA that becomes much smaller than the primary transcript and contains all coding sequences (Figure 10.10).

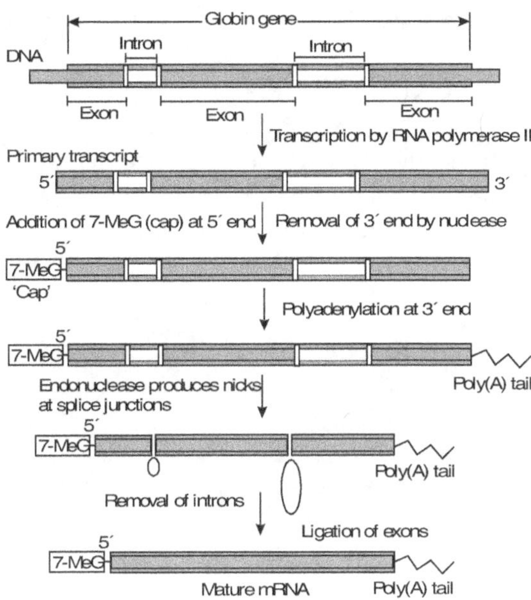

Figure 10.10   Overall steps involved in the post-transcriptional modification of globin mRNA

GENE SILENCING BY RNA INTERFERENCE (RNAi)

It is a biological phenomenon resulting in post-transcriptional silencing of a targeted gene by introduction of double-stranded RNA (dsRNA). RNAi has the ability to knock down the expression of a targeted gene by 75% or more in a simple, effective, specific, and reproducible manner and has tremendous use in both science and medicine. RNAi was first noticed in plants by biologists who are attempting to boost the colour of petunias by manipulating pigment-producing genes. Instead of gaining an expected deep purple colour, the flowers turned white. Researchers conducted anti-sense experiments in the nematode *Caenorhabditis elegans* and noted the effects of gene silencing. Antisense RNA is complementary to mRNA and can form duplex (dsRNA) and block the protein synthesis. The presence of dsRNA longer than 30 nucleotides gives rise to an anti-viral response in mammalian cells, which triggers a non-specific shutdown of protein synthesis and causes degradation of mRNA. **Small (short) interfering RNA (siRNA)** less than 23 nucleotides in length does not trigger such non-specific effects and have been successfully used for silencing genes similar to the long dsRNAs in other systems. RNAi has tremendous applications in combating viral diseases, cancer, ophthalmic diseases, and metabolic disorders like diabetes that offer validated targets and clearly defined clinical endpoints and in diseases where current therapies are not very effective.

RIBOZYME (CATALYTIC RNA)

It is another class of modified RNA that has the ability to catalyse certain chemical reactions. Ribozymes function as true catalysts enhancing the rate of chemical reactions without any net change to themselves. The catalytic ability of ribozymes arises due to their three-dimensional structures (Figure 10.11) which is able to generate in them the substrate specific building site, similar to enzymes.

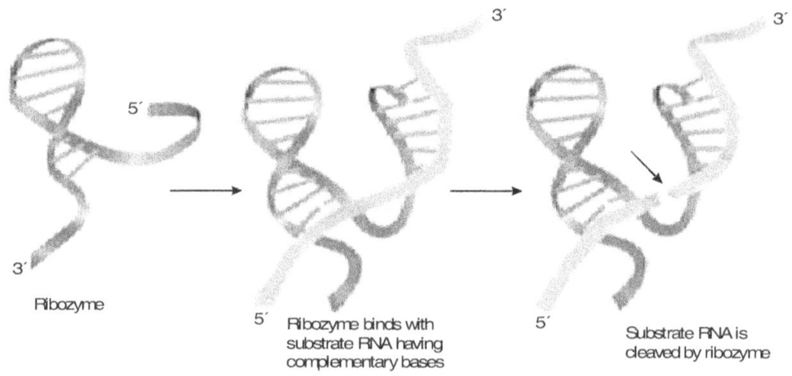

Figure 10.11   Role of ribozyme in breaking down the substrate RNA

Ribozymes have two properties that make them potentially important agents in fighting against certain chronic diseases in humans, animals, and plants.

1. They contain stretches of nucleotides that allow them to base-pair with a complementary RNA.

2. They have a catalytic site (as mentioned above) that is able to cleave the backbone of the complementary RNA.

With these properties, ribozymes are ideal subjects for manipulation by genetic engineers because they can be custom-designed to recognize and destroy any specific RNA target. The custom-designed ribozymes are used in the fight against viral diseases. Recent experiments carried out by Wongstaal and Hempel of University of California and Northern Illinois University respectively, have demonstrated the feasibility of using ribozymes in the fight against diseases caused by RNA viruses. In their experiments, they isolated T lymphocytes from patients with AIDS and transfected the cells with a genetically engineered DNA sequence that encodes a ribozyme capable of recognizing and cleaving HIV mRNA. It is hoped that transfected cells will prove resistant to the virus and preserve the patient's crippled immune system. The ultimate goal in this type of gene therapy is to introduce the gene for an RNA destroying ribozyme into all cells of the body that are infected by the virus, rather than simply those that circulate in the bloodstream.

Ribozymes may also prove useful in the fight against certain types of human cancers. The mutation in an oncogene that encodes a protein involved in regulative cell division, results in human cancer. Since the malignancy in these cells is due to synthesis of an RNA that is different from that produced by normal cell, it should be feasible to reverse the malignant state of the cell by destroying the altered RNA with the help of ribozyme. To be effective as a treatment for cancer, the gene encoding the ribozyme would have to be introduced into all the cells of the tumour, which requires the development of a vector that is capable of delivering the gene to the body's internal tissues. Ribozymes are also used to engineer plants for virus resistance. Transgenic tobacco plants expressing ribozymes against TMV showed some resistance to TMV infection.

SUMMARY

- *Francis Crick (1956) proposed the central dogma of molecular biology indicating that the flow of genetic information from DNA to the protein takes place via RNA. It involves firstly transcription during which mRNA is synthesized from sense strand of DNA and secondly translation in which the codons of mRNA are deduced to sequence the amino acids.*

- *Temin and Baltimore provided a new dimension to central dogma due to their discovery of enzyme called RNA-dependant DNA polymerase or reverse transcriptase in RNA viruses that synthesize DNA duplex from a single-stranded RNA template.*

- *RNA is a single-stranded nucleic acid with ribose sugar instead of deoxyribose and thymine replaced by uracil in its ribonucleotides. Usually the number of purines in RNA is not equal to number of pyrimidines.*

- *Crick et al. proposed the messenger hypothesis suggesting that messenger RNA (mRNA) forms as a complementary copy of one strand of DNA of a particular gene, and the adapter hypothesis indicating that the role of tRNA is that of an adapter molecule that binds a specific amino acid with one region and recognizes the sequence of nucleotides with another region.*

SUMMARY

- Bacterial RNA polymerase is a complex, multimeric protein having six subunits with two identical alpha (α), and each one of beta (β), beta dash (β'), omega (ω), and sigma (σ) chains.

- Single RNA polymerase accomplishes transcription in bacteria whereas there are three RNA polymerases (I, II, and III) in eukaryotic cells to perform the same function. RNA polymerase I present in nucleolus synthesizes the rRNAs. RNA polymerase II located in the nucleoplasm makes hnRNA that modifies to form mRNA and RNA polymerase III present in the nucleoplasm synthesizes tRNAs.

- The promoter gene of RNA polymerase II contains three distinct regions that include TATA box, CAT box, and GC box, each of which is identified by its specific base sequences. Enhancers and silencers often regulate transcription in eukaryotes.

- Ribosomal RNAs are stable or insoluble molecules that are synthesized from rDNA sequences present in the nucleolus in the presence of RNA polymerase I. During mitosis, nucleoli disappear and are represented as nucleolar organizers on a chromosome. Along with proteins, rRNAs form ribosomes in prokaryotic (70S) and eukaryotic cells (80S) and provide the sites for protein synthesis.

- RNA polymerase III present in the nucleoplasm synthesizes about 60 different species of transfer RNAs (tRNA) or soluble RNAs by transcribing DNA sequences that are repeated a number of times within the genome. Transfer RNAs pick up activated amino acids and transfer them to the site of protein synthesis.

- The mRNA precursors called hnRNAs (primary transcripts) are synthesized by RNA polymerase II from nuclear genes with the help of general transcription factors. The primary transcript is modified to form mRNA that carries the message in the form of codons, which are decoded to form a specific sequence of amino acids in the protein.

- The post-transcriptional modifications of hnRNA to form mRNA include addition of 'cap' (7-methyl guanosine) at the 5' end, addition of poly(A) (adenylic acid) tail at 3' end, and removal of intervening sequences or introns (RNA splicing).

- Genes can be silenced or the expression of a targeted gene can be knocked down by RNA interference (RNAi). It is a biological phenomenon resulting in post-transcriptional silencing of a targeted gene by introduction of double-stranded RNA (dsRNA) that has tremendous applications in science and medicine.

- Ribozymes are modified RNAs, which have the ability to catalyse certain chemical reactions. They can be custom-designed to recognize and destroy any specific RNA target specifically. They are used in the treatment of viral diseases and human cancer.

REVIEW QUESTIONS

1. Explain the concept of gene. Describe the one-gene-one enzyme hypothesis and its modification.

2. What is the central dogma of molecular biology? How does genetic information flow from genotype to phenotype?

3. What is Teminism? How did the phenomenon modify the central dogma of molecular biology?

4. In what respect does RNA differ from DNA?

5. What is transcription? Explain the composition of bacterial RNA polymerase.

6. Describe the process of transcription in *E. coli*.

7. Explain three different RNA polymerases mediating transcription in eukaryotes.

8. Describe the promoters, enhancers, and silencers in eukaryotic genome.

9. Explain the contrasting characters between the three different classes of RNAs.

10. What are rDNA sequences? Explain its role in transcription.

11. Describe the post-transcriptional modifications in hnRNA to become mRNA.

12. Write short notes on:

    i. Introns

    ii. Ribozymes

    iii. Small interfering RNA (siRNA)

    iv. Short nuclear RNA

    v. Primary transcript

    vi. RNA splicing

    vii. Sigma factor

    viii. Prokaryotic ribosomes

    ix. Rho factor

    x. Poly(A) tail

    xi. Reverse transcriptase

    xii. 7-methyl guanosine

    xiii. TATA box

    xiv. Transcription factors

TRANSLATION—THE SYNTHESIS OF PROTEIN

INTRODUCTION

Proteins are the end products of most information pathways. A typical cell requires thousands of different proteins at any given moment. The functions of proteins can be described at three levels. **Phenotypic function** describes the effect of a protein on the entire organism. For example, the loss of the protein may lead to slower growth of the organism, and altered development pattern, or even death. **Cellular function** is a description of the network of interactions engaged by a protein at the cellular level. Interactions with other proteins in the cell can help define the kinds of metabolic processes in which the protein participates. Finally, **molecular function** refers to the precise biochemical activity of a protein, including details such as the reactions that an enzyme catalyses or the ligands a receptor binds. Proteins must be synthesized in response to the cell's current needs, targeted to their appropriate cellular locations, and degraded when no longer needed. Proteins are macromolecules with different structural complexity and integrity and are made up of polypeptide chains of several amino acids joined by peptide bonds.

Protein synthesis is the most complex biosynthetic process that involves several to thousands of copies of many different proteins, enzymes, and RNA molecules. In a prokaryotic cell, a single mRNA molecule may code for one or several polypeptide chains. If the mRNA carries the coded language, which is transcribed from the template strand of DNA and codes for only one polypeptide, then such mRNA is called **monocistronic** (the "cistron" indicates the functional unit of a gene); if it codes for two or more different polypeptides, the mRNA is referred to as **polycistronic**. In eukaryotes, most mRNAs are monocistronic. On the translation of mRNA, a protein is synthesized that undergoes post-translational modifications and may perform the function as an enzyme, a hormone, a structural component of cell membrane or cell organelle, an antibody, carrier protein such as haemoglobin or myoglobin, a storage protein such as albumin or globulin, contractile protein such as actin and myosin, or receptor proteins. Despite the great complexity of protein synthesis, proteins are made at exceedingly high rates. A polypeptide of 100 amino acid residues is synthesized in about 5 seconds in an *Escherichia coli* cell at 37°C. The minimum length of an mRNA is set by the length of polypeptide chain for which it codes. For example, a polypeptide chain of 50 amino acid residues requires an mRNA coding sequence of at least 150 nucleotides, because a triplet of nucleotides codes each amino acid. To understand the relationship between the genetic code of DNA, codon of mRNA, and the sequence and number of amino acids in a protein, several researchers put their efforts.

GENETIC CODE

How do transcription and translation produce a specific and functional protein product? This was a major challenge faced by the molecular biologists. Once the structure of DNA had been described, the investigators directed their efforts to find out the relation between sequence of amino acids in a polypeptide and the sequence of nucleotides in the DNA of a gene. It was widely assumed that the information for protein synthesis in the nucleotide sequence in the DNA was present in some type of **genetic code**. The coded message of DNA is called **cryptogram**. This genetic code specifies which amino acids will be used to build a protein. During transcription, the genetic information of DNA is handed over to the complementary mRNA molecule as a series of sequential, non-overlapping three-letter "words" called codons. The sequence of codons (triplets of nucleotides) on mRNA along the chain specifies the sequence of particular amino acids in a polypeptide chain. Our present knowledge of the genetic code is based on the experimental output obtained by several investigators.

The physicist George Gamow (1954), for the first time proposed the model of genetic code and suggested that each amino acid in a polypeptide was encoded by three sequential nucleotides (codon). Each codon is complementary to the corresponding triplet in the DNA molecule from which it was transcribed. Gamow reasoned that it would require at least three nucleotides for each amino acid to have its own unique codon. It was a well-established fact that the genetic code of DNA made up of four-letter alphabets (A, T, G, and C corresponding to deoxyribonucleotides that contain adenine, thymine, guanine and cytosine respectively) codons on mRNA containing A, U (for uracil), G, and C. Consider that each nucleotide will code for one amino acid. It would be insufficient to code for all 20 different amino acids by only 4 nucleotides ($4^1 = 4$ codons such as A, U, G, and C). If it were assumed that each code is formed of two alphabets, i.e., 2 nucleotides (any two of the 4 nucleotides are combined), by permutation and combination, it would produce 16 codons ($4^2 = 16$ combinations as shown in Figure 11.1) that would produce ambiguity in determination of all the 20 amino acids. But as per Gamow's model, there must be at least 3 (triplet) successive nucleotides for the specification of 20 different amino acids. Thus, there are 64 codons ($4^3 = 4 \times 4 \times 4 = 64$ combinations as shown in Figure 11.2) possible that are more than the number of amino acids. Obviously then each amino acid will have more than one code indicating the degeneracy of codes. The triplet nature of the code was soon verified by a number of genetic researchers.

Second letter in code

		A	U	G	C
First letter in code	A	AA	AU	AG	AC
	U	UA	UU	UG	UC
	G	GA	GU	GG	GC
	C	CA	CU	CG	CC

Figure 11.1   Sixteen possible doublet codons that are insufficient to code 20 amino acids

Francis Crick reasoned out on how the genetic information encoded in 4-letter language of nucleic acids could be translated into the 20-letter language of proteins. A small nucleic acid, i.e., tRNA acts as an adapter that "translates" the nucleotide sequence of mRNA into amino acid sequence of a polypeptide. Hoagland and Zamecnik in their experiments found that amino acids were "activated" in the presence of ATP before they bind to tRNA molecules. Robert Holley discovered and characterized the enzyme **aminoacyl-tRNA synthetase** that catalyses the binding reaction of activated amino acid to the tRNA molecule.

Second letter in code

First letter in code		U	C	A	G	Third letter in code
	U	UUU, UUC Phe; UUA, UUG Leu	UCU, UCC, UCA, UCG Ser	UAU, UAC Tyr; UAA, UAG Termination codons	UGU, UGC Cys; UGA - Termination codon; UGG - Trp	U C A G
	C	CUU, CUC, CUA, CUG Leu	CCU, CCC, CCA, CCG Pro	CAU, CAC His; CAA, CAG Gln	CGU, CGC, CGA, CGG Arg	U C A G
	A	AUU, AUC, AUA Ileu; AUG-Met Initiation codon	ACU, ACC, ACA, ACG Thr	AAU, AAC Asn; AAA, AAG Lys	AGU, AGC Ser; AGA, AGA Arg	U C A G
	G	GUU, GUC, GUA, GUG Val	GCU, GCC, GCA, GCG Ala	GAU, GAC Asp; GAA, GAG Glu	GGU, GGC, GGA, GGG Gly	U C A G

Figure 11.2 The genetic code. Out of sixty-four possible codons, 61 codons specify amino acids, AUG acts as initiation codon and codes for methionine, while UAA, UAG, and UGA are termination codons that stop the process of protein synthesis.

CRACKING OF GENETIC CODE USING ARTIFICIAL mRNA

In 1955, Manago and Ochoa made successful attempts for the synthesis of polynucleotides (mRNA) containing only a single type of nucleotide (U, C, A, or G) that is repeated several times. For example, a polynucleotide containing several units of adenine is called poly(A) strand. Similarly they artificially synthesized poly(U), poly(C), and poly(G) mRNA strands that were utilized in deciphering the genetic code. In similar attempt, Nirenberg and Mathhaei reported the first breakthrough to codon assignment when they used *in vitro* system for the synthesis of a polynucleotide using an artificially synthesized mRNA molecule containing only one type of nucleotide (such mRNA is called homopolymer). They incubated synthetic polyuridylate, poly(U), with an *E.coli* extract, GTP, ATP, ribosomes, tRNAs, aminoacyl-tRNA synthetase enzymes and a mixture of 20 amino acids in 20 different tubes, each tube containing different radioactively labelled amino acids. Because poly(U) mRNA is made up of many successive UUU triplets, it should promote synthesis of a polypeptide containing only one amino acid encoded by triplet UUU. The radioactive polypeptide was indeed formed in only one of the 20 tubes, the one containing radioactive phenylalanine. Nirenberg and Mathhaei came to the conclusion that the triplet UUU encodes phenylalanine. In the same approach, it was found that poly(C) encodes the polypeptide containing proline and poly(A) encodes a polypeptide containing lysine. Whereas poly(G) message was found non-functional *in vitro* system, because it spontaneously attains secondary structure that cannot be bound by ribosomes.

Nirenberg and Philip Leder (1964) achieved another breakthrough in providing ribosome-binding technique. They made use of the finding that aminoacyl-tRNA molecule specifically binds to ribosome–mRNA complex. This binding does not require the presence of a long mRNA molecule; in fact, even trinucleotide or minimessenger could promote specific binding of appropriate tRNAs, so these experiments could be carried out with chemically synthesized small oligonucleotides. Using this technique, researchers determined the amino acid corresponding to each triplet and thus it was possible to decipher 61 out of the possible 64 codons.

PROPERTIES OF GENETIC CODE

On the basis of genetic studies, several key properties of genetic code were established that are as follows:

The Code is a Triplet Codon

A codon is a triplet of nucleotides present on mRNA that codes for a specific amino acid. Translation of this message takes place in such a way that these nucleotide triplets are read in a successive and non-overlapping fashion. The first codon of mRNA establishes the "reading frame" in which the new codon begins every three nucleotide residues. The protein-synthesizing machinery reads the codons from initiation to termination in a non-overlapping manner and there is no punctuation between codons for successive amino acid residues. In principle, any single-stranded DNA or mRNA sequence has three possible reading frames, and on translation of each reading frame different sequence of amino acids is formed, but the correct reading frame with initiation codon is likely to encode a given protein (Figure 11.3).

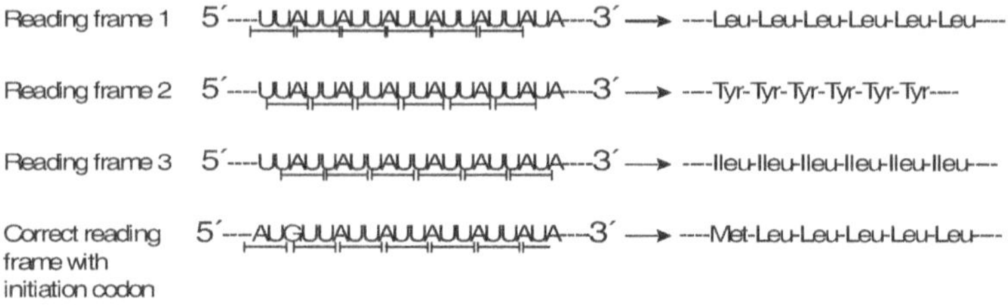

Reading frame 1   5′---UUAUUAUUAUUAUUAUUAUA---3′ ⟶ ---Leu-Leu-Leu-Leu-Leu-Leu---

Reading frame 2   5′----UUAUUAUUAUUAUUAUUAUA---3′ ⟶ ---Tyr-Tyr-Tyr-Tyr-Tyr-Tyr----

Reading frame 3   5′----UUAUUAUUAUUAUUAUUAUA---3′ ⟶ ---Ileu-Ileu-Ileu-Ileu-Ileu-Ileu---

Correct reading frame with initiation codon   5′---AUGUUAUUAUUAUUAUUAUA---3′ ⟶ ----Met-Leu-Leu-Leu-Leu-Leu---

Figure 11.3   Three possible reading frames that encode different amino acids in a polypeptide. Last one is correct reading frame that starts with initiation codon.

Evidently, during encoding the message on mRNA, due to deletion, addition, or deletion and addition mutations (see chapter 9) the reading frame is changed and leads to synthesis of aberrant protein. Such alterations in the sequence of nucleotides in DNA or mRNA are referred to as "frameshift mutations". Sometimes the long sequence of codons may be interrupted by stop codon due to any one of mutations mentioned above that cause synthesis of truncated polypeptide chain.

The Code is Non-overlapping

During translation, the codons on mRNA are "read" sequentially in a group of three nucleotides without overlapping. Thus a base from a first codon does not participate to form the next codon indicating that each codon is an independent unit. Usually, nine bases of the three codons are translated into three amino acids suggesting that each codon is an independent unit of three bases as shown in Figure 11.4. Thus each nucleotide along an mRNA is a part of one, and only one, codon.

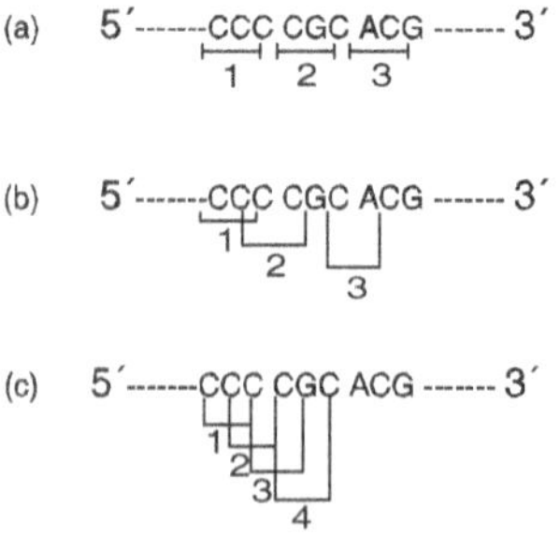

(a)   5′-------CCC CGC ACG ------- 3′
         1   2   3

(b)   5′-------CCC CGC ACG ------ 3′
         1   2   3

(c)   5′-------CCC CGC ACG ------ 3′
         1   2   3   4

Figure 11.4   (a) The non-overlapping codons, (b) Overlapping codons due to one base, and (c) Overlapping of codons due to two bases

Mutational studies revealed that the code is non-overlapping. In tobacco mosaic virus the change of one base in the DNA due to mutation would alter the composition of protein by a single amino acid proving that the code is non-overlapping. Similar base substitution mutations that resulted in single amino acid alterations were reported in insulin, tryptophan synthetase, alkaline phosphatase, and haemoglobin.

The Code is Commaless

There is no comma or punctuation between successive codons. Once the first codon is translated into an amino acid, the second amino acid will automatically be coded by the next three bases and there is no base reserved for comma or punctuation mark. This clearly indicates that the genetic code is commaless. Khorana and his associates clearly indicated that there is no delimiting point between one codon and the other. They used artificially synthesized long polynucleotide chains to produce polypeptides with repeating amino acids. For example, the repeating sequence ACACAC contains the codons for threonine (ACA) and histidine (CAC). The polypeptide synthesized on this artificial messenger RNA contained equal amount of threonine and histidine. This clearly indicated that triplets of bases are commaless.

The Code is Degenerate

As mentioned in Figure 11.2, there are 64 possible codons, of which 61 code for amino acids. Since there are 20 different amino acids to constitute the protein, it is obvious that more than one codon may specify the same amino acid or in other words, each amino acid must have more than one codon. This is called **degeneracy** or multiplicity of the code that is a striking feature of the genetic code (Table 11.1). For example, of the 20 amino acids, tryptophan and methionine have a single codon each; all 18 amino acids have more than one codon. Thus, nine amino acids, namely phenylalanine, tyrosine, histidine, glutamine, asparagine, lysine, aspartic acid, glutamic acid and cysteine have two codons each. Isoleucine has three codons. The amino acids namely, valine, proline, threonine, alanine, and glycine have four codons each. The amino acids leucine, arginine and serine have six codons each.

The degeneracy of the code may be of partial or complete type. It is called **partial degeneracy** when first two bases of codons are identical but third base differs, for example, UAU and UAC code for tyrosine. The **complete degeneracy** occurs when any one of the four bases can take third position in a codon and still it codes for the same amino acid, for example, ACU, ACC, ACA, and ACG code for threonine.

The multiplicity or degeneracy of code has biological significance since it protects the organisms from mutations. The degeneracy of code permits essentially the same enzyme or other proteins to be specified by microorganisms varying widely in their DNA base composition. Thus, by one or other way, degeneracy provides a mechanism to decrease mutational hazards.

Table 11.1   Degeneracy of genetic code

Amino acid	Abbreviation	No. of codons	Amino acid	Abbreviation	No. of codons
Methionine	Met	1	Cysteine	Cys	2
Tryptophan	Trp	1	Isoleucine	Ileu	3
Phenylalanine	Phe	2	Valine	Val	4
Tyrosine	Tyr	2	Proline	Pro	4
Histidine	His	2	Threonine	Thr	4
Glutamine	Gln	2	Alanine	Ala	4
Asparagine	Asn	2	Glycine	Gly	4
Lysine	Lys	2	Leucine	Leu	6
Aspartic acid	Asp	2	Arginine	Arg	6
Glutamic acid	Glu	2	Serine	Ser	6

The Code has Polarity

Every codon on mRNA is decoded in a fixed direction, i.e., in the $5' \rightarrow 3'$ direction. The particular gene sequence always specifies the amino acid sequence in all the situations unless it is altered due to mutation. This is possible only when the sequence of codons "read" from a fixed initiation point and terminate at a fixed end point in a precise direction. Any change in the direction leads to synthesis of aberrant protein (Figure 11.5). This clearly indicates that codes have polarity and the message on mRNA is decoded in $5' \rightarrow 3'$ direction.

Codon   5′----AUG CUU UGC AAC GCA AGG---3′

Polypeptide synthesized ⟶ Met----Leu----Ser----Asn----Ala----Arg

In 5′⟶3′ direction   Val----Phe----Arg----Gln----Thr----Gly⟵ Polypeptide synthesized
                                                                                in reverse direction

Figure 11.5   The sequence of codons is decoded in 5′ to 3′ direction to synthesize essential protein. If it is decoded in a reverse direction the non-functional protein is synthesized.

Some Codes Serve for Initiation of Protein Synthesis

In most of the organisms, the process of protein synthesis is initiated by AUG, which acts as **start codon** or **initiation codon** and in rare cases GUG serves as initiation codon that codes for valine. It has been found that the chain initiation by GUG is not as efficient as AUG possibly because the former has less affinity for formyl-methionyl-tRNA. In eukaryotes, the amino acid methionine is placed in the first position in a growing polypeptide chain whereas in prokaryotes, in *N*-formyl-methionine (f-Met) occupies the first position in a polypeptide chain. The methionyl or f-Met-tRNA complex specifically binds to the initiation codon on mRNA from where the process begins and continues up to the termination codon.

Some Codes Act as Stop Codons

Of the 64 codons, three codons UAG, UAA, and UGA specifically provide the signal to end the process of protein synthesis. Thus, these codons act as **stop** or **termination** or **nonsense** codons. They do not code for any of the amino acids. These codons are not read by any tRNA molecules, but they are decoded by some specific proteins, called releasing factors, e.g. RF-1, RF-2, and RF-3 in prokaryotes and RF in eukaryotes.

The termination codons are also named after colours such as **amber** (brown) for UAG, **ochre** (yellow red or pale yellow) for UAA, and **opal** (milky white). Of the three termination codons, UAG was the first stop codon to be discovered by Sidney Brenner (1965). It was named **amber** after a German graduate student named Bernstein (means amber in German) who helped in the discovery of a class of mutations. The presence of more than one termination codon might be a safety measure, in case the first codon could not stop the protein synthesis.

The Code is Universal

The same genetic code is found valid for all organisms ranging from bacteria to man. Cracking of genetic code was first carried out in microorganisms but later on it has been clearly proved that the codes are same in their amino acid specification from prokaryotes to eukaryotes. Hence the code is said to be universal. In 1967, Marshall, Caskey, and Nirenberg have shown in their experiments with diverse organisms including *E.coli* (bacterium), *Xenopus laevis* (amphibian), and guinea pig (mammal) use the same code for their aminoacyl-tRNAs.

Additional evidence for universality of codes can be obtained from gene mutation experiments. The base substitution mutation results in replacement of one amino acid by the other during protein synthesis. Studies of this type revealed that gene mutation involving the substitution of a single base affected the same amino acid in TMV, *E.coli*, and also in man.

RELATIONSHIP BETWEEN GENETIC CODE, CODON, ANTICODON AND AMINO ACID

The genetic information on DNA is in the form of genetic codes that are made up of four types of deoxyribonucleotides (represented simply as A, T, G, and C). This information is transcribed into codons on mRNA during the transcription.

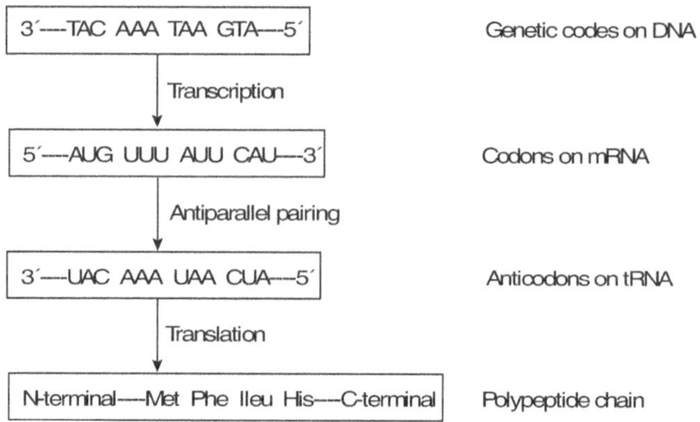

Figure 11.6   Relationship between code, codons, and anticodons

The codons on mRNA are represented simply as A, U, G, and C. The triplet code of DNA is complementary to codon of mRNA, i.e., DNA codes run in 3′ → 5′ direction and mRNA codons run in 5′ → 3′ direction. The three bases present on anticodon loop of tRNA form the anticodon that pairs with the codon of mRNA during protein synthesis indicating that anticodon on tRNA is complementary to codon of mRNA and their pairing is antiparallel. Very often, the anticodon is written in 3′ → 5′ direction. The tRNA molecule brings activated amino acid at the site of protein synthesis (i.e., ribosome) facing its N-terminal (NH$_2$ group) end in the growing polypeptide chain and proceeds in

NH$_2$ → COOH direction. For example, the code TAC on DNA is transcribed to AUG as codon of mRNA. The tRNA with anticodon UAC brings methionine during translation of the message and places this amino acid in the first position facing its N-terminal in the beginning (Figure 11.6).

CRICK'S WOBBLE HYPOTHESIS

The hypothesis is based on the degeneracy of the genetic code or codons. Protein synthesis involves only 20 amino acids while there are 64 possible codons, of which any one of three codons provides the signal to stop the process. Thus, more than one codon can code for the same amino acid indicating the degeneracy of codons. The degeneracy of the genetic code may be of two types: the **first** and **second base degeneracy**, where codons with different bases in the first two positions may encode the same amino acid. In third base degeneracy, codons with different bases in the third position may encode the same amino acid. A group of codons that code for the same amino acid is called a codon family and the members are referred to as synonymous codons. The maximum size of a codon family is six and the minimum size is one (*see* Table 11.1).

There are several codons which differ at the third base position (at the 3′ end) and code for the same amino acid. For example, valine is coded by the triplets GUU, GUC, GUA, and GUG. These 4 codons form a codon family and each codon of this family is called synonymous codon. The codons for most amino acids can be symbolized by XYA/G or XYU/C. The first two letters of each codon are primarily responsible for determination of a specific amino acid while the third position of the base is least important in the specification of the amino acid.

As mentioned above, the codons of mRNA are complementary to anticodons of tRNA and they form pairing in an antiparallel direction, i.e., the first base of the codon in mRNA (read in 5′ → 3′ direction) pairs with the third base of anticodon. Since the third base of the codon and first base of anticodon do not significantly involve in specification of amino acid, this position is called wobble (means non-specific). The third base degeneracy is explained by the wobble hypothesis of Francis Crick (1966). This hypothesis allows a single tRNA species to be recognized by several different codons as shown in Figure 11.7.

The wobble hypothesis suggests that the pairing in the third base is ambiguous, i.e., the normal bases become less discriminating in the wobble position, and in some cases can only recognize the type of base (purine or pyrimidine) rather than the specific base in the opposite strand. Thus a single tRNA can bind with more than one mRNA codon differing in only the third base. For example, the anticodon UCG of tRNA read codons AGC and AGU of mRNA indicating that the first base G in anticodon of tRNA can

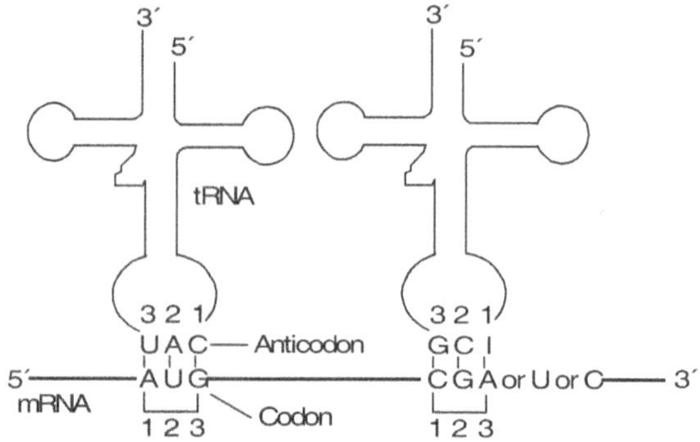

Figure 11.7 Base-pairing between anticodon and codon according to wobble hypothesis

pair with C or U, the third base in codon of mRNA. In other cases, the presence of inosine (I) in the first place in anticodon can interact with A, U, or C present at third position in codon of mRNA. The first two bases in codon of mRNA always form strong Watson–Crick base-pairing with the corresponding bases in anticodon of tRNA and confer most of the specificity. The pairing between third base of codon and first base of anticodon is loose and hence it permits rapid dissociation of tRNA from its codon during protein synthesis. If all three bases of codon are engaged in strong Watson–Crick base-pairing with three bases of anticodon, the dissociation of tRNA would become very slow and this would severely hamper the rate of protein synthesis. The interaction between codon and anticodon is neatly balanced for the accuracy and speed of protein synthesis.

EXCEPTIONS TO UNIVERSALITY OF GENETIC CODE

It has been mentioned above that the genetic code is universal and ever since it originated it has remained unaltered. However, the genetic study in yeast mitochondria and in Mycoplasma carried out during 1980s revealed that genetic codes in these organisms are rather different from others. In yeast mitochondria UGA codes for tryptophan while in nuclear genes in the same organism UGA is a stop codon that terminates the protein synthesis. This however is not to say that in all mitochondria of all organisms the UGA codes for tryptophan. For example, in maize cytochrome oxidase, tryptophan is coded by CGC. This code in nuclear genes codes for arginine. Similarly in 1968, a different genetic code was shown to be working in the ciliate member of Mycoplasma (*Mycoplasma capricolum*). In this organism the codes UAA and UAG do not serve for termination of protein synthesis instead they specify glutamine in the growing polypeptide.

MECHANISM OF TRANSLATION

The RNA-directed protein biosynthesis is the most complex biochemical process in which the message of mRNA is decoded with the help of tRNA molecules that bring the activated amino acids to the site provided by ribosomes. The ribosomes are viewed as protein-synthesizing machines, which bring all necessary components together for protein synthesis and also provide the enzymes for catalysing the peptide bond formation between activated amino acids brought by tRNA molecules. Prokaryotic and eukaryotic ribosomes are quite similar in design and in their functions (see chapter 2 and 10). Each of these is composed of a large and a small subunit. Eukaryotic ribosomes (80S) are roughly half the weight of RNA. The small subunit (40S) consists of one ribosomal RNA (18S rRNA) and about 33 different proteins designated as S_1, S_2, $\cdots$ S_{33}. The large subunit (60S) is made up of three different rRNAs (28S, 5.8S, and 5S) and 49 different ribosomal proteins designated as L_1, L_2, $\cdots$ L_{49}. Most of the ribosomal proteins are rich in basic amino acids and their molecular weight ranges between 7000 to 32,000 daltons. The prokaryotic ribosomes are slightly smaller than their eukaryotic counterparts. The prokaryotic ribosome (70S) consists of a large subunit (50S) with 23S, 5S rRNAs and 35 ribosomal proteins and a small subunit (30S) with 16S rRNA and 21 specific proteins. Mitochondria and chloroplasts also contain ribosomes, some of which are similar to those of prokaryotic ribosomes. Different proteins and rRNAs in ribosomal subunits are held together by ionic and hydrophobic forces and not by covalent bonds. If these forces are disrupted by detergents, proteins and rRNAs separate from each other. If detergent is removed, the entire complex structure self-assembles.

Prokaryotic and eukaryotic ribosomes do not produce just one kind of protein. A ribosome can use any mRNA and all species charged of tRNAs, and thus can be used to make many different polypeptide molecules. Thus the ribosome serves as a molecular workbench where the process of protein synthesis is accomplished. Its structure helps it to hold the mRNA and tRNA molecules with activated amino acids in the right positions, thus allowing the growing polypeptide to be assembled efficiently. Several domains of the ribosome have particular significant functions during translation of the message of mRNA. The small subunit contains a binding site for mRNA and two major sites for tRNA. The **A-site** (aminoacyl-tRNA site) is one into which the aminoacyl-tRNAs enter the ribosome–mRNA complex, whereas the **P-site** is one from which tRNA donates amino acid to the growing polypeptide. The initiator tRNA (tRNA^{Met}) probably appears initially in the P-site of the ribosome. Bacterial ribosomes have a third **E-site** (exit site) to which spent tRNAs are dispatched prior to ejection (Figure 11.8). The large subunit possesses a peptidyl transferase domain that catalyses the peptide bond formation and a GTP domain that is required for translocation of ribosomes along the mRNA.

Regardless of the location of ribosomes (whether they are present in cytoplasm or found attached to endoplasmic reticulum or may be present in mitochondria or plastid), their function is to facilitate the translation of a copy of genetic information encoded in mRNA into specific sequence of amino acids in a polypeptide chain. In an overall view, the process of protein synthesis is completed in initiation, elongation and termination.

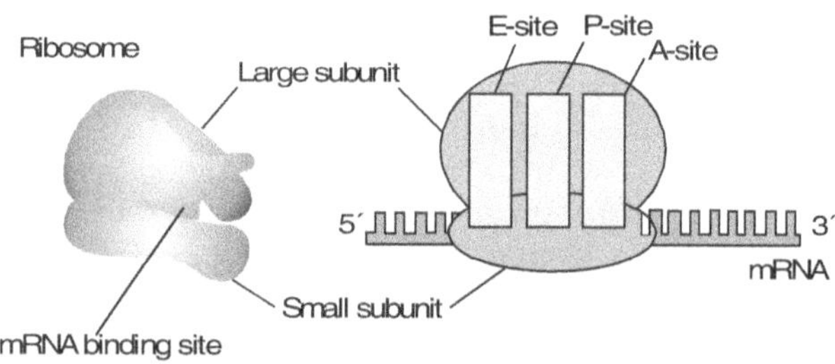

Figure 11.8 Two subunits of ribosomes showing their RNA binding sites

Activation of Amino Acids and Formation of Aminoacyl-tRNA

It is the initial step in the protein biosynthesis in which amino acids are brought to the ribosome by their appropriate adapter tRNA molecules. In the cytoplasm, each of the 20 amino acids is present in an inactive state and before of its attachment to its specific tRNA, it is activated by a specific activating enzyme called aminoacyl-tRNA synthetase in the presence of ATP. Thus each amino acid has its appropriate synthetase enzyme, the molecular weight of which ranges from 100,000 to 240,000 daltons and made up of single or more polypeptides and all of them require Mg^{++} ions for their activity. Aminoacyl-tRNA synthetase catalyses the following two-step reaction:

$$AA + ATP \underset{\text{Aminoacyl-tRNA synthetase}}{\rightleftharpoons} AA{\sim}AMP + PPi$$

$$\text{Aminoacyl adenylate complex} \qquad \text{Pyrophosphate}$$

$$AA{\sim}AMP + tRNA \underset{\text{Aminoacyl-tRNA synthetase}}{\rightleftharpoons} \text{Aminoacyl-tRNA} + AMP$$

It is a critically important process during protein synthesis in which each tRNA molecule is attached to the "correct" (cognate) amino acid. Each amino acid covalently attaches to the 3′ end of its cognate tRNA in two-step reactions that are mediated by the enzyme aminoacyl-tRNA synthetase. As a result, AMP (adenosine monophosphate) and enzyme are released and a final product, aminoacyl-tRNA, is formed that moves towards the site of protein synthesis, i.e., ribosomes.

Thus, each amino acid is recognized by a specific aminoacyl-tRNA synthetase, which is capable of "charging" all the tRNAs that are appropriate for that amino acid. For example, methionine attaches to its tRNA (tRNA^{Met}) by methionyl-tRNA synthetase to form methionyl-tRNA (simply met-tRNA) and similarly, cysteine joins to its tRNA (tRNA^{cys}) by cysteinyl-tRNA synthetase to form cysteinyl-tRNA (or simply cys-tRNA) (Figure 11.9).

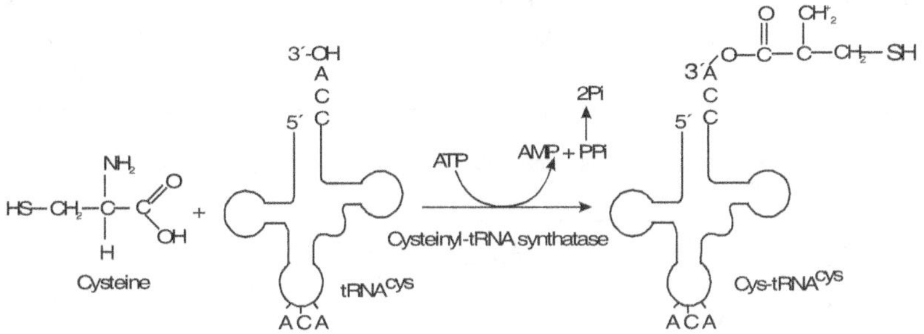

Figure 11.9 Activation of cysteine and formation of Cys-tRNA^{Cys} in the presence of ATP and cysteinyl-tRNA synthetase. The tRNA molecule has CCA base sequence at 3'end, where cysteine attaches by ester link forming charged Cys-tRNA^{Cys}.

In the case of prokaryotes, the first tRNA is acylated with modified amino acid N-formyl methionine (fMet) and its tRNA is often referred to as tRNA^{fMet}. Both tRNA^{fMet} and tRNA^{Met} recognize the codon AUG, but only tRNA^{fMet} is used for initiation. The tRNA^{fMet} molecule is first charged with methionine, and an enzyme (found only in prokaryotes) adds a formyl group to the amino group of methionine. In eukaryotes also, the initiating tRNA molecule is charged with methionine, it is not formylated. In all prokaryotes, N-terminal (–NH$_2$ end) of the nascent polypeptide is marked by the presence of N-formyl methionine and all eukaryotic proteins have methionine at amino terminus. However, these amino acids are frequently removed by the action of hydrolytic enzymes during post-translational processing of protein.

Formation of Initiation Complex

It is of vital importance that a ribosome begins its translation of an mRNA at exactly the correct point. The mRNAs have 5' and 3' polarity and coding regions lie in between. Initiation of protein synthesis involves the formation of a complex between the small subunit of ribosomes, initiator tRNA and its cognate mRNA codon.

The process of translation, or protein synthesis, in prokaryotes (for example in *E.coli*) begins with the association of 30S ribosomal subunit, an mRNA molecule, fMet-tRNA^{fMet}, three protein molecules called initiation factors (IFs such as IF1, IF2, and IF3, properties of which are mentioned in Table 11.2), and guanosine 5′-triphosphate (GTP). The initiation site is the codon AUG (less frequently GUG); however, AUG codon can occur anywhere in the mRNA since it codes for methionine, and for a lengthy period, biochemists were wondering how the ribosome initiated only at the first AUG and not one of the others further down the mRNA. Why wasn't there a special codon always used for initiation? The answer to this confusing situation came from the facts that there are two different tRNAs, both are specific for methionine and they have same anticodon (UAC), but one tRNA is used for initiation and the other exclusively for addition of methionine in the elongating polypeptide.

Table 11.2 Properties of protein factors required for initiation of translation in *E.coli*

Initiation factor	Mol.wt.	Functions
IF1	9,000	Prevents premature binding of tRNAs to A-site. Stimulates activity IF2 and IF3.
IF2	1,20,000	Facilitates binding of fMet-tRNA^{fMet} to 30S ribosome.
IF3	22,000	Binds to 30S subunit; prevents premature association of 50S subunit; enhances specificity of P-site for fMet-tRNA^{fMet}.

Translation of mRNA begins with the formation of initiation complex, which consists of a charged tRNA bearing first amino acid of the polypeptide chain and a small ribosomal (30S) subunit, both bound to a complementary ribosome recognition sequence on mRNA (Figure 11.10). This sequence of bases on mRNA is known as the **Shine–Dalgarno sequence**, which is complementary to a section of 16S rRNA present in the small ribosomal subunit. This sequence is "upstream" (towards the 5′ end) of the actual start codon that begins translation. In the presence of initiation factors, mRNA, fMet-tRNA^{fMet}, and GTP, a complex of these with a 30S subunit is formed and IF3 is released. The charged tRNA, bearing the anticodon UAC and bringing the first *N*-formyl methionine, is positioned at the P-site on the subunit with its anticodon paired with AUG codon. Once IF3 is released, the complex then associates with a 50S subunit.

The event is accompanied by the hydrolysis of GTP and release of GDP and Pi and of IF1 and IF2. As a consequence, a complete 70S ribosome is positioned on mRNA in P-site with the donor fMet-tRNA^{fMet} anticodon base-paired with initiation AUG codon. The A-site being vacant, awaits the second charged tRNA to deliver the specific amino acid. This brings the completion of the initiation process.

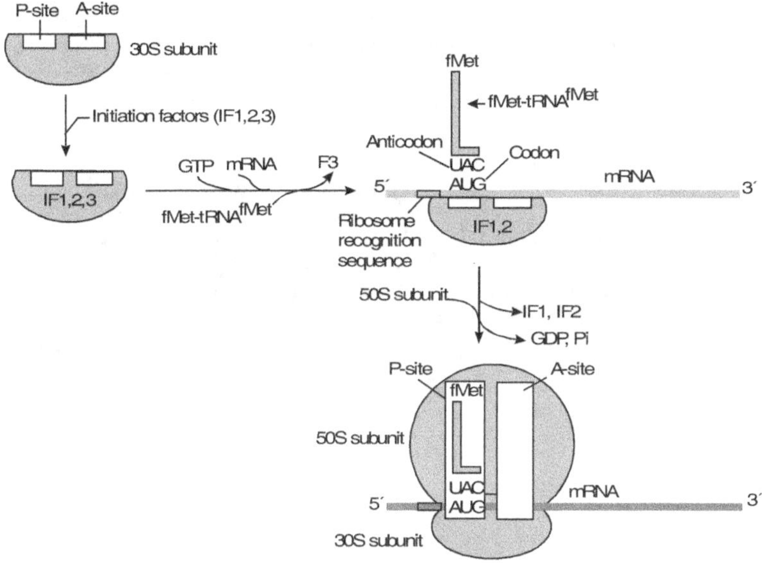

Figure 11.10   Steps in the formation of initiation complex during protein synthesis in prokaryotes

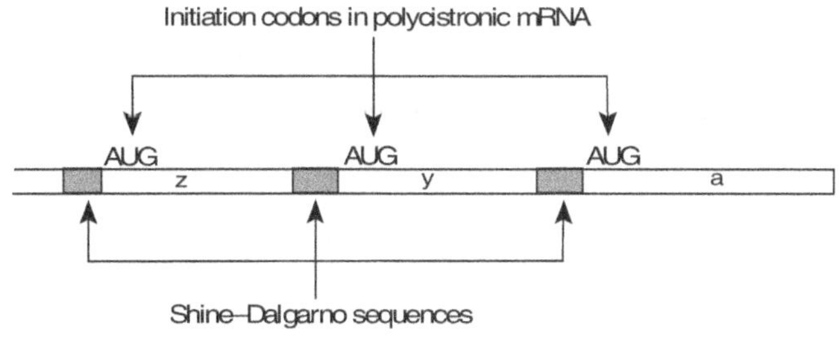

Figure 11.11   Bacterial polycistronic mRNA showing three coding regions, z, y, and a with their independent initiation codons

Most of the bacterial mRNA molecules are polycistronic (i.e., coding for more than one protein), the *lac* mRNA is one such example (*see* chapter 12). In this, there are three regions in the one mRNA molecule coding for three different proteins. In this situation, each coding region has a Shine–Dalgarno sequence adjacent to it so that each can be transitionally initiated independently as shown in Figure 11.11.

Table 11.3   Properties of eukaryotic initiation factors (eIFs) involved in the formation of initiation complex.

Factor	Structure and Mol. wt.	Functions
eIF1,	Monomer; 15,000	Helps in mRNA binding.
eIF2	Trimeric; 1,28,000	Facilitates binding of initiating Met-tRNA^{Met} to 40S ribosomal subunit.
eIF2B	Monomer; 38,000	Binds to Met-tRNA^{Met}.
eIF3	Multimer; 7,50,000	Assist mRNA binding.
eIF4A	Multimer; 15,000	RNA helicase activity removes secondary structure in mRNA to permit binding to 40S subunit.
eIF4B	Monomer; 15,000	Facilitates scanning of mRNA to locate the first AUG.
eIF4E	Monomer; 15,000	Binds to the 5´ cap of mRNA; part of eIF4F complex.
eIFG	Monomer; 15,000	Binds to eIF4E and to poly(A) binding protein; part of the eIF4F complex.
eIF5	Monomer, 1,50,000	Promotes dissociation of several other initiation factors from 40S subunit and assist in association of 60S subunit to form 80S initiation complex.
eIF6	Monomer; 23,000	Helps in dissociation of inactive 80S ribosome into 40S and 60S subunits.

As compared to prokaryotes, in eukaryotes the process of protein synthesis follows the same general pattern so far its initiation complex is concerned except for the number of initiation factors involved. In eukaryotic cells (e.g. reticulocytes) about ten initiation factors have been identified (Table 11.3) and are referred to as eIF1, eIF2, eIF2B, eIF3, eIF4A, eIF4B, eIF4E, eIF4G, eIF5, and eIF6 (a prefix "e" indicates the eukaryotic origin for each initiation factor).

Mechanism of Elongation of Polypeptide

Once the initiation complex is formed, the initiator fMet-tRNA^{fMet}, occupies the P-site, and the A-site remains vacant (Figure 11.12a). Only in the initiation process does the P-site accept tRNA charged with an amino acid, in this case N-formyl methionine, while all subsequent aminoacyl-tRNAs enter the A-site. The second aminoacyl-tRNA is complexed with the elongation factor EF-Tu, carrying a molecule of GTP bound to it. The EF-Tu ensures the correct hydrogen bonding between codon of mRNA and anticodon of tRNA and helps in proper positioning of tRNA in A-site. This process requires energy that is provided by the hydrolysis of GTP. Once this activity is performed, the EF-Tu dissociates from the ribosome, and in the cytoplasm it is subsequently regenerated to its active form by another elongation factor, the EF-Ts. At this moment, both sites of ribosome are occupied by tRNAs, each of which carries an amino acid. As shown in the Figure 11.12(b), the P-site is filled by fMet-tRNA^{fMet} and the A-site is occupied by Phe-tRNA^{Phe}.

The aminoacyl groups on the two tRNA molecules on the P-site and A-site are in the vicinity of the ribosomal enzyme, **peptidyl transferase**, which transfers the first amino acid (N-formyl methionine) to the free amino (NH$_2$) group of the second amino acid (in this it is phenylalanine). Thus, the first amino acid is placed "on the top of" the second amino acid by forming a peptide bond (—CO—NH—) between the carboxyl group of the first amino acid and amino group of the second amino acid, resulting in the formation of a **dipeptide**. Now the A-site is filled with dipeptide-carrying tRNA and P-site has discharged tRNA^{Met} (Figure 11.12c).

Once dipeptide is formed, for continuous synthesis of polypeptide the next charged tRNA must enter the A-site of ribosome and to achieve this objective the ribosome moves one codon further along the mRNA. This process is called translocation of ribosome in which discharged tRNA^{fMet} is ejected from the E-site (not shown in the figure), the third exit site on 50S subunit and later on this tRNA is to be re-used. The dipeptide-carrying tRNA is shifted from A-site to P-site constituting the **transpeptidation** reaction. The movement of the ribosome relative to mRNA requires an elongation factor called EF-G (that is also called **translocase**) and hydrolysis of GTP. As a result, the P-site is occupied by dipeptidyl-tRNA and A-site is vacated, which is to be filled by a third charged tRNA bringing the amino acid to elongate the polypeptide chain (Figure 11.12d).

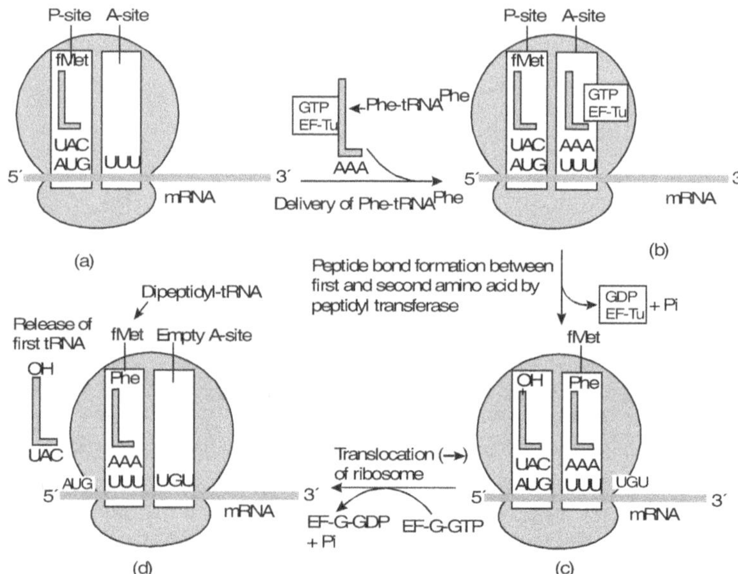

Figure 11.12   Stages in the process of elongation of polypeptide. (a) Binding of fMet-tRNA^{fMet} to P-site, (b) Entry of second charged tRNA at A-site, (c) Peptide bond formation between first and second amino acid, and (d) translocation of ribosome.

The elongation of polypeptide continues as ribosomes move exactly three nucleotides along the mRNA molecule in 5′ to 3′ direction. The peptidyl-tRNA molecule in P-site transfers its burden of growing polypeptide to the next succeeding tRNA, followed by translocation of ribosome, release of discharged tRNA, and the filling of A-site by incoming new tRNA with activated amino acid and correct anticodon complementary to codon of mRNA. In prokaryotes, each cycle requires one-twentieth of a second under optimal conditions. Thus an average-sized protein of 400 amino acids requires only 20 seconds for its synthesis.

Termination of Protein Synthesis

The polypeptide chain elongates as the ribosome moves codon by codon on mRNA and it continues until it encounters any one of the three termination or stop codons for which no tRNA with anticodons complementary to codons exists; these are UAG, UAA, and UGA. When the ribosome reaches one of the stop codons, specific cytoplasmic **protein-releasing factors** (RFs), which bind to the A-site on the ribosome. The RF1 identifies termination codon UAA and UAG, RF2 recognizes codon UGA, and RF3 seems to promote the action of RF1 and RF2. Overall, releasing factors cause release of the nascent polypeptide from tRNA by altering the peptidyl transferase such that it hydrolyses the ester bond between the –COOH of the protein and the –OH of the 3′ terminal nucleotide of tRNA. Finally the ribosome detaches from mRNA and dissociates into subunits that are reused for the next round of initiation.

POLYSOME FORMATION

As mentioned earlier and in chapter 2, it takes about 20 seconds for a ribosome to synthesize an average protein in *E.coli*. If only a single ribosome at a time moved along the mRNA, it could be possible to synthesize the protein at the rate of one molecule per 20 seconds. However, as soon as an initiated ribosome has got underway and has moved along 30 codons, another initiation complex can occur so that one ribosome after another moves on to the mRNA, each independently synthesizing a protein molecule. Thus, several ribosomes can assemble on the mRNA molecule to form **polyribosome** or simply **polysome** to synthesize multiple copies of protein in a metabolically active cell (Figure 11.13).

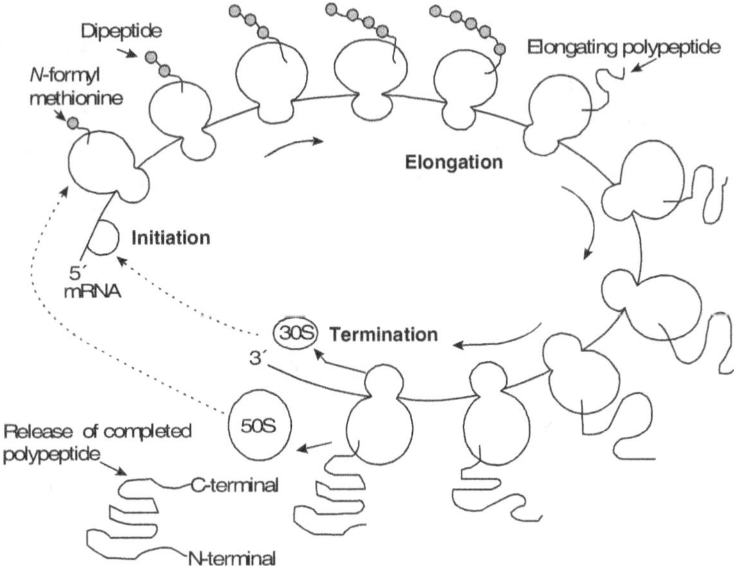

Figure 11.13   Diagrammatic representation of polyribosome formation. Several ribosomes move along the codons on mRNA and independently synthesize a protein molecule.

The number of ribosomes associated in a polysome depends on the size of the protein being synthesized, for example, in rapidly growing *E. coli* cell, there may be an association of as many as 15,000 ribosomes per mRNA molecule, whereas in the reticulocytes (immature red blood cells which have lost the nucleus), only four to six ribosomes are attached to the messenger RNA that synthesizes only one protein (haemoglobin). In HeLa tumour cell, where many proteins are being synthesized, the number of ribosomes per mRNA varies from few to 30 or even more.

PROTEIN SYNTHESIS IN EUKARYOTES

The process of polypeptide synthesis in eukaryotes is the same in essentials, but important differences exist. Eukaryotic ribosomes are larger (80S, with 60S and 40S subunits) and have extra rRNA and proteins. Methionine is always the first amino acid but it is not formylated. About ten initiation factors are involved in the formation of initiation complex during protein synthesis in eukaryotes. Similarly, the factors required for elongation of polypeptide are also different; the counterpart of EF-Tu in eukaryotes is called EF1α; that of EF-G is called EF2. In eukaryotic system, only one releasing factor is known i.e., eRF1 in contrast to three releasing factors (RF1, RF2, and RF3) in prokaryotes. Furthermore, in prokaryotes there is coupled transcription–translation that is not possible in eukaryotes. In *E.coli*, there is no nuclear membrane to separate DNA and ribosomes. As a result, on one hand, an mRNA molecule is transcribing on a DNA template and on the other hand, translation is initiated at its 5′ terminus. Such coupled activity speeds up the protein synthesis that does not occur in eukaryotes, because the mRNA is synthesized and processed in the nucleus and later on it is transported to the cytoplasm through the nuclear pore where the ribosomes form initiation complex to synthesize the desired protein. In addition, eukaryotic mRNAs are **monocistronic** (code for a single polypeptide) in contrast to prokaryotic mRNAs that are **polycistronic** with Shine–Dalgarno sequences provided for the initiation of translation of each cistron.

POST-TRANSLATIONAL MODIFICATION OF PROTEIN

Many proteins are covalently modified, either before their detachment from ribosome or after their synthesis has been completed. Since most of the modifications in the polypeptide take place after the translation, they are referred to as post-translational modifications that mainly include trimming or proteolysis, phosphorylation, glycosylation, and hydroxylation (Figure 11.14). A functional protein is not necessarily the same as the polypeptide chain that is released from the ribosome. Especially in eukaryotic cells, the polypeptide may need to be moved far from the site of synthesis in the cytoplasm, destined to an organelle, or even secreted from the cells. In addition, the polypeptide is often modified by the addition of new chemical groups that have functional significance.

The nascent proteins are either sent to the nucleus, mitochondria, plastids, or peroxisomes, or remain in the cytoplasm where they perform specific functions. Polypeptides may be retained in ER or sent to lysosome via Golgi complex or they may be secreted from the cell via vesicles, which fuse with plasma membrane and release their content by the process of exocytosis. During **proteolysis**, the lager and functionally inactive polypeptides are cleaved to form active molecules in the presence of specialized endoproteases. Proteases are essential in some viruses, including HIV, because the large viral polypeptides cannot fold properly unless it is cut. Certain drugs used to treat AIDS work by inhibiting the HIV protease, thereby preventing the formation of proteins needed for viral reproduction. **Phosphorylation** is the addition of phosphate groups to proteins in the presence of protein kinases. The charged phosphate groups alter the conformation of targeted proteins that may increase or decrease its activity. **Glycosylation** is addition of sugars ("sugar coating") to proteins by the enzymes present in ER and the Golgi apparatus. Glycosylation is essential for addressing proteins to lysosomes, important in conformation and recognition functions of proteins at cell surface and also helps in stabilizing proteins stored in storage vacuoles in plant seeds.

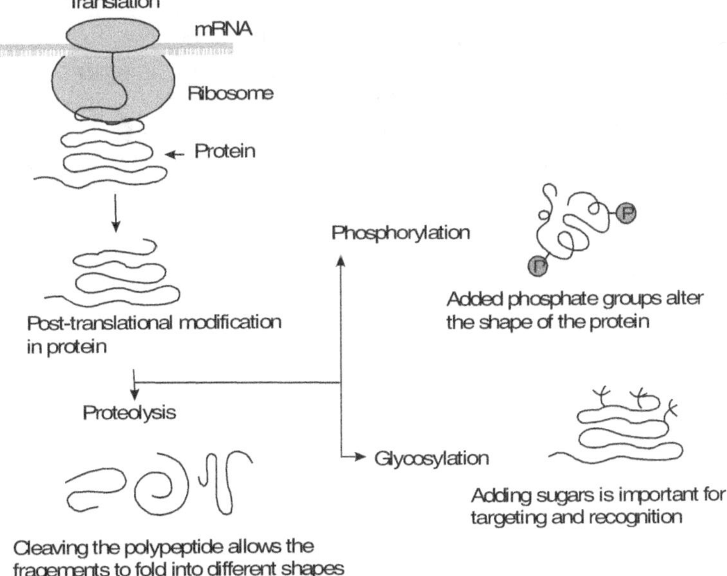

Figure 11.14   Various methods of post-translational modifications in protein

INHIBITORS OF PROTEIN SYNTHESIS

The accuracy of translation depends upon the interaction between the codons of mRNA and anticodons of tRNA. It is evident that only about 1 in 10,000 amino acids incorporated incorrectly. Several bacterial diseases are treated by antibiotics, which are considered as chemical missiles since they are responsible for placing incorrect aminoacid in the growing polypeptide during protein synthesis. As a result, aberrant and non-functional protein is formed. Antibiotics also cause premature polypeptide chain termination or even completely prevent the biosynthesis of protein. As a consequence the growth and reproduction of bacteria is checked and disease is cured. The antibiotics **streptomycin** and **neomycin** bind to particular protein (called S3 and S5) in the small subunit of ribosome and induces a conformational change in the subunit so that mRNA-tRNA initiation complex at P-site and aminoacyl-tRNA at A-site become much less tightly bound to ribosome. As a result, these antibiotics in prokaryotes prevent transition from initiation complex to chain-elongating ribosome. Since these antibiotics do not bind to eukaryotic ribosomes, they have no effect on translation of host cell's mRNAs. Another antibiotic **chloramphenicol** blocks the peptidyl transferase reaction of 50S ribosomal subunit in prokaryotes. **Cycloheximide** inhibits the activity of peptidyl transferase activity of 60S ribosomal subunit in eukaryotes. **Puromycin** in prokaryotes and also in eukaryotes causes the premature release of nascent polypeptides. **Tetracyclines** are a widely used class of antibiotics in clinical medicine to cure bacterial diseases like pneumonia, listeria, gonorrhoea, and cholera. In addition, these drugs are effective against chlamydia, another widespread venereal disease. Tetracyclines can selectively kill bacterial infection without inhibiting host's protein synthesis because microbes have membrane protein that actively transports tetracycline into cell, but mammalian cells lack this protein. Bacteria become resistant to tetracycline only when they have a defect in the gene that is responsible for transport of protein. Since several antibiotics have inhibitory effect on protein biosynthesis, they are used to study the various steps involved in the process.

SUMMARY

- Thousands of different proteins are present in a typical cell where they are formed during translation of mRNA, and perform phenotypic, cellular, or molecular functions.

- Genetic code is in the nucleotide sequence in the DNA that contains the information for protein synthesis.

- George Gamow (1954) proposed the model of genetic code and suggested that each amino acid in a polypeptide was encoded by three sequential nucleotides.

- In 1955, Manago and Ochoa cracked the genetic code using artificial mRNA.

- Genetic code is a triplet codon, which is universal, commaless, and shows linear arrangement and degeneracy. Some codes serve for initiation while few for termination of protein synthesis.

- Wobble hypothesis suggests that the pairing in the third base in a codon on mRNA and first base in anticodon of tRNA is ambiguous.

- Protein synthesis in prokaryotes involves formation of initiation complex, elongation of polypeptide chain and termination of process, which are mediated by several factors including initation factors, elongation factors, releasing factors, ATP, GTP, enzymes, and three types of RNAs.

- In eukaryotes, protein synthesis follows the same path but with some differences including the size of the ribosomes, number of initiation factors, form of mRNA, speed of addition of amino acids in a growing polypeptide, etc.

- Once protein is synthesized, it undergoes post-translational modifications that include trimming or proteolysis, phosphorylation, glycosylation, and hydroxylation, which are essential to perform their assigned functions.

- Antibiotics such as streptomycin, neomycin, chloramphenicol, puromycin, tetracyclines and others prevent protein synthesis by one or the other way.

REVIEW QUESTIONS

1. Explain the genetic code dictionary and describe how it was cracked.

2. What do you understand by genetic code? Give the properties of the genetic code.

3. Comment on the universality of the genetic code.

4. Discuss Crick's wobble hypothesis.

5. Why is multiple system of coding known as degenerate system? What are reasons for degeneracy of the code?

6. Give an account of the biosynthesis of protein.

7. Describe the mechanism of translation during protein synthesis in a prokaryotic cell.

8. Explain the role of aminoacyl-tRNA synthetase in the formation of initiation complex.

9. How does protein synthesis differ in prokaryotes and eukaryotes?

10. Enlist the factors essential for initiation, elongation and termination of polypeptide synthesis in *E.coli* with their significance.

11. Explain the polyribosome with its importance.

12. What are post-translational modifications of protein?

13. Describe the action of various antibiotics on protein synthesis.

14. Write short notes on:

    i. Cryptogram
    ii. Initiation codon
    iii. Non-overlapping codons
    iv. Termination codons
    v. RNA-binding sites on ribosomes
    vi. Polarity of anticodon and codon
    vii. Shine–Dalgarno sequence
    viii. fMet-tRNA^{fMet}
    ix. Translocation
    x. Peptidyl transferase
    xi. Sugar coating of proteins
    xii. Effect of tetracycline on protein synthesis
    xiii. Polycistronic mRNA

REGULATION OF BACTERIAL GENE EXPRESSION

INTRODUCTION

Bacteria harbour in dynamic environmental conditions ranging from the human intestine to polluted rivers, ponds, etc. where they are exposed to different metabolites and molecules. In such a wide range of ecological conditions, bacteria have to adapt rapidly and correctly, which is an indispensable phenomenon for their survival. This could be possible because of their ability to "switched on" and "switched off" the expression of specific sets of genes in response to the specific demand of the environment. As mentioned earlier, in *E.coli* there is a single circular DNA molecule 1100 μm length bearing 4,406 genes that are composed of 4.6 × 10^6 base pairs. As per the demand of the environmental condition, *E.coli* responds by regulating the expression of only few of the genes at a given time, based on the cellular requirement. The expression of particular genes means the specific genes are "switched on". It further indicates the transcription and translation of DNA to synthesize proteins, which are essential for growth and multiplication of microbes in a given environment. Their expression is "switched off" when the products of these genes are no longer required by bacteria. The protein synthesizing machinery of the cell consumes considerable energy. It is highly essential for organisms for the optimal use of the available energy by "switching off" the expression of genes when their products are not needed. Consider for an instant that *E.coli* "turned on" all its 4,406 genes to produce proteins that are not needed; soon it will become energy-deficient and unable to survive.

METHODS FOR REGULATION OF PROKARYOTIC GENE EXPRESSION

Prokaryotes can conserve energy and utilize their available resources in the most efficient way by making proteins only when they are required. The protein content of a bacterium can alter rapidly when conditions warrant. A prokaryotic cell could shut off the supply of unwanted protein by either of the following ways:

- ❖ Block the mRNA synthesis for a particular protein. This is transcriptional control mediated by repressor proteins.

- ❖ Hydrolyse the mRNA once it has been formed and prior to its translation. This is post-transcriptional control.

- ❖ Prevent translation of mRNA at the ribosome. This is translational control.

- ❖ Hydrolyse the protein after it is made. This is possible during post-translational modification.

- ❖ Inhibit the function of protein.

These methods would all have to be selective, affecting some genes and proteins and not others. In addition, they would all have to respond to some biochemical signal. Selective inhibition of transcription is far more efficient than transcribing the gene, translating the mRNA, and then degrading or inhibiting the protein. While examples of all five mechanisms for regulating protein levels are found in nature, however, extensive data indicate that prokaryotes generally use transcriptional regulation as the most important mode of control of gene expression. Regulatory fine-tuning at translational levels is clearly important in the overall control of metabolic processes in living organisms. The regulatory mechanisms with the largest effects on phenotype, however, have been shown to act at the level of transcription.

COMPONENTS OF BACTERIAL OPERON

The bacterial genome is organized so that genes with a common function (like synthesis of enzymes involved in biosynthetic or degradative pathway) are often grouped together in units that are referred to as **operons**. The genes of the operon are transcribed into a polycistronic mRNA that can be activated or repressed as a unit at the level of transcription with the help of an inducer, regulator, and other associated factors. The clusters of related genes code for the enzymes needed to synthesize amino acids, purines and pyrimidines, fatty acids, and vitamins, as well as the genes controlling the catabolism of most known carbon and nitrogen sources.

The **inducer** is the substance that induces or activates the expression of genes so that in its presence, the genes transcribe and translate to synthesize enzymes, which then metabolize the inducer (that can act as carbon source to produce energy). This phenomenon is known as **induction** and the enzymes synthesized are called inducible enzymes. For example, *E.coli* growing in the presence of lactose as a carbon source, lactose acts as inducer and activates the genes to synthesize the enzyme β-galactosidase, which hydrolytically cleaves lactose into glucose and galactose. The opposite response is shown by many enzymes involved in cellular biosynthesis. For example, *E.coli* growing in medium without any amino acids contain all the enzymes necessary for biosynthesis of the 20 necessary amino acids. When, on the other hand, the growth medium contains these amino acids, their biosynthetic enzymes are almost entirely missing. Biosynthetic enzymes whose amount is reduced by the presence of their end products (e.g. histidine is the end product of the histidine biosynthetic enzymes) are called **repressible enzymes**. Those end-product metabolites whose introduction into growth medium specifically decreases the amount of specific enzymes are known as **co-repressors**. The inductive and repressive responses are equally useful to bacteria. When enzymes are needed to transform a specific food molecule or to synthesize a necessary constituent, they are present; when they are unnecessary, the enzymes are effectively absent.

The amount of enzymes in *E. coli* as a function of the amount of inducer present in the growth medium as an adaptive response is shown in Figure 12.1. There are a number of proteins within the cell whose amount does not seem to be influenced by the external environment. Such proteins are called **constitutive proteins**. In *E. coli*, for example, the amount of enzymes controlling the degradation of glucose do not radically change when glucose is either removed or added to the growth medium.

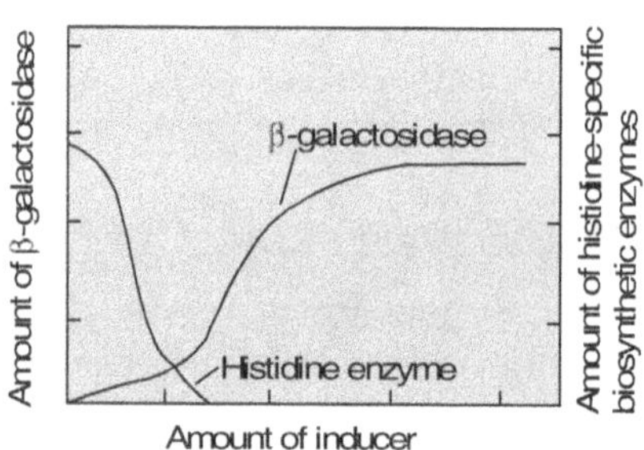

Figure 12.1   Effect of inducer on enzyme level in *E.coli*

The components of an operon are basically classified into categories of genes, 1) **Control genes** that include regulator, promoter and operator genes, and 2) **Structural genes** that transcribe to form polycistronic mRNA which on translation synthesize enzymes (Figure 12.2).

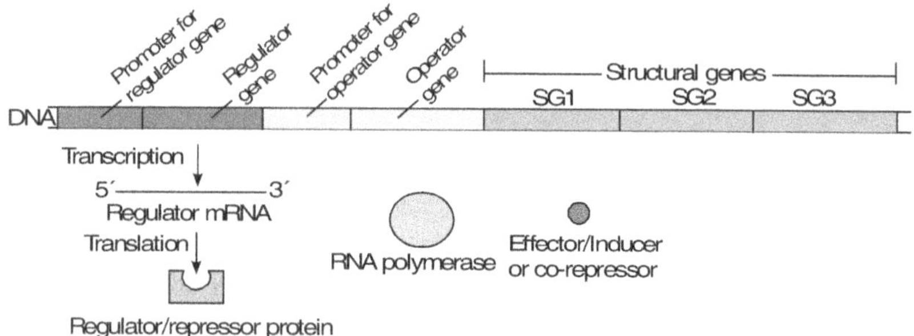

Figure 12.2   Components of operon in a bacterium

The controlling genes are primarily responsible for controlling the structural genes. The product of regulator gene is **regulator protein** (repressor protein or activator) that determines when many inducible and repressible enzymes are to be made. The regulatory protein affects the expression of one or more specific genes. They are of two fundamental types, **positive regulators**, and **negative regulators**, distinguished by the contrasting effect of mutations in the genes that encode them. Negative regulation is by means of repressor protein, which binds to the DNA (specifically to operator gene) and prevents transcription of structural genes. In positive regulation, the activator binds to adjacent promoter (site for binding RNA polymerase) and activates the process of transcription.

THE *LAC* OPERON—AN INDUCIBLE OPERON

In 1961, Jacob and Monad of Pasteur Institute in Paris proposed the operon model on the basis of their studies of the *lac* operon in *E.coli* in order to explain the induction or repression of enzyme synthesis. The *lac* operon is a group of genes that contain the information for assembling the enzymes required for the catabolism of lactose, the predominant sugar (disaccharide) present in milk. It contains a promoter, an operator, and three structural genes, z, y, and a, coding for enzymes, β-galactosidase, β-galactoside transacetylase, and *lac* permease. An enzyme β-galactosidase is the product of structural gene z, which contains 3510 base pairs and catalyses the hydrolysis of lactose to glucose and galactose. The structural gene y contains 780 base pairs and determines the structure of an enzyme *lac* permease, which is a carrier protein present in plasma membrane and facilitates the entrance of lactose into the cell. The third structural gene a comprises 825 base pairs and is responsible for the synthesis of the enzyme β-galactoside transacetylase, which is indirectly involved in the utilization of lactose. It transfers an acetyl group from acetyl-CoA to β-galactoside.

When *E.coli* is grown on the medium that does not contain lactose or other β-galactosides, the levels of the three enzymes are extremely low. Thus the bacterial cell does not waste the energy in the synthesis of these unneeded enzymes. If, however, the environment changes such that lactose is supplied in the medium, the bacteria promptly begin to synthesize all three enzymes needed to catabolize the lactose present in the medium. This is a rapid, simultaneous, and coordinated response shown by the bacteria, as there is a change in the environment. Thus, lactose acts as an **inducer** of the operon and induces the synthesis of the enzymes needed for its catabolism. Lactose can induce the synthesis of 3,000 molecules of β-galactosidase per cell. This phenomenon is known as **enzyme induction**.

The regulatory mechanism of the *lac* operon consists of a regulator gene (i) that is located close to the operon but separated from that of the operator gene (o). The *i* gene codes for a repressor protein comprising 360 amino acids. In high concentration, inducer (lactose) binds to repressor protein forming repressor–inducer complex. As a consequence, repressor becomes inactive and cannot bind to operator. Without the repressor in the way, RNA polymerase attaches to the promoter (P) and transcribes the structural genes, z, y, and a. It leads into production of polycistronic mRNA that subsequently translates to manufacture all three lactose-digesting enzymes (Figure 12.3).

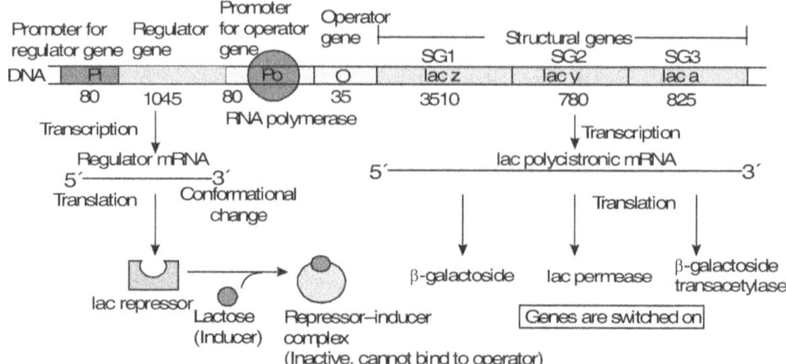

Figure 12.3 An inducible lactose operon

As the lactose is catabolized in presence of β-galactosidase, β-galactoside transacetylase, and lac permease, the concentration of lactose in the medium decreases. This causes bound inducer molecule to dissociate from the repressor–inducer complex. As a consequence, repressor protein becomes active and then regains its ability to bind the operator, thereby preventing transcription of structural genes (Figure 12.4). Thus, when inducer concentration is low, the operon is repressed, preventing the synthesis of unneeded enzymes, to conserve energy by switching off the genes. This phenomenon is known as enzyme repression.

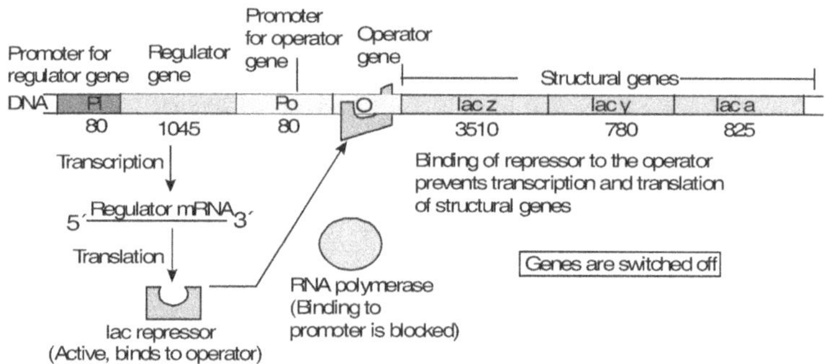

Figure 12.4 *Lac* operon. In the absence of lactose, the structural genes are switched off.

The *lac* repressor protein exerts its influence by **negative control**, since the interaction of the DNA with this protein inhibits gene expression. The striking view of how regulatory proteins, like repressors, bind specific DNA sites has been given by the X-ray crystallography of *E.coli* **catabolite gene activator protein** (CAP), λ Cro protein, λ repressor, and the repressor of phage 434. As protein crystallographers had expected, the active binding unit is a dimer of two identical globular polypeptide chains oriented oppositely in space to give an overall shape with a twofold axis of symmetry matching that of the binding site in DNA. In the case of λ repressor, for example, the amino terminus of each chain folds into a domain that binds to one half of the operator (Figure 12.5).

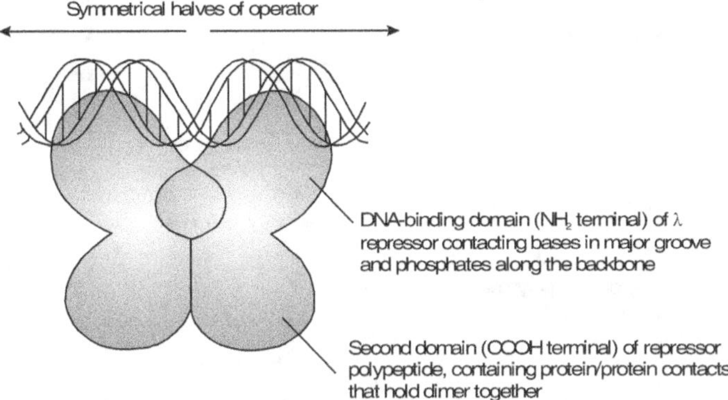

Figure 12.5   Diagrammatic representation of binding of λ *repressor* to operator DNA

The *lac* operon is also under positive control, as was discovered during an early investigation of a phenomenon called the glucose effect. If the bacterial cells are supplied with glucose, as well as a variety of other types of substrates such as lactose or galactose, the cells metabolize the glucose in the medium for production of energy and ignore the other compound. The glucose in the medium acts to suppress the production of various catabolic enzymes such as β-galactosidase, which are needed to degrade these other substrates. In 1965, a surprising finding was made. Cyclic AMP, previously thought to be involved only in eukaryotic metabolism, was detected in *E.coli*. It was found that the concentration of cAMP in the cells was related to the presence of glucose in the medium: the higher the glucose concentration, the lower the cAMP concentration. Furthermore, when cAMP was added to the medium in the presence of glucose, the cells suddenly synthesized all of the catabolic enzymes that were normally absent.

THE *TRP* OPERON—A REPRESSIBLE OPERON

The *trp* (tryptophan) operon in *E.coli* is probably the best-known repressible operon in which the repressor is unable to bind to the operator DNA by itself. Instead, the repressor is active as a DNA-binding protein only when complexed with a co-repressor, such as tryptophan (Figure 12.6a). The *trp* operon repressor is the product of gene *trp*R, which is not closely linked the *trp* operon. In the

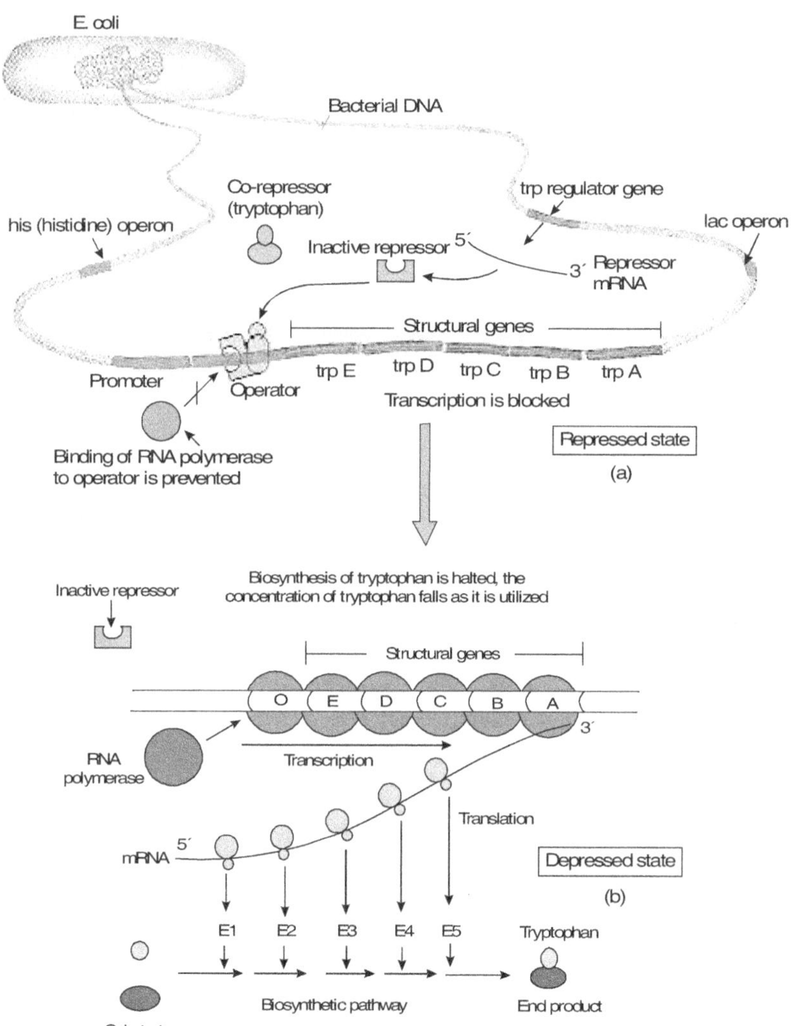

Figure 12.6 (a) *E.coli* DNA showing the tryptophan (*trp*) operon in the repressed state in which repressor–co-repressor (tryptophan) complex binds to operator DNA preventing the binding of RNA polymerase thereby blocking the transcription of *trp* structural genes. (b) When the concentration of tryptophan is low, the repressor proteins lack a co-repressor and therefore fail to attach to the operator. As a consequence, structural genes are transcribed and translated to synthesize enzymes needed for biosynthesis of tryptophan. As the biosynthetic pathway continues, tryptophan accumulates to high-enough concentration to again repress the operon.

absence of tryptophan, the operator site is open to binding by RNA polymerase, which transcribes the structural genes (*trpE, trpD, trpC, trpB, trpA*) of *trp* operon. As a result, enzymes needed for biosynthesis of tryptophan are produced (Figure 12.6b). When tryptophan becomes available, the enzymes of tryptophan synthetic pathway are no longer needed, and the tryptophan–repressor complex blocks transcription.

SUMMARY

- The single circular DNA of E. coli has 1100 μm length with 4,406 genes that are composed of 4.6×10^6 base pairs. Only few of these genes are expressed at a given time as per the cellular requirement.

- The expressions of genes indicate that the specific genes are "switched on" to transcribe and translate DNA for synthesis of proteins essential for growth and reproduction of microbes, while the rest of the genes are "switched off" to conserve the energy of the cell.

- Bacteria could shut off the supply of unwanted protein by either transcriptional control mediated by repressor proteins, post-transcriptional control, translational control, post-translational modification, or by preventing the function of protein.

- In prokaryotes, genes are organized into regulatory units called operons. Operons are clusters of structural genes that typically code for different enzymes in the same metabolic pathway.

- All the structural genes of the operon are transcribed into a single mRNA, their expression is regulated in a coordinated manner.

- The lac operon is an inducible operon in which the gene expression is controlled by a key metabolic compound such as inducer (e.g. lactose) that attaches to a repressor protein. This event causes the RNA polymerase to bind to promoter and turn on the structural genes.

- When tryptophan is present in enough quantity there is no need of its biosynthesis, hence the repressor along with co-repressor (typtophan) bind to operator DNA and thereby present and thereby prevent transcription. But when there is absence of tryptophan RNA polymerase attaches to the operator site and transcribes the structural genes to synthesize enzymes needed for biosynthesis of tryptophan.

REVIEW QUESTIONS

1. Explains various methods of regulation of gene expression in prokaryotes.

2. What is operon? Describe different components of an operon.

3. Give an illustrated account of *lac* operon in *E.coli*.

4. Compare and contrast the role of inducer and co-repressor in lactose and tryptophan operons in *E.coli*.

5. If you were to find a mutant of *E.coli* that produced continuous enzymes β-galactosidase and *lac* permease, how might you explain how this happened?

6. Comment on how transcriptional level control conserve energy in prokaryotes.

7. What are some of the consequences you might expect in bacterial cell as a result of presence of glucose in medium in addition to other carbon sources?

8. Distinguish between the structural genes of *lac* and *trp* operons.

9. How are the functions of *lac* repressor and *trp* repressor are different for each other?

10. Explain the organization of *trp* operon in relation to the position of His and *lac* operons in *E.coli*.

11. Write short notes on:

    i. Inducible enzymes

    ii. Catabolite gene activator protein

    iii. Role of cAMP in gene regulation

    iv. Positive control over operon

    v. Co-repressor

    vi. Enzyme repression

    vii. Regulator genes

    viii. Constitutive proteins

    ix. Role of enzymes in lactose metabolism

    x. Biosynthesis of tryptophan

Appendix

NOBEL PRIZE WINNERS OF PHYSIOLOGY/ MEDICINE AND CHEMISTRY

Year	Scientist/s	Research
2006	Andrew Z. Fire and Craig C. Mello	RNA interface-gene silencing by double-stranded RNA.
2005	Barry J. Marshall and J. Robin Warren	The role of bacterium *Helicobacter pylori* in gastritis and peptic ulcer disease.
2004	Richard Axel and Linda B. Buck	Discoveries of odorant receptors and the organization of olfactory system.
2003	Paul C. Lauterbur, and Sir P. Mansfield	Magnetic resonance imaging.
2002	Sydney Brenner, H. Robert and John E. Sulston	Genetic regulation of organ development and programmed cell death.
2001	Leland H. Hartwell, R. Timothy and P.M. Nurse	Key regulators of the cell cycle.
2000	Arvid Carlsson, Paul Greengard and Eric Kandel	Signal transduction in the nervous system.
1999	Gunter Blobel	Proteins have intrinsic signals that govern their transport and localization in the cell.
1998	Robert F. Furchgott, Louis J. Ignarro and Ferid Murad	Nitric oxide as a signalling molecule in the cardiovascular system.
1997	Stanley B. Prsuiner	Prions, a new biological principle of infection.
1996	P.C.Doherty and R.F. Zinkernagel	Specificity of the cell-mediated immune defence.
1995	E.B. Lewis, C.N. Volhard and E.F. Wieschaus	The genetic control of early embryonic development.

(Contd.)

NOBEL PRIZE WINNERS OF PHYSIOLOGY/ MEDICINE AND CHEMISTRY

Year	Scientist/s	Research
1994	Alfred G. Gilman and M. Rodbell	G-proteins and the role of these proteins in signal transduction in cells.
1993	R. J. Roberts and P. A. Sharp	Split genes and RNA processing.
	K. Mullis	Polymerase chain reaction.
	M. Smith	Site-directed mutagenesis.
1992	E. H. Fischer and E. G. Krebs	Reversible protein phosphorylation as a biological regulatory mechanism.
1991	E. Neher and B. Sakmann	The function of single ion channels in cells.
1990	J. E. Murray and E.D. Thomas	Organ and cell transplantation in the treatment of human disease.
1989	J.M. Bishop and H. E. Varmus	The cellular origin of retroviral oncogenes.
	T.R. Cech, S. Altman	Ribozyme
1988	Sir J. W. Black, G. B. Elion and G. H. Hitchings	Important principles for drug treatment.
	J. Deisenhofer, R. Huber and H. Michel	Bacterial photosynthetic reaction centre.
1987	Susumu Tonegawa	The genetic principle for generation of antibody diversity.
1986	Stanley C. and Rita Levi-Montalcini	Growth factors.
1985	M.S. Brown and J.L. Goldstein	The regulation of cholesterol metabolism.
1984	N.K.Jerne, G.J.F. Kohaler and Cesar Milstein	The specificity in development and control of the immune system and the discovery of the principle for production of monoclonal antibodies.
1983	Barbara Mc Clintock	Mobile genetic elements.
1982	John R. Vane	Prostaglandins and related biological active substances.
	A. Klug	Structure of virus and components of life.

(Contd.)

NOBEL PRIZE WINNERS OF PHYSIOLOGY/ MEDICINE AND CHEMISTRY

Year	Scientist/s	Research
1981	David H. Huble and T.N. Wiesel	Information processing in the visual system.
	Roger W. Sperry	Functional specialization of cerebral hemisphere.
1980	Baruj B., J. Dausset and G.D. Snell	Genetically determined structures on the cell surface that regulate immunological reactions.
1979	A.M. Cormack and Sir G.N. Hounsfield	Computer-assisted tomography (CAT) scan.
1978	Warner A., Daniel N. and Hamilton O. Smith	Restriction enzymes and their application to problems of molecular genetics.
1977	R. Guillemin and Andrew V. Schally	Peptide hormone production of the brain.
	Rosalyn Yalow	Radioimmunoassays of peptide hormones.
1976	B.S. Blumberg and D.C. Gajdusk	New mechanisms for the origin and dissemination of infectious diseases.
1975	D. Baltmore, R. Dulbecco and H.M. Temin	The interaction between tumour viruses and the genetic material of the cell.
1974	A. Claude, C. De Duve and G.E. Palade	The structural and functional organization of the cell.
1973	K.V. Frisch and Nikolaas T.	Organization and elicitation of individual and social behaviour patterns.
1972	G.M. Edelman and R.R. Porter	The chemical structure of antibodies.
1971	E.W.Jr. Sutherland	The mechanisms of the action of hormones.
1970	Sir B.Katz U.V. Euler and Julius Axelford	The humoral transmitters in the nerve terminals and their mechanism of release and inactivation.
1969	Max Delbruck, A. D. Hershey and S. E. Luria	The replication mechanism and the genetic structure of viruses.
1968	H. G. Khorana, R. W. Holley and M. W. Nirenberg	Interpretation of the genetic code and its function in protein synthesis.

(Contd.)

NOBEL PRIZE WINNERS OF PHYSIOLOGY/ MEDICINE AND CHEMISTRY

Year	Scientist/s	Research
1967	R. Granit, H.K. Hartline and George W.	The primary physiological and chemical visual processes in the eye.
1966	Preyton Rous	Tumour-inducing viruses.
	C. B. Huggins	Hormonal treatment of prostatic cancer.
1965	F. Jacob, Andre L. and J. Monod	Genetic control of enzyme and virus synthesis.
1964	K. Bloch and F. Lynen	The mechanism and regulation of cholesterol and fatty acid metabolism.
1963	Sir John C. E., Sir A.L. Hodegkin and Sir. A. F. Huxley	The ionic mechanisms involved in excitation and inhibition in the peripheral and central portions of the nerve cell membrane.
1962	Francis H.C. Crick, J.D. Watson and M.H.F. Wilkins	Molecular structure of nucleic acids and its significance for information transfer in living material.
1961	G. V. Bekesy	The physical mechanism of stimulation within the cochlea.
1960	Sir F.M.Burent and Sir Peter B. Medawar	Acquired immunological tolerance.
1959	S.Ochoa and A. Kornberg	The mechanisms in the biological synthesis of ribonucleic acid and deoxyribonucleic acid.
1958	G.W. Beadle and E.L. Tatum	Genes act by regulating definite chemical events.
	Joshua Lederberg	Genetic recombination and the organization of the genetic material of bacteria.
1957	Daniel Bovet	Synthetic compounds that inhibit the action of certain body substances, and especially their action on the vascular system and the skeletal muscles.

Glossary

A (aminoacyl) site Site into which aminoacyl-tRNAs enter the ribosome–mRNA complex.

Acellular Lacking cellular organization; not having a delimiting cytoplasmic membrane; organizational description of viruses, viroids and prions.

Acquired immunity An ability of an individual to produce specific antibodies in response to antigens to which the body has been previously exposed based on the development of memory response.

Acrocentric A chromosome or chromatid that has its centromere near the end.

Adenine One of the four nitrogen bases that make up the letters ATGC in DNA code.

Adenosine triphosphate (ATP) An energy-rich compound present in biological systems, composed of Adenosine and three phosphate groups. The free energy released from the hydrolysis of the ATP is utilized to carry out energy-dependent processes in the biological systems.

AIDS An infectious disease syndrome caused by HIV retrovirus, characterized by loss of normal immune response, followed by various opportunistic infections.

Allele One of the many possible forms of the gene; different alleles produce variation in the inherited characteristics.

Allosteric enzyme Enzyme with a binding and catalytic site for the substrate and a different site where a modulator (allosteric effector) acts.

Amino acid The building blocks of protein.

Amniocentesis Puncture of uterine wall with needle for the purpose of obtaining amniotic fluid, which can be analysed to determine whether the foetus has a genetic abnormality.

Amplification Production of additional copies of a chromosomal sequence.

Anaphase The stage of mitosis or meiosis during which the daughter chromosomes pass from the equatorial plate to opposite poles of the cell (towards the ends of the spindle). Anaphase is followed by metaphase and precedes telophase.

Antibiotic One of the many natural organic substances (or their synthetic analogues) secreted by undergrowth or microbes that are toxic to other species, retard or prevent their growth and presumably function as defence mechanisms.

Antibodies Proteins produced by the immune system in order to neutralize potentially harmful foreign molecules (antigens).

Antigen A substance that can be recognized by the body as alien and that may induce an immune response by antibody.

Antisense therapy The *in vitro* treatment of a genetic disease by blocking translation of a protein with a DNA or RNA sequence that is complementary to a specific mRNA.

Apoptosis Programmed cell death in which the cell responds to certain signals by initiating a normal response that leads to the death of the cell.

Autoimmunity A situation in which the body produces an immune response against it own organs or tissues.

Autoradiography A record or photograph prepared by labelling a substance such as DNA with a radioactive material such as tritiated thymidine and allowing the image produced by decay radiations to develop on a film over a period of time. Thus, it is the technique by which emissions from radioactively labelled microscopic specimens can be visualized by exposure of photographic emulsion to the labelled tissue.

Autosome Any chromosome that is not a sex chromosome; human somatic cells have 22 pairs of autosomes and one pair of sex chromosomes.

Bacteriophage A virus that which infects the bacteria and replicates in bacteria. In genetic engineering it is used to introduce genes into bacterial cells.

Bacterium A structurally simple single cell with no nucleus. Its role in biotechnology is indispensable.

Barr body Also called sex chromatin that is derived form X chromosome and present only in genetic female.

Basal body Small granule to which a cilium or flagellum is attached.

Base analog Unnatural purine or pyrimidine base, differing slightly from normal base, that can be incorporated into nucleic acid.

Base pair (bp) A pair of molecules (either A and T or G and C) from one rung in the DNA ladder. Pairing is also possible in RNA under some circumstances.

B lymphocyte White blood cells that are able to produce specific antibodies.

Carcinogenic A substance capable of inducing cancer in an organism.

cDNA Complementary DNA; DNA strand formed from RNA using the enzyme reverse transcriptase.

Cellulose Complex polysaccharide forming the cell wall.

Centriole An organelle in many animal cells that appears to be involved in the formation of the spindle during mitosis.

Chromatid In mitosis or meiosis, one of the two identical strands resulting from self-duplication of a chromosome.

Chromatin The complex of de-oxyribonucleic acid and protein (histone) that constitutes the chromosome.

Chromocentre Body produced by fusion of the heterochromatic region of the chromosomes in polytene tissues (e.g. salivary glands) in Dipteran insects.

Chromonema An optically single thread forming an axial structure within each chromosome.

Chromosomal aberration The structural or numerical alterations in chromosomes.

Chromosomal banding Staining of chromosomes in such a way that light and dark areas occur along the length of the chromosome.

Cistron/gene The functional unit of genetic inheritance; a segment of genetic nucleic acid that codes for a specific protein.

Clone A collection of genetically identical cells or organisms derived from a common ancestor.

Codon A group of three nucleotides of mRNA coding for an amino acid.

Colchicine An alkaloid derived from the autumn crocus that is used as an agent to arrest spindle formation and interrupt mitosis.

Conjugate protein Protein linked covalently or non-covalently to substances other than amino acids, such as metals, nucleic acids, lipids, and carbohydrates.

Conjugation Union of sex cells (gametes) or unicellular organisms during fertilization in *E.coli*, a one-way transfer of genetic material from donor ("male") to a recipient ("female").

Constitutive enzyme/protein An enzyme that is produced in fixed quantities regardless of need.

Covalent bond The type of chemical bond in which electron pairs are shared between two atoms.

Crossing over A process in which chromosomes exchange material through the breakage and reunion of their DNA molecules.

Cyclic adenosine monophosphate (cyclic AMP, or cAMP) It is a second messenger that is produced from ATP by an enzyme, adenylate cyclase, that

is attached to the inside of the cell membrane.

Cytokinesis Cytoplasmic division and other changes exclusive of nuclear division that are a part of mitosis or meiosis.

Cytoplasm The protoplasm of a cell outside the nucleus in which cell organelles reside. All living parts of the cell except the nucleus.

Cytosine A pyrimidine base found in RNA and DNA.

Dalton The mass of a hydrogen atom.

Deficiency/Deletion Absence of a segment of a chromosome, reducing the number of loci.

Degeneracy An aspect of the genetic code in which a particular amino acid is encoded by more than one codon.

Dehydrogenase An enzyme that removes hydrogen from the substrate.

Denaturation Loss of native structure of a macromolecule (e.g. protein) that is usually accompanied by loss of biological activity.

Denaturation of DNA: Conversion of double-stranded DNA into single-stranded state usually due to heating.

De novo Arising from very simple precursors.

Diakinesis A stage of prophase I in meiosis in which the bivalents are shortened and thickened.

Diploid An organism or cell with two sets of chromosomes (2n) or two gametes. Somatic cells of higher plants and animals are ordinarily in diploid condition.

Disulphide bond A covalent bond formed between –SH groups of two cysteine residues.

DNA Deoxyribonucleic acid; the genetic material that stores, expresses, and inherits the genetical characters.

DNase An enzyme that hydrolyses DNA.

DNA cloning A broad-base technique for producing large quantities of a specific DNA segment.

DNA polymerase Enzyme responsible for the synthesis of new DNA strand.

DNA profiling The process of identifying the genetic make-up of an individual.

DNA tumour viruses Viruses capable of infecting vertebrate cells, transforming them into cancer cells. These viruses have DNA in mature virus particle.

Down's syndrome (Mongolism or 21 Trisomy) A genetic disorder produced as a result of non-disjunction of autosome number 21. Persons with such abnormality are usually mentally retarded.

Embryo transfer Implantation of embryos from donor animals or generated by *in vitro* fertilization into the uteri of recipient animals.

Endomitosis Duplication of chromosomes without division of nucleus, resulting in increased chromosome number within a cell.

Endonuclease An enzyme that catalyses the cleavage of DNA, normally it cuts DNA at specific sites.

Endotoxin Toxic substances found as part of some bacterial cells.

Enhancer A substance that increases a chemical activity or physiological process. It may be a DNA segment that influences the expression of a gene.

Enzyme A class of protein that catalyses biological reactions.

Equatorial plate The figure formed by the chromosomes at the centre of the spindle in mitosis.

Euchromatin Less condensed and fairly stained genetic material that consists of active genes.

Eukaryotes Cellular organisms having a membrane-bound nucleus in which the genome of the cell is stored. They include algae, fungi, protozoa, plants and animals.

Extrachromosomal The structure that is not a part of the chromosome, e.g. plasmids in bacteria.

F factor A bacterial plasmid that confers the ability to function as a genetic donor ("male") in conjugation; the fertility factor in bacteria.

Fibroblast Flattened connective tissue cell.

Flagellum A whip-like organelle of locomotion in certain cells; locomotor structures in flagellated protozoa.

Frameshift A mutation that upsets the reading frame of an mRNA, by either inserting or deleting nucleotides.

Fungi A group of diverse, widespread, unicellular and multicellular eukaryotes lacking chlorophyll and usually bearing spores and often filaments.

Gametogenesis The process of formation of gametes—sperm and ovum.

Gene The fundamental physical and functional unit of heredity. A gene is a sequence of DNA that codes for protein, enzyme or RNA.

Gene amplification Process by which a gene is replicated selectively.

Gene expression The process in which the gene is activated and the DNA code of that gene is translated into specific protein.

Genetically modified organisms (GMO) An organism whose genetic make-up has been changed by any method including natural processes and genetic engineering.

Genetic code The DNA sequence of a gene, which determines the sequence of amino acids in a protein or enzyme.

Genetic engineering The technology used to isolate genes from an organism, manipulate them in the laboratory and insert them into another cell system for industrial, medical and research purposes.

Genetic mapping Determination of relative positions of genes in DNA or RNA.

Genome A complete set (n) of chromosomes (hence, of genes) inherited as a unit from one parent.

Genotype The genetic make-up of an organism or individual.

Glycolysis The catabolic pathway in which the glucose molecule is broken down into two molecules of pyruvic acid.

Guanine One of the four nitrogen bases found in nucleic acid.

Haploid An organism or cell having only one set (n) of chromosomes of one genome.

HeLa cell Cell line originally derived from uterine carcinoma cells of a patient named Henrietta Lacks.

Helix Any structure with a spiral shape. The DNA model is in the form of a double helix.

Heterochromatin Chromatin staining darkly even during interphase, often containing repetitive DNA with few genes.

Heterozygous Having two different alleles for a given trait.

Hfr High-frequency recombination strain of *E. coli*; in such a strain, the F-factor is integrated into the bacterial chromosome.

Histones Group of proteins rich in basic amino acids associated with eukaryotic genes.

Homologous chromosomes Chromosomes that occur in pairs and are generally similar in size and shape, in a pair one is paternal and other is maternal chromosome.

Homozygous Having two identical alleles for a given (gene in the diploid cell) trait.

Inducer A substance of low molecular weight that inactivates a repressor by combining with it, thereby stimulating gene expression.

Inoculum A material introduced onto or into a host or medium.

***In situ* hybridization** Technique in which DNA is kept in place while it is allowed to react with particular preparation of labelled DNA or RNA.

Interphase The stage in the cell cycle when the cell is not dividing: the metabolic stage during which DNA replication occurs. The stage is following telophase of one division and extending to the beginning of prophase in the next division.

Intron Intervening sequence of DNA located within a gene, that is not included in the mature mRNA.

In vitro Literally means "in glass", e.g. a test tube. Experimentation on organisms or portion thereof in glassware of culture container under an artificial condition.

In vivo Experimentation or organisms under natural conditions within intact living organisms.

Karyotype The chromosome constitution of a cell or an individual; chromosomes arranged in order of length and according to the position of centromere.

Kinetosome Granular body at the base of a flagellum or a cilium.

Klinefelter's syndrome The individual has the genotype 44A + XXY and is characterized by mental retardation, underdevelopment of genitalia, and presence of feminine physical characteristics.

Lactose A disaccharide of glucose and galactose found in milk.

Lamella A double-membrane structure, plate, or vesicle that is formed by two membranes lying parallel to each other.

Lampbrush chromosome Large diplotene chromosomes present in oocyte nuclei, particularly conspicuous in amphibians.

Ligase Enzyme that joins cut ends of DNA strands.

Locus The normal positions of gene on a chromosome.

LTR (Long terminal repeat) A DNA sequence present at each of the ends of a retrotransposon.

Lysis Bursting of a cell by the destruction of the cell membrane following infection by a virus.

Lysogenic bacteria Bacteria harbouring temperate bacteriophage.

Meiosis The process of cell division by which the diploid number of chromosomes of parent cell is reduced to half and results in the formation of haploid gametes in animals or spores in plants; important source of variation through recombination.

Messenger RNA (mRNA) The type of RNA that carries information for protein synthesis from DNA to the ribosomes.

Metabolism Sum total of all chemical processes in living cells by which energy is provided and used.

Metabolite Product of biochemical activity.

Metacentric chromosome A chromosome with centromere near the middle and two arms of equal length.

Metaphase The stage of cell division in which chromosomes are most discrete and arranged in an equatorial plate; stage following prophase and preceding anaphase.

Metastasis The spread of cancer cells to previously unaffected organs.

Microbes Microscopic organisms that may be pathogenic or non-pathogenic.

Microtubules Hollow filaments in the cytoplasm making up a part of the locomotor apparatus of a motile cell; component of the mitotic spindle.

Mitochondria Organelles in the cytoplasm of plant and animal cells where oxidative phosphorylation takes place to produce ATP.

Mitosis The process of cell division to divide the diploid parent cell to produce two genetically identical daughter cells.

Mutagen An environmental agent, either physical or chemical, that is capable of inducing mutation.

Mutant A cell or individual organism that shows a change brought about by mutation; an altered gene.

Mutation The stable change in gene or chromosome that can cause a disorder or the inherited susceptibility to a disorder.

Nucleic acid A macromolecule composed of pentose sugar, nitrogen base, and phosphoric acid; DNA and RNA.

Nucleolar organizer A chromosomal segment containing genes that control the synthesis of ribosomal RNA, located at the secondary constriction of some chromosomes.

Nucleolus An RNA-rich, spherical sac in the nucleus of metabolic cells associated with the nucleolar organizer; storage place for ribosomes and ribosome precursors.

Nucleoprotein Conjugated protein composed of nucleic acid and protein; the material of which the chromosomes are made.

Nucleosome Spherical subunit of eukaryotic chromatin that are composed of a core particle consisting of an octamer of histones and surrounded by DNA (146 bp).

Nucleotide A unit of DNA and RNA molecule made up of pentose sugar, nitrogen base, and phosphoric acid.

Nucleus The part of a eukaryotic cell that contains the chromosomes and separated from the cytoplasm by a membrane.

Oncogene Gene thought to be capable of producing cancer.

Oocyte The egg mother cell that undergoes oogenesis to produce ovum.

Oogenesis The process of formation of egg or ovum in animals.

Operator A part of an operator that controls the activity of structural gene/s by binding a regulatory protein.

Operon A group of genes making up a regulatory or control unit. The unit includes an operator, a promoter, and structural genes.

Organelle Specialized part of a cell with a particular function/s.

Origin of replication (*ori*) The nucleotide sequence at which DNA synthesis is initiated.

Pachytene A stage during prophase I of meiosis in which chromosomes are visible as long, paired threads.

Pathogen An organism that causes a disease.

pH The standard measure of relative acidity; it is the negative logarithm of the hydrogen ion concentration in any solution.

Phenotype The external or physical make-up of an individual.

Photon A packet of light energy.

Phagocytosis Ingestion of particulate material by cells either as a means of feeding or defence.

Plasma cell Cells that are able to synthesize a specific antibody and secondary B cells.

Plasmid An autonomous, self-replicating, extrachromosomal DNA found in bacteria playing significant role in genetic engineering.

Primer A short nucleotide sequence with a reactive 3´ OH that can initiate DNA synthesis along a template.

Prokaryote Organisms that lack a true nucleus.

Promoter A nucleotide sequence to which RNA polymerase binds and initiates transcription.

Protooncogene A normal cellular gene that can be changed to an oncogene by mutation.

Prototroph An organism such as bacterium that will grow on a minimal medium.

Provirus A viral chromosome that has integrated into a host genome.

Poly(A) Long polyadenylic acid segment added posttranscriptionally in many eukaryotic mRNA at the 3′ end.

Polymer Large molecules made up of repeating structural subunits.

Polynucleotides A linear sequence of linked nucleotides in DNA or RNA.

Polytene chromosomes Giant chromosomes produced by interphase replication without division and consisting of many identical chromatids arranged side by side.

Prion An infectious agent associated with certain mammalian degenerative diseases that is composed solely of protein; derived from proteinacious infectious particle.

Protein Structurally and functionally diverse group of polymers built of amino acid monomers.

Purine Type of nitrogen bases found in nucleic acid having two fused rings consisting of nitrogen and carbon atoms (e.g. adenine, guanine).

Pyrimidine Type of nitrogen bases found in nucleic acid having single heterocyclic ring (e.g. cytosine, thymine, uracil).

Recombinant DNA Genetic material with novel gene sequences produced by cross overs, chromosome reassortment, by other natural means or through genetic engineering.

Regulator gene A gene that controls the rate of expression of another gene/s.

Renaturation The restoration of a molecule to its native form. In relation to DNA, it refers to the formation of a double-stranded helix from complementary single-stranded molecules.

Replication A duplication process of DNA.

Replicon A unit of replication.

Repressible enzyme An enzyme whose synthesis is diminished by a regulatory molecule.

Restriction enzymes DNA endonucleases that can recognize specific short sequences of usually unmethylated DNA and that cut the duplex DNA.

Retrotransposon A transposable element that resembles the integrated form of a retrovirus.

Reverse transcriptase An enzyme that catalyses the synthesis of DNA using an RNA template.

RNA tumour viruses Virus capable of infecting vertebrate cells, transforming them into cancer cells. RNA viruses have RNA in the mature virus particle.

Satellite DNA A component of the genome that can be isolated from the rest of the DNA by centrifugation; usually it consists of short, highly repetitive sequences.

Sexduction The incorporation of bacterial genes into F factors and their subsequent transfer by conjugation to a recipient cell.

Silencer A DNA sequence that helps to reduce or shut off the expression of a nearby gene.

Site-directed mutagenesis A technique to modify a gene in a predetermined way so as to produce a protein with a specifically altered amino acid sequence.

Somatic Refers to the vegetative or non-sexual stages of life cycle.

Southern blotting Procedure for transferring denatured DNA from an agarose gel to a nitrocellulose paper where it can be hydrolysed with a complementary nucleic acid.

Spermatogenesis The process of formation of spermatozoa in testes.

Spermiogenesis Formation of sperm from spermatids.

Splicing Gene manipulation, the objective of which is to attach one DNA molecule to another. RNA splicing is the removal of introns from precursors (hnRNA).

Stem cell An undifferentiated cell from which specialized cells develop.

Structural genes A gene that specifies the synthesis of a polypeptide.

Synapsis The pairing of homologous chromosomes in meiotic prophase.

Syndrome A group of symptoms that occur together and represent a particular disease.

Telophase The stage in mitotic and meiotic division in which the chromosomes are assembled at the poles of the division spindle.

Temperate phase A phase (virus) that invades but may not destroy (lyse) the host (bacterial cell); however, it may continue into the lytic cycle.

Template A pattern or mould. DNA stores coded information and acts as a model or template from which information is transcribed into messenger RNA.

Tetrad The four cells arising from the second meiotic division in plants (pollen tetrads) or fungi (ascopores). The term is also used to identify the quadruple group of chromatids that is formed by the association of duplicated homologous chromosomes during meiosis.

Thymine One of the four nitrogen bases that make-up the letters ATGC in DNA.

Transcription Formation of an RNA from DNA template with the help of RNA polymerase.

Transcription factors Auxiliary proteins that help in the operation of eukaryotic polymerases.

Transduction Genetic recombination in bacteria mediated by bacteriophage.

Transfection The process that introduces DNA into cultured cells.

Transformed Normal cells that have been converted into cancer cells by treatment with a carcinogenic material, radiation, or an infective tumour virus.

Transgenic A term applied to organisms that have been altered by introducing DNA molecule into them.

Transition A mutation caused by the substitution of one purine by another purine or one pyrimidine by another pyrimidine in DNA or RNA.

Translation Synthesis of protein in the cytoplasm using the information coded on mRNA.

Translocation Movement of ribosome relative to mRNA during elongation of polypeptide. The term is also applied to the change in the position of a segment of chromosome to another part of the same chromosome or to a different chromosome.

Transposons DNA segment that can move from one position in a DNA molecule to another.

Transversion A mutation caused by the substitution of a purine by a pyrimidine or a pyrimidine by a purine in DNA or RNA.

Tissue culture A process where individual cells, or clumps of plant or animal tissue are grown *in vitro*.

Tumour-suppressor gene Gene that encodes protein that retains cell growth and prevents cells from becoming malignant.

Uracil One of the four nitrogen bases that make up the letters AUGC found in RNA.

Vaccine A preparation that contains an antigen made up of disease-causing organisms in a dead or weakened state. It is used to boost immunity against the given disease, and can be created using recombinant DNA technology.

Vector An entity, such as plasmids, viruses or bacteriophages, used to carry new DNA into an organism. In parasitology, vectors like mosquitoes, flies, etc. are carriers of pathogenic organisms.

Virulent phase A virus that destroys the host (bacterial) cell.

Virus A submicroscopic organism consisting of DNA or RNA core surrounded by protein coat capable of replicating the cells of the host organism.

Wobble hypothesis Hypothesis to explain how one tRNA may recognize two codons. The first bases of mRNA codon and anticodon pairs properly, but the third base in the anticodon has some play (to wobble) that permits it to pair with more than one base.

X-chromosome A sex-chromosome-associated determination of sex. In most animals, the female has two and male has one X chromosome.

X-ray crystallography/diffraction Scattering of X-ray by the atoms of a crystal in a manner such that the resulting diffraction pattern provides information about the structure and/or identity of the substance.

Y-chromosome A sex chromosome associated with sex chromosome. In animal species, male has Y chromosome as a partner with X chromosome.

Z-DNA The type of DNA that exhibits conformational flexibility. "Z" indicates the zig-zagged path of sugar–phosphate backbone of the structure.

Zygonema The stage in prophase I of meiosis in which synapsis of chromosomes takes place.

REFERENCES

Adams, R.L.P., *et al.*1986. *The Biochemistry of the Nucleic Acids*. Chapman and Hall.

Adolph, K.W. (Ed.) 1988. *Chromosomes and Chromatin*. CRC Press, Boca Raton, Fl.

Albert, B.D., Bray, J. Lewis, M. Raff, K. Robertis and J.D. Watson. 1989. *Molecular Biology of the Cell*, 2nd edn. Garland Publishing, New York.

Aliva, J. 1992. *"Microtubule Functions"*, *Life Science*. 50.

Balows, A. (Ed.) 1992. *The Prokaryotes*. Vol. 4 Springer-Verlag, NY.

Baserga, R. 1985. *Biology of Cell Reproduction*. Harvard University Press, Cambridge.

Bershadsky, A., and Vasiliev, J. 1988. *The Cytoskeleton*. Plenum, New York.

Beermann, W. 1972. *Developmental Studies on Giant Chromosomes*. Springer-Verlag, New York.

Bonner, J., and Varner, J. (Eds.). *Plant Biochemistry*. Academic Press, New York.

Bray, D.1992. *Cell Movement*. Garland Publishing, New York.

Brachet, J. 1985. The *Cell Cycle: Molecular Cytology*, Vol.1. Academic Press, NY.

Brachet, J. 1985. *Cell Interaction: Molecular Cytology*, Vol.2. Academic Press, NY.

Bretscher, M. 1984. *Endocytosis: Relation to Capping and Cell Locomotion*. *Science*. 224.

Brett, E.T., and Hillman, J.R. (Eds.). 1985. *Biochemistry of Plant Cell Walls*. Cambridge Univ. Press, Cambridge.

British Council, 2003. *DNA and after*. 50 Years of UK Excellence published by Foreign and Commonwealth Office, London.

Brock T.D., *et al.* 1990. *Biology of Microorganisms*, 6th edn. Prentice-Hall.

Busch, H. (Ed.) 1983. *The Cell Nucleus*, Academic Press, NY.

Carafoili, E. and Roman, I. 1980. Mitochondria and disease. *Mole. Aspects Med* 3.

Data, D.B. 1987. *Comprehensive Introduction to Membrane Biochemistry*, Madison: Floral Press.

Davis B.D., *et al.* 1990. *Microbiology*, 4th edn. Lippincott.

David, E. S. 1993. *Cell Biology, Organelle Structure and Function*. Jones and Bartlett Publishers, Boston.

Davis, R.W., D. Botsein, and J.R. Roth. 1980. *Advanced Bacterial Genetics*. Cold Spring Harbour Laboratory Press, Cold Spring Harbor, N.Y.

Darlington, C.D. and L. Cour.1976. *The Banding of Chromosomes*, 6th edn. Halsted, New York.

Darnell, J., H. Lodish, and D. Baltimore. 1986. *Molecular Cell Biology*. Scientific American Books, New York.

DeDuve, C. 1991 *Blueprint for a Cell: The Nature and Origin of Life*. Neil Patterson Pub.

De Grouchy J. and C. Tureieau. 1977. *Clinical Atlas of Human Chromosomes*. Wiley, New York.

DE Robertis, E.D.P., and E.M.F.DE Robertis, Jr. 1980. *Cell and Molecular Biology*, 7th edn. Saunders College/Holt, Rine-hart and Winston, Philadelphia.

Drake, J.W. 1982. *The Molecular Basis of Mutation*. Holden-Day, San Francisco.

Dulbecco, R., and Ginsberg, H.S. 1991. *Virology*, 3rd edn. Lippincott.

Freifelder, D. 1983. *Molecular Biology, A Comprehensive Introduction to Prokaryotes and Eukaryotes*. Sci. Books Internationals, Boston.

Gardner, E.J., M.J. Simmons, and D.P. Snustad. 1991. *Principles of Genetics*, (8th edn.). John Wiley and Sons Inc., NewYork.

Gennis, R.B.1990. *Biomembranes*. Springer-Verlag, New York.

Gelard, K. 1996. *Cell and Molecular Biology*, John Wiley and Sons. Inc., New York.

Gesteland, R.F., and Atkins, J.F. (Eds.). 1993. *The RNA World*. Cold Spring Harbor Lab. Press.

Gregory, R.P.F. 1989. *Biochemistry of Photosynthesis*, 3rd edn. Wiley, New York.

Hober, J.K. 1984. *Chloroplasts*. Plenum, New York.

Holtzman, E., 1989. *Lysosomes*. Plenum, New York

Hyams, J.S., and Lloyd, C.W. 1994. *Microtubules*. Wiley-Liss, NY.

John, B.1990. *Meiosis*. Cambridge University Press, Cambridge, UK.

Le Beau, N.M. and J.D. Rowley. 1986. "Chromosomal Abnormalities in Leukemia and Lymphoma: Clinical and Biological Significance". *Adv.Human Genet*. 15: 1–54.

Lehninger, A.L 1982. *Principles of Biochemistry*. Worth, New York.

Leviene, A. 1992. *Viruses*. Scientific American Library.

Lewin, B. 1994. *Genes V,* Cell Press, Cambridge, MA/Oxford Univ. Press, Oxford.

McCarty, M. 1985. *The Transforming Principle*. Norton.

McKnight, S.L., and Yamamoto, K. 1992. *Transcriptional Control*. Cold Spring Harbor Lab. Press.

Morrow, J. 1983. *Eukaryotic Cell Genetics*, Academic Press, New York.

Moore, A., and Beechey, R. 1987. *Plant Mitochondria*, Plenum, New York.

Neidhardt, F.C. (editor in chief), J.I. Ingraham, K.B. Low, B. Magasanik, M. Schechter, and H.E. Umbarger, (Eds.) 1987. *E.coli and S. typhimurium, Cellular and Molecular Biology*. American Society for Microbiology, Washington, DC.

Pasternak, C.A. 1989. *"Membrane Transport and Disease"*. *Mol. Cell Biochem.* 91.

Priest, J.H. 1977. *Medicinal Cytogenetics and Cell Culture*, 2nd edn. Lea and Febier, New York.

Ptashne, M.1987. *A Genetic Switch: Gene Control and Phage* λ. Cell Press, Cambridge, MA.

Singer, M. and Berg, P. 1991. *Genes and Genomes*, Uni. Sci. Books, California and Blackwell Scientific Publishers, Oxford.

Sparks, R.S. *et al.*1977. *Molecular Human Cytogenetics*. Academic Press, New York.

Spirin, A.S. 1986. *Ribosome Structure and Protein Synthesis*, Benjamin-Cumming, Menlo Park, CA.

Stainer, R.Y., *et al.* 1986. *The Microbial World*, 5th edn. Prentice-Hall.

Stryer, L. 1988. *Biochemistry*. Wiley, New York.

Sotaniemi, E. (Ed).1987. *Enzyme Induction in Man*, Taylor and Francis, Philadelphia.

Thurman, E., and Susman, M.1993. *Human Chromosomes*. 3rd edn. Springer-Verlag.

Turpin, R. and J. Lejeune. 1969. *Human Afflictions and Chromosomal Aberrations*. Pergamon, Oxford, England.

Van Holde, K. 1989. Chromatin. Springer-Verlag, New York.

Verma, R.S. (Ed). 1989. *Heterochromatin: Molecular and Structural Aspects*, Cambridge Univ. Press. Cambridge.

Von Figura, K., and Hasilik, A. 1086. *"Lysosomal Enzymes and their Receptors"*, *Ann Rev. Biochem.* 55.

Ucker, D.S. 1991. "Death by suicide: One way to go in mammalian cellular development?" *New Bio.* 3.

Watson, J.D. and J.Tooze. 1981. *The DNA Story*. Freeman, San Francisco.

Watson, J.D., N.H. Hopkins, J.W. Roberts, J.A. Steitz, and A.M. Weiner. 1987. *Molecular Biology of the gene*, 4th edn. Benjamin, Menlo Park. CA.

Whaley, W.G. 1975. *The Golgi Apparatus*. Springer-Verlag, New York.

William, H.E. and Daphne, C.E.1997. *Biochemistry and Molecular Biology*. Oxford Univ. Press, Oxford.

Wolfee, A.P. 1992. *Chromatin Structure and Function*, Academic Press.

Zubey, G.1987. *Biochemistry*, Addison-Wesley, Reading, MA.